Current Topics in Membranes, Volume 63

Membrane Protein Crystallization

Current Topics in Membranes, Volume 63

Series Editors

Dale J. Benos
Department of Physiology and Biophysics
University of Alabama
Birmingham, Alabama

Sidney A. Simon
Department of Neurobiology
Duke University Medical Centre
Durham, North Carolina

Current Topics in Membranes, Volume 63

Membrane Protein Crystallization

Edited by

Larry DeLucas
Center for Biophysical Sciences and Engineering
University of Alabama at Birmingham
Birmingham, Alabama

ELSEVIER

AMSTERDAM • BOSTON • HEIDELBERG • LONDON
NEW YORK • OXFORD • PARIS • SAN DIEGO
SAN FRANCISCO • SINGAPORE • SYDNEY • TOKYO
Academic Press is an imprint of Elsevier

Academic Press is an imprint of Elsevier
30 Corporate Drive, Suite 400, Burlington, MA 01803, USA
525 B Street, Suite 1900, San Diego, CA 92101-4495, USA

First edition 2009

ISBN: 978-0-12-374987-1
ISSN: 1063-5823

For information on all Academic Press publications visit our website at elsevierdirect.com

Printed and bound in the United Kingdom
Transferred to Digital Printing, 2011

To my former mentor, Dr F.L. (Bud) Suddath (1942–1992).

Contents

Contributors

Numbers in parentheses indicate the pages on which the authors' contributions begin.

Jeff Abramson (109) Department of Physiology, Division of Molecular Medicine, David Geffen School of Medicine, University of California, Los Angeles, California 90095

Konstantinos Beis (269) Imperial College London, South Kensington Campus, London SW7 2AZ, United Kingdom

James U. Bowie (109) Department of Chemistry and Biochemistry, UCLA-DOE center for Genomics and Proteomics, Molecular Biology Institute, University of California, Los Angeles, California 90095

M. Caffrey (83) Department of Chemical and Environmental Sciences, University of Limerick, Limerick, Ireland

Richard J. Cogdell (127) Division of Molecular and Cellular Biology, Faculty of Biological Life Sciences, Glasgow Biomedical Research Centre, University of Glasgow, 120 University Place, Glasgow G12 8TA, UK

Larry DeLucas (151) Center for Biophysical Sciences and Engineering, University of Alabama at Birmingham, Birmingham, Alabama 35294-4400

Volker A. Erdmann (25) Institut für Chemie und Biochemie, Freie Universität Berlin, Thielallee 63, D-14195 Berlin, Germany

Salem Faham (109) Department of Physiology, Division of Molecular Medicine, David Geffen School of Medicine, University of California, Los Angeles, California 90095

Petra Fromme (127, 191) Department of Chemistry and Biochemistry, Arizona State University, Tempe, Arizona 85287-1604

Mads Gabrielsen (127) Division of Molecular and Cellular Biology, Faculty of Biological Life Sciences, Glasgow Biomedical Research Centre, University of Glasgow, 120 University Place, Glasgow G12 8TA, UK

Alastair T. Gardiner (127) Division of Molecular and Cellular Biology, Faculty of Biological Life Sciences, Glasgow Biomedical Research Centre, University of Glasgow, 120 University Place, Glasgow G12 8TA, UK

Cory Gerdts (179) deCODE biostructures, woodridge, Illinois 60517

Michael Gerrits (25) RiNA GmbH, Takustrasse 3, D-14195 Berlin, Germany

Ingo Grotjohann (191) Department of Chemistry and Biochemistry, Arizona State University, Tempe, Arizona 85287-1604

Deborah K. Hanson (51) Biosciences Division, Argonne National Laboratory, Argonne, Illinois 60439

D. J. Hart (83) Department of Chemistry, The Ohio State University, Columbus, Ohio 43210

Charles Henry (151) Department of Chemistry, Colorado State University, Fort Collins, Colorado 80523-1872

Tina M. Iverson (229) Department of Biochemistry, Vanderbilt University Medical Center, Nashville, Tennessee 37232-6600; Department of Pharmacology, Vanderbilt University Medical Center, Nashville, Tennessee 37232-6600

David H. Johnson (153) Center for Biophysical Sciences and Engineering, University of Alabama at Birmingham, Birmingham, Alabama 35294-4400

Stefan Kubick (25) RiNA GmbH, Takustrasse 3, D-14195 Berlin, Germany

Philip D. Laible (51) Biosciences Division, Argonne National Laboratory, Argonne, Illinois 60439

Hubing Lou (269) Centre for Biomolecular Sciences, The University of St Andrews, Fife KY16 9ST, United Kingdom

J. Lyons (83) Department of Chemical and Environmental Sciences, University of Limerick, Limerick, Ireland

Alex McPherson (5) Department of Molecular Biology, University of California-Irvine, Irvine, California 92697

Helmut Merk (25) RiNA GmbH, Takustrasse 3, D-14195 Berlin, Germany

Donna L. Mielke (51) Biosciences Division, Argonne National Laboratory, Argonne, Illinois 60439

James H. Naismith (269) Centre for Biomolecular Sciences, The University of St Andrews, Fife KY16 9ST, United Kingdom

Peter Nollert (179) Emerald BioSystems, Inc., Bainbridge Island, Washington 98110

Robert W. Payne (151) Department of Chemistry, Colorado State University, Fort Collins, Colorado 80523-1872

T. Smyth (83) Department of Chemical and Environmental Sciences, University of Limerick, Limerick, Ireland

Wolfgang Stiege (25) RiNA GmbH, Takustrasse 3, D-14195 Berlin, Germany

Mikio Tanabe (229) Department of Pharmacology, Vanderbilt University Medical Center, Nashville, Tennessee 37232-6600

Rachna Ujwal (109) Department of Physiology, Division of Molecular Medicine, David Geffen School of Medicine, University of California, Los Angeles, California 90095

Gregg Whited (151) Genencor International, Inc., Palo Alto, California 94304

W. William Wilson (151) Department of Chemistry, Mississippi State University, Mississippi State, Mississippi 39762

Preface

Membrane protein biology has become increasingly important as scientists use a variety of new techniques to investigate systems biology. Estimated to represent more than one third of the genomes from human and most other species, membrane proteins play critical roles in cell functions. Some of these roles include ion, metabolite and macromolecular transport, signal processing, electron transport, oxidative phosphorylation, muscle contraction, and interactions with a variety of cell regulatory elements. The experimental determination rate for new macromolecular structures (via X-ray crystallographic techniques) has grown exponentially in the past decade with more than 35,000 structures currently deposited in the protein data bank (PDB). However, membrane proteins represent less than 1% of deposited structures, with the majority determined from prokaryotic species. The inability to express membrane proteins in quantities required for crystallographic studies combined with added difficulties encountered with the production of high quality three-dimensional protein crystals have emerged as the major bottlenecks preventing high throughput determination of membrane protein structures.

The purpose of this book is to familiarize the membrane biologist with the general theory and experimental approaches for membrane protein crystallization. The book begins with a comprehensive explanation of general theory and experimental approaches of crystallization followed by a detailed description of two of the more recent and successful membrane protein expression systems (eukaryotic cell-free expression and a novel photosynthetic bacterial expression system). Traditional expression systems for both prokaryotic and eukaryotic membrane proteins are summarized in subsequent chapters that address a variety of membrane protein crystallization techniques. Structural themes and biological implications resulting from X-ray crystallographic structures for specific membrane protein classes are also reviewed.

The main goal of this book is to provide membrane biologists with sufficient knowledge about membrane protein crystallization to enable them to perform crystallization studies within their own laboratories. However, I believe the book contains a wealth of new information that experienced membrane protein crystallographers will also find useful.

I want to express my appreciation to the contributors for devoting a significant amount of time addressing their subject matter in such a way that readers not familiar with protein crystallization methods would understand and be able to transfer this knowledge into the laboratory setting. It is my hope that this book has succeeded in placing these contributions in a broader perspective, providing insight into the multiple approaches used to enhance membrane proteins crystallization success rates.

Previous Volumes in Series

Current Topics in Membranes and Transport

Volume 23 Genes and Membranes: Transport Proteins and Receptors* (1985)
Edited by Edward A. Adelberg and Carolyn W. Slayman

Volume 24 Membrane Protein Biosynthesis and Turnover (1985)
Edited by Philip A. Knauf and John S. Cook

Volume 25 Regulation of Calcium Transport across Muscle Membranes (1985)
Edited by Adil E. Shamoo

Volume 26 Na^+–H^+ Exchange, Intracellular pH, and Cell Function* (1986)
Edited by Peter S. Aronson and Walter F. Boron

Volume 27 The Role of Membranes in Cell Growth and Differentiation (1986)
Edited by Lazaro J. Mandel and Dale J. Benos

Volume 28 Potassium Transport: Physiology and Pathophysiology* (1987)
Edited by Gerhard Giebisch

Volume 29 Membrane Structure and Function (1987)
Edited by Richard D. Klausner, Christoph Kempf, and Joe van Renswoude

Volume 30 Cell Volume Control: Fundamental and Comparative Aspects in Animal Cells (1987)
Edited by R. Gilles, Arnost Kleinzeller, and L. Bolis

Volume 31 Molecular Neurobiology: Endocrine Approaches (1987)
Edited by Jerome F. Strauss, III, and Donald W. Pfaff

**Part of the series from the Yale Department of Cellular and Molecular Physiology.*

Volume 32 Membrane Fusion in Fertilization, Cellular Transport, and Viral Infection (1988)
Edited by Nejat Düzgünes and Felix Bronner

Volume 33 Molecular Biology of Ionic Channels* (1988)
Edited by William S. Agnew, Toni Claudio, and Frederick J. Sigworth

Volume 34 Cellular and Molecular Biology of Sodium Transport* (1989)
Edited by Stanley G. Schultz

Volume 35 Mechanisms of Leukocyte Activation (1990)
Edited by Sergio Grinstein and Ori D. Rotstein

Volume 36 Protein–Membrane Interactions* (1990)
Edited by Toni Claudio

Volume 37 Channels and Noise in Epithelial Tissues (1990)
Edited by Sandy I. Helman and Willy Van Driessche

Current Topics in Membranes

Volume 38 Ordering the Membrane Cytoskeleton Trilayer* (1991)
Edited by Mark S. Mooseker and Jon S. Morrow

Volume 39 Developmental Biology of Membrane Transport Systems (1991)
Edited by Dale J. Benos

Volume 40 Cell Lipids (1994)
Edited by Dick Hoekstra

Volume 41 Cell Biology and Membrane Transport Processes* (1994)
Edited by Michael Caplan

Volume 42 Chloride Channels (1994)
Edited by William B. Guggino

Volume 43 Membrane Protein–Cytoskeleton Interactions (1996)
Edited by W. James Nelson

Volume 44 Lipid Polymorphism and Membrane Properties (1997)
Edited by Richard Epand

Volume 45 The Eye's Aqueous Humor: From Secretion to Glaucoma (1998)
Edited by Mortimer M. Civan

Volume 46 Potassium Ion Channels: Molecular Structure, Function, and Diseases (1999)
Edited by Yoshihisa Kurachi, Lily Yeh Jan, and Michel Lazdunski

Volume 47 AmilorideSensitive Sodium Channels: Physiology and Functional Diversity (1999)
Edited by Dale J. Benos

Volume 48 Membrane Permeability: 100 Years since Ernest Overton (1999)
Edited by David W. Deamer, Arnost Kleinzeller, and Douglas M. Fambrough

Volume 49 Gap Junctions: Molecular Basis of Cell Communication in Health and Disease
Edited by Camillo Peracchia

Volume 50 Gastrointestinal Transport: Molecular Physiology
Edited by Kim E. Barrett and Mark Donowitz

Volume 51 Aquaporins
Edited by Stefan Hohmann, Søren Nielsen and Peter Agre

Volume 52 Peptide–Lipid Interactions
Edited by Sidney A. Simon and Thomas J. McIntosh

Volume 53 CalciumActivated Chloride Channels
Edited by Catherine Mary Fuller

Volume 54 Extracellular Nucleotides and Nucleosides: Release, Receptors, and Physiological and Pathophysiological Effects
Edited by Erik M. Schwiebert

Volume 55 Chemokines, Chemokine Receptors, and Disease
Edited by Lisa M. Schwiebert

Volume 56 Basement Membrances: Cell and Molecular Biology
Edited by Nicholas A. Kefalides and Jacques P. Borel

Volume 57 The Nociceptive Membrane
Edited by Uhtaek Oh

Volume 58 Mechanosensitive Ion Channels, Part A
Edited by Owen P. Hamill

Volume 59 Mechanosensitive Ion Channels, Part B
Edited by Owen P. Hamill

Volume 60 Computational Modelling of Membrane Bilayers
Edited by Scott E. Feller

Volume 61 Free Radical Effects on Membranes
Edited by Sadis Matalon

Volume 62 The Eye's Aqueous Humor
Edited by Mortimer M. Civan

Introduction

Macromolecular crystallography has resulted in structures for approximately 40,000 proteins with an increasing number of new structures determined each year. This is due to a number of advancements in methodology including: (a) expression systems that yield milligram quantities of homogeneous protein, (b) use of molecular biology to engineer protein constructs more likely to crystallize, (c) high-throughput nanocrystallization robots capable of preparing hundreds of experiments/hour using 10–200 nanoliters of protein solution for each experiment, (d) high-throughput crystal imaging systems, (e) dramatic increases in the speed and quality of X-ray data collection, (f) automated X-ray data collection systems, (g) novel methods to obtain initial phase information required for a structural solution, and (h) new and improved software programs for processing raw X-ray data, producing initial electron density maps, and model fitting and refinement (Fogg & Wilkinson, 2008; Hui & Edwards, 2003; Manjasetty, Turnbull, Panjikar, Bussow, & Chance, 2008; Puri et al., 2006; Sauder et al., 2008; Stevens, 2007; Tickle, Sharff, Vinkovic, Yon, & Jhoti, 2004). The improvements in these technologies, most of which began more than 15 years ago, resulted in the establishment of structural genomics groups around the world, with a goal of performing high-throughput structural genomics on a genome-wide scale (Burley et al., 1999; Goulding et al., 2004; Meng et al., 2008; Montelione & Anderson, 1999; O'Toole, Grabowski, Otwinowski, Minor, & Cygler, 2004). These groups have coordinated their efforts in that target information is openly shared and results (from initial cloning and expression to structure determination) are immediately posted on each structural genomics center's website with crystallization and structures deposited in the Protein Data Bank (PDB) (Kouranov et al., 2006). The new structural information is likely to play a major role, supporting our knowledge of biological mechanisms for individual molecules as well as entire cellular systems, that is, systems biology (Hendrickson, 2007; Kambach, 2007; Service, 2006).

In spite of these major advances in technology, the crystallization of macromolecules continues to present a major challenge, particularly for membrane proteins. The current state-of-the-art for protein crystallization employs extreme brute force, using high-throughput technologies that rapidly prepare 2000 solution conditions using a milligram of purified protein. Although the structural genomics era has led to the development of a number of high-throughput crystallization robotic systems, it is clear that robotics alone does not dramatically improve success rates. Thus far, the overwhelming majority of the structures resulting from the structural genomics effort have been for prokaryotic aqueous proteins. However, novel eukaryotic expression technologies combined with knowledge of protein structural factors that hinder crystallization have enabled the transition to more complex eukaryotic aqueous proteins and protein complexes. Similarly, momentum is growing, as advances in membrane protein expression, purification, and crystallization have resulted in recent structure-determination breakthroughs for several different membrane protein families from prokaryotic and to a lesser extent, eukaryotic organisms.

This book introduces the noncrystallographer, membrane biologist to the fundamental theory and experimental aspects of protein crystallization (chapter 1) with the majority of the chapters focused on membrane proteins. Chapters 2 and 3 discuss common protein overexpression problems and two novel approaches that have already made significant impacts on expression of prokaryotic membrane proteins, also showing promise for eukaryotic membrane proteins. Chapter 4, 5, and 6 describe different approaches used to crystallize membrane proteins via lipidic- and detergent-based systems. This is followed in chapter 7 by a review of a fluidic crystallization device capable of screening thousands of crystallization conditions using less than a milligram of protein, and chapter 8 describes a novel diagnostic approach that may be useful for optimizing solution conditions for membrane protein solubility, stability, and crystallization. The final three chapters review crystallization methodology and address specific classes of membrane proteins, providing a compilation of the crystallization methods and conditions that yielded crystals and protein structures. After reading this volume, the noncrystallographer membrane biologist should have a sufficient theoretical and experimental background in membrane protein crystallization to independently initiate crystallization efforts in their own laboratories. Although the chapters do not provide a detailed explanation of all approaches used to coax membrane proteins to crystallize, those not described in detail are mentioned and appropriate references provided.

References

Burley, S. K., Almo, S. C., Bonanno, J. B., Capel, M., Chance, M. R., Gaasterland, T., et al. (1999). Structural genomics: Beyond the human genome project. *Nature Genetetics, 23*, 151–157.

Fogg, M. J., & Wilkinson, A. J. (2008). Higher-throughput approaches to crystallization and crystal structure determination. *Biochemical Society Transactions, 36*, 771–775.

Goulding, C. W., Apostol, M., Anderson, D. H., Gill, H. S., Smith, C. V., Kuo, M. R., et al. (2002). The TB structural genomics consortium: Providing a structural foundation for drug discovery. *Current Drug Targets. Infectious Disorders, 2*, 121–141.

Hendrickson, W. A. (2007). Impact of structures from the protein structure initiative. *Structure, 15*, 1528–1529.

Hui, R., & Edwards, A. (2003). High-throughput protein crystallization. *Journal of Structural Biology, 142*, 154–161.

Kambach, C. (2007). Pipelines, robots, crystals and biology: What use high throughput solving structures of challenging targets. *Current Protein and Peptide Science, 8*, 205–217.

Kouranov, A., Xie, L., de la Cruz, J., Chen, L., Westbrook, J., Bourne, P. E., et al. (2006). The RCSB PDB information portal for structural genomics. *Nucleic Acids Research, 34*, D302–305.

Manjasetty, B. A., Turnbull, A. P., Panjikar, S., Bussow, K., & Chance, M. R. (2008). Automated technologies and novel techniques to accelerate protein crystallography for structural genomics. *Proteomics, 6*, 612–625.

Meng, W., Forwood, J. K., Guncar, G., Robin, G., Cowieson, N. P., Listwan, P., et al. (2008). Overview of the pipeline for structural and functional characterization of macrophage proteins at the University of Queensland. *Methods in Molecular Biology, 426*, 577–587.

Montelione, G. T., & Anderson, S. (1999). Structural genomics: Keystone for a human proteome project. *Nature Structural and Molecular Biology, 6*, 11–20.

O'Toole, N., Grabowski, M., Otwinowski, Z., Minor, W., & Cygler, M. (2004). The structural genomics experimental pipeline: Insights from global target lists. *Proteins: Structure, Function and Bioinformatics, 56*, 201–210.

Puri, M., Robin, G., Cowieson, N., Forwood, J. K., Listwan, P., Hu, S.-H., et al. (2006). Focusing in on structural genomics: The University of Queensland structural biology pipeline. *Biomolecular Engineering, 23*, 281–289.

Sauder, M. J., Rutter, M. E., Bain, K., Rooney, I., Gheyi, T., Atwell, S., et al. (2008). High throughput protein production and crystallization at NYSGXRC. *Methods in Molecular Biology, 426*, 561–575.

Service, R. F. (2006). The impact of structural genomics: Expectations and outcomes. *Science, 287*, 1954–1956.

Stevens, R. C. (2007). Generation of protein structures for the 21st century. *Structure, 15*, 1517–1519.

Tickle, I., Sharff, A., Vinkovic, M., Yon, J., & Jhoti, H. (2004). High-throughput protein crystallography and drug discovery. *Chemical Society Reviews, 33*, 558–565.

CHAPTER 1

Introduction to the Crystallization of Biological Macromolecules

Alex McPherson

Department of Molecular Biology, University of California-Irvine, Irvine, California 92697

I. OVERVIEW

Biological macromolecules, which include proteins, nucleic acids, and their complexes, can be crystallized by a wide variety of techniques involving a broad range of reagents. The objective in all cases is to produce supersaturated mother liquors. The crystallization trials may in turn be carried out under different physical conditions such as temperature. The most commonly employed approaches for discovering successful crystallization conditions for specific macromolecules, and the factors that influence them are summarized here. In addition, some of the classical ideas from crystallization science and protein science, such as solubility in salts as a function of pH, are described and discussed in terms of their practical application, and many of the difficulties that are commonly encountered are addressed. While the methodologies have been developed principally for protein crystallization, they are equally applicable to large biological assemblies such as viruses and ribosomal particles.

Current Topics in Membranes, Volume 63 1063-5823/09 $35.00

DOI: 10.1016/S1063-5823(09)63001-5

II. INTRODUCTION

Crystallization has emerged as the rate-limiting step in macromolecular crystallography and has become a major barrier to advances in structural biology. Macromolecules are intricate physical-chemical systems whose properties vary as a function of environmental conditions such as temperature, pH, ionic strength, contaminants, and solvent composition, to name only a few. They are structurally dynamic, often microheterogeneous, aggregating systems, and the macromolecules' conformations are often sensitive to the presence of a spectrum of ligands. Superimposed on this is the limited extent of our current understanding of macromolecular crystallization phenomena and the forces that promote and maintain protein and nucleic acid crystals.

Macromolecular crystallization is a matter of searching, as systematically as possible, the ranges of the individual parameters that impact upon crystal formation, finding a set, or multiple sets of these factors that yield some kind of crystals, and then optimizing the variables to obtain the best possible crystals for X-ray analysis. This is done, most simply, by conducting an extensive series, or establishing a vast array, of crystallization trials, evaluating the results, and using information obtained to improve matters in successive rounds of trials. Because the number of variables is large, and its range is broad, intelligence and intuition in designing and evaluating the individual and collective trials becomes essential.

III. THE REQUIREMENT FOR SUPERSATURATION

In a saturated solution, including one saturated with respect to protein, two states exist in equilibrium, the solid phase, and one consisting of molecules free in solution. At equilibrium, no net increase in the proportion of solid phase can accrue because it would be counter balanced by an equivalent dissolution. Thus, crystals do not grow from a saturated solution. The system must be in a nonequlibrium, or supersaturated state to provide the thermodynamic impetus for crystallization.

When the objective is to grow crystals of any compound, an undersaturated solution of the molecule must be transformed or brought into the supersaturated state, whereby its return to equilibrium forces exclusion of solute molecules into the solid state, the crystal. If, from an undersaturated solution, for example, solvent is gradually withdrawn by evaporation, temperature is lowered or raised appropriately, or some other property of the system is altered, then the solubility limit may be exceeded and the solution will become supersaturated. If a solid phase is already present, or introduced, then strict saturation will be reestablished as molecules leave the solvent, join the solid phase, and equilibrium is regained.

If no solid is present as conditions are changed, then solute will not immediately partition into two phases, and the solution will remain in the supersaturated state. The solid state does not necessarily develop spontaneously as the saturation limit is exceeded because energy, or an improbable event, analogous to the activation energy and transition state of a chemical reaction, is required to create the ordered second phase, the stable nucleus of a crystal or a precipitate. Thus a kinetic, or energy barrier allows conditions to proceed further from equilibrium, into the zone of supersaturation. On a phase diagram, like that seen in Fig. 1, the line indicative of saturation is also a boundary that marks the requirement for energy-requiring events to occur in order for a second phase to be established, the formation of the ordered nucleus of a crystal, or the nonspecific aggregate that characterizes a precipitate.

Once a stable nucleus has formed in a supersaturated solution, it will continue to grow until the system regains equilibrium. So long as nonequilibrium prevails and some degree of supersaturation exists to drive events, a crystal will grow, or the precipitate continues to form. It is important to understand the significance of the term "stable nucleus." Many aggregates or nuclei spontaneously form

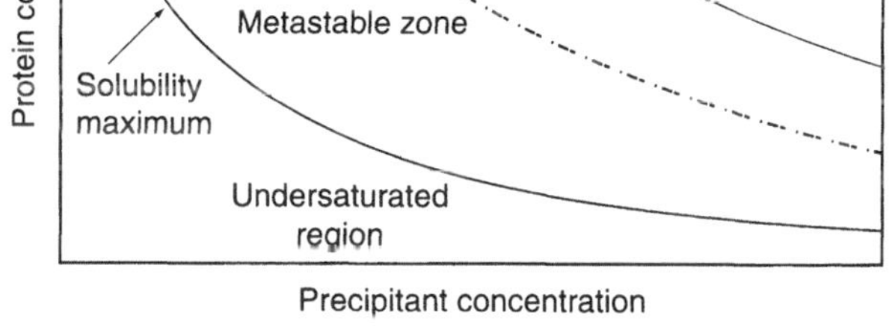

FIGURE 1 The phase diagram for crystallization. It consists of three regions: the undersaturated, the supersaturated, and the equilibrium line which separates them. This line denotes the maximum solubility, and the concentration of solute at which the solid state is in equilibrium with solute molecules in solution. The supersaturated region is divided into a labile region where crystals may nucleate and grow, and a metastable region where crystals are unlikely to nucleate, but if present, can grow.

once supersaturation is achieved, but most are, in general, not "stable." Instead of continuing to develop, they redissolve as rapidly as they form and their constituent molecules return to solution. A stable nucleus is an ordered molecular aggregate of such size and physical coherence that it will enlist new molecules into its growing surfaces faster than others are lost into solution; that is, it will continue to grow so long as the system is supersaturated.

In classical theories describing crystal growth of conventional molecules, the region of supersaturation that pertains above saturation is further divided into what are termed the metastable region and the labile region, as shown in Fig. 1. By definition, stable nuclei cannot form in the metastable region just beyond saturation because the probability is too low. If, however, a stable crystal nucleus or solid is already present in the metastable region, then it can and will continue to grow. The labile region of greater supersaturation is discriminated from the metastable in that stable nuclei can spontaneously form because the probability of nucleation is a function of the degree of supersaturation. Further, because the nuclei are stable they will accumulate molecules and thus deplete the liquid phase of solute until the system passes through the metastable and ultimately reaches the saturated state.

An important point, shown graphically in Fig. 1, is that there are two regions above saturation, one of which can support crystal growth but not formation of stable nuclei, and the other which can yield nuclei as well as support growth. Now the rates of nucleation and crystal growth are both a function of the distance of the solution from the equilibrium position, saturation. Thus a nucleus that forms far from equilibrium and well into the labile region will initially grow very rapidly and, as the solution is depleted of nutrient, move back toward the metastable state. It will grow slower and slower. The nearer the system is to the metastable state when a stable nucleus initially forms, the slower it will proceed to mature.

It might appear that the best approach for obtaining crystals is to press the system as far into the labile region, supersaturation, as possible. There, the probability of nuclei formation is greatest, the rate of growth is greatest, and the likelihood of crystals is maximized. As the labile region is penetrated further, however, the probability of spontaneous and uncontrolled nucleation is also enhanced. Thus, crystallization from solutions in the labile region far from the metastable state frequently results in extensive and uncontrolled "showers" of crystals. By virtue of their number, none is favored and, in general, none will grow to a size suitable for X-ray diffraction studies. In addition, when crystallization is initiated at high supersaturation, then initial growth is extremely rapid. Rapid growth is frequently associated with the occurrence of defects, dislocations, and the incorporation of impurities. Hence, crystals produced from highly saturated solutions tend to be numerous, small, and afflicted with growth defects. In addition, because the nucleation of precipitate, also a solid state, is favored by high supersaturation, the likelihood of its formation, rather than crystals, is promoted.

In terms of the phase diagram, ideal crystal growth would begin with nuclei formed in the labile region, but just beyond the metastable. Here, the growth would occur slowly, the solution, by depletion, would return to the metastable state where no more stable nuclei could form, and the few nuclei that had established themselves would continue to grow to maturity at a pace free of defect formation. Thus, in growing crystals for X-ray diffraction analysis, one attempts, by either dehydration or alteration of physical conditions, to transport the solution into a labile, supersaturated sate, but one as close as possible to the metastable phase.

A. Physical Chemistry of Crystallization

The natural inclination of any system proceeding toward equilibrium is to maximize the extent of disorder, or entropy by freeing individual constituents from physical and chemical constraint. At the same time, there is a thermodynamic requirement to minimize the free energy (or Gibbs free energy) of the system. This is achieved by the formation of chemical bonds and interactions that generally provide negative free energy. Clearly, the assembly of molecules into a fixed lattice severely reduces their mobility and freedom, yet crystals do form and grow.

It follows, then, that crystal nucleation and growth must be dominated by noncovalent chemical and physical bonds arising in the crystalline state that either cannot be formed in solution or are stronger than those that can. These bonds are, in fact, what hold crystals together. They are the energetically favorable intermolecular interactions that drive crystal growth in spite of the resistance to molecular constraint. From this, it is clear that if one wishes to enhance the likelihood of crystal nuclei formation and growth, then one must do whatever is possible to ensure the greatest number of most stable interactions between the molecules in the solid state.

One may ask why molecules should arrange themselves into perfectly ordered and periodic crystal lattices, when they could equally well form random and disordered aggregates, which we commonly refer to as precipitate. The answer is the same as for why solute molecules leave the solution phase at all: to form the greatest number of most stable bonds, to minimize the free energy, or enthalpy of the system. While precipitates represent, in general, a low-energy solid state in equilibrium with a solution phase, crystals, not precipitates, are the states of lowest free energy. A frequently noted phenomenon has been the formation of precipitate followed by its slow dissolution concomitant with the formation and growth of crystals. The converse is not observed. This is one empirical demonstration that crystals represent more favorable energy states.

B. Proteins Create Unique Problems for Crystallization

In principle, the crystallization of a protein, nucleic acid, or virus is little different than the crystallization of conventional small molecules. Crystallization requires the gradual creation of a supersaturated solution of the macromolecule followed by spontaneous formation of crystal growth centers or nuclei. Once growth has commenced, emphasis shifts to maintenance of virtually invariant conditions so as to sustain continued ordered addition of single molecules, or perhaps ordered aggregates, to surfaces of the developing crystal.

The perplexing difficulties that arise in the crystallization of macromolecules, in comparison with conventional small molecules, stem from the greater complexity, lability, and dynamic properties of proteins and nucleic acids. The description offered above of labile and metastable regions of supersaturation are still applicable to macromolecules, but it must now be borne in mind that as conditions are adjusted to transport the solution away from equilibrium by alteration of its physical and chemical properties, the very nature of the solute molecules is changing as well. As temperature, pH, pressure, or solvation is changed, so may be the conformation, charge state, or size of the solute macromolecules. In addition, proteins and nucleic acids are very sensitive to their environment and if exposed to sufficiently severe conditions may denature, degrade, or randomize in a manner that ultimately precludes any hope of their forming crystals. They must be constantly maintained in a thoroughly hydrated state at or near physiological pH and temperature.

Macromolecular crystals are composed of approximately 50% solvent on average, though this may vary over 25–90% depending on the particular macromolecule (Matthews, 1968; McPherson, 1989, 1999). The protein or nucleic acid occupies the remaining volume. The entire crystal is permeated with a network of interstitial spaces through which solvent and other small molecules may freely diffuse.

In proportion to molecular mass, the number of bonds (salt bridges, hydrogen bonds, hydrophobic interactions) that a conventional molecule forms in a crystal with its neighbors far exceeds the few exhibited by crystalline macromolecules. Since these contacts provide the lattice interactions that maintain the integrity of the crystal, this largely explains the difference in properties between crystals of salts or small molecules, and macromolecules, as well as why it is so difficult to grow protein and nucleic acid crystals.

Crystals of small molecules exhibit firm lattice forces, which are highly ordered, generally physically hard and brittle, easy to manipulate, usually can be exposed to air, have strong optical properties, and diffract X-rays intensely. Macromolecular crystals are by comparison usually more limited in size, are very soft and crush easily, disintegrate if allowed to dehydrate, exhibit weak optical properties, and diffract X-rays poorly. Macromolecular crystals are temperature sensitive and undergo extensive damage after prolonged exposure to radiation. In some cases, many crystals must be analyzed for a structure determination to be successful. The extent of the diffraction pattern from a crystal is directly correlated with its degree of internal order. The level of detail to which atomic positions can be determined by a crystal structure analysis corresponds closely with the degree of crystalline order. While conventional molecular crystals often diffract almost to their theoretical limit of resolution, protein crystals by comparison are characterized by diffraction patterns of limited extent.

The liquid channels and solvent cavities that characterize macromolecular crystals are primarily responsible for the limited resolution of the diffraction patterns. Because of the relatively large spaces between adjacent molecules and the consequent weak lattice forces, every molecule in the crystal may not occupy exactly equivalent orientations and positions in the crystal, but they may vary slightly from lattice point to lattice point. Furthermore, because of their structural complexity and their potential for conformational dynamics, protein molecules in a crystal may exhibit slight variations in the course of their polypeptide chains or the dispositions of side groups.

C. Strategic Considerations

The strategy employed to bring about crystallization is to guide the system very slowly toward a state of reduced solubility by modifying the properties of the solvent. This is accomplished by increasing the concentration of precipitating agents, or by altering some physical property such as pH (Aschaffenburg et al., 1965; Bailey, 1940, 1942; Cohn & Ferry, 1950; McPherson, 1982; Northrop et al., 1948). In this way, a limited degree of supersaturation may be achieved. In very concentrated solutions the macromolecules may aggregate as an amorphous precipitate. This result is to be avoided if possible and is indicative that supersaturation has proceeded too extensively or too swiftly. One must endeavor to approach very slowly the point of inadequate solvation and thereby allow the macromolecules sufficient opportunity to order themselves in a crystalline lattice.

For a specific protein, the precipitation points, or solubility minima, are usually dependent on the pH, temperature, the chemical composition of the precipitant, and the properties of both the protein and the solvent. As shown in Fig. 2, at very low ionic strength a phenomenon known as "salting-in" occurs in which the solubility of the protein rises as the ionic strength increases from zero. The physical effect that diminishes solubility at very low ionic strength is the absence of ions essential for satisfying the electrostatic requirements of the protein molecules. As the ions are removed, and in this region of low ionic strength cations are most important (Cohn & Ferry, 1950; Czok & Buecher, 1960), the protein molecules seek to balance their electrostatic requirements through interactions among themselves. Thus, they tend to aggregate and separate from solution. Alternatively, one may say that the chemical activity of the protein is reduced at very low ionic strength. The salting-in effect, when applied in the direction of reduced ionic strength, can itself be used as a crystallization tool. In practice, one extensively dialyzes a protein that is soluble at moderate ionic strength against distilled water. Many proteins have been crystallized by this means.

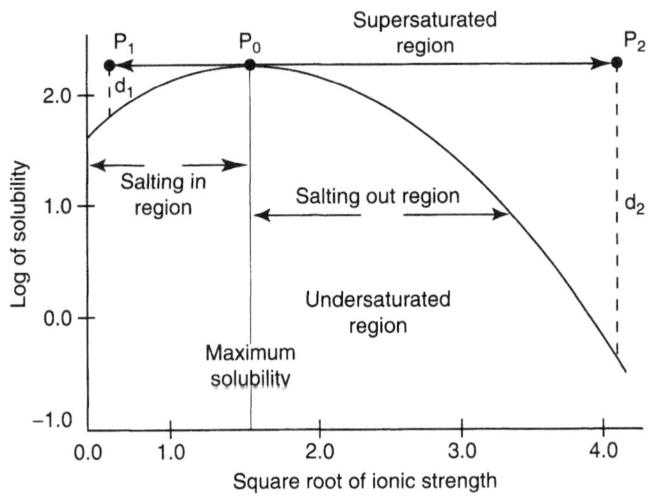

FIGURE 2 The solubility of a protein is indicated by the curve, which has a maximum in this case at about 1.5 M ionic strength. To the left of the maximum, at lower ionic strength, is the "salting-in" region, and to the right, at higher ionic strength, is the "salting-out" region. Supersaturation may be attained by bringing the protein solution to point P_0, clarifying the solution of solid, and then removing ions, thereby moving the solution into the salting-in region to point P_1. Supersaturation may be brought about by adding salt and transforming the protein solution into the salting-out region to point P_2.

As ionic strength in Fig. 2 is increased, the solution again reaches a point where the solute molecules begin to separate from solvent and preferentially form self-interactions among themselves that result in crystals or precipitate. The explanation for this "salting-out" phenomenon is that the salt ions and macromolecules compete for the attention of solvent molecules, that is, water. Both the salt ions and the protein molecules require hydration layers to maintain their solubility. When competition between ions and proteins becomes sufficiently intense, the protein molecules begin to self associate in order to satisfy, by intermolecular interactions, their electrostatic requirements. Thus, dehydration, or the elimination and perturbation of solvent layers around protein molecules, induces insolubility.

Just as proteins may be driven from solution at constant pH and temperature by the addition or removal of ions, they can similarly be crystallized or precipitated at constant ionic strength by changes in pH or temperature, as illustrated in Fig. 3. This is because the electrostatic character of the macromolecule, its surface features, or its conformation may change as a function of pH, temperature, and other variables as well. By virtue of its ability to inhabit a range of states, proteins may exhibit a number of different solubility

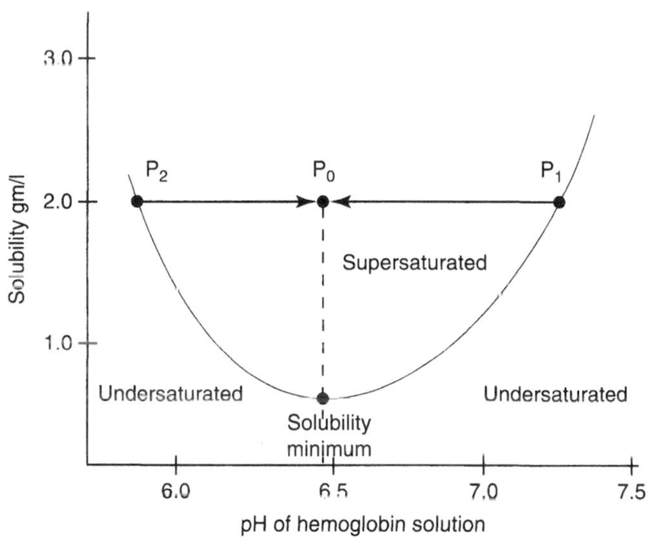

FIGURE 3 The solubility of hemoglobin as a function of pH at constant ionic strength and temperature is shown by the curve. Hemoglobin displays a sharp solubility minimum at about pH 6.5, but is freely soluble at both lower and higher pH. A supersaturated hemoglobin solution (or that of many other proteins) may be created by making a saturated protein solution at high (point P_1) or low (point P_2) pH, clarifying the solution by filtration or centrifugation, and then gradually altering the pH so that the solution is transformed to the point P_0.

minima as a function of the variables, and each of these minima may afford the opportunity for crystal formation. Thus, we may distinguish the separation of protein from solution according to methods based on variation of precipitant at constant pH and temperature from those based on alteration of pH, temperature, or some other variable at constant precipitant concentration. The principles described here for salting-out with a true salt are not appreciably different if polymeric precipitating agents are used. In practice, proteins may equally well be crystallized from solution by increasing the poly (ethylene glycol) concentration at constant pH and temperature, or at constant poly(ethyleneglycol) concentration, by variation of pH or temperature (McPherson, 1976a,b, 1985, 1999).

The most common approach to crystallizing macromolecules, be they proteins or nucleic acids, is to alter gradually the characteristics of a highly concentrated protein solution to achieve a condition of limited supersaturation. As discussed above, this may be achieved by modifying some physical property such as pH or temperature, or through equilibration with precipitating agents. The precipitating agent may be a salt such as ammonium sulfate, an organic solvent such as ethanol or methylpentanediol, or a highly soluble synthetic polymer such as poly(ethyleneglycol). The three types of precipitants act by slightly different mechanisms, though all share some common properties.

Polymers such as poly(ethyleneglycol) also serve to dehydrate proteins in solution as do salts, and they alter somewhat the dielectric properties in a manner similar to organic solvents. They produce, however, an additional important effect. Poly(ethyleneglycol) perturbs the natural structure of the solvent and creates a more complex network having both water and itself as structural elements. The underlying basis for the solvent exclusion effect is that polymeric precipitants, such as PEG, are not like proteins, but lack any fixed or consistent conformation. They writhe and twist randomly in solution, exhibit a large hydrodynamic radius, and occupy far more space than they otherwise deserve. This results in less solvent available space for the other macromolecules, which then segregate, aggregate, and ultimately form a solid state, often crystals. The various approaches to creating supersaturated solutions of biological macromolecules currently in use are summarized in Table I.

D. The Two Stages: Screening and Optimization

There are generally two phases in the pursuit of protein crystals for an X-ray diffraction investigation, and these are (1) the identification of chemical, biochemical, and physical conditions that yield some crystalline material, though it may be entirely inadequate for X-ray diffraction, and (2) the

TABLE I

Strategies for Creating Supersaturation

1.	Direct mixing to immediately create a supersaturated condition (batch method)
2.	Alter temperature
3.	Alter salt concentration (salting-in or -out)
4.	Alter pH
5.	Add a ligand that changes the solubility of the macromolecule
6.	Alteration of the dielectric of the medium
7.	Direct removal of water (evaporation)
8.	Addition of a polymer that produces volume exclusion
9.	Addition of a cross bridging agent
10.	Concentration of the macromolecule
11.	Removal of a solubilizing agent

systematic alteration of those initial conditions by incremental amounts to obtain optimal crystalline samples for diffraction analysis. The first of these bears the greater risk, as some proteins simply refuse to form crystals, and any clues as to why are elusive or absent. The latter, however, often proves to be more demanding, time consuming, and frustrating.

There are basically two approaches to screening for crystallization conditions. The first is a systematic variation of what are believed to be the most important variables, precipitant type and concentration, pH, temperature, etc. Figure 4 illustrates one such strategy for a systematic grid search. The second is what we might term a shotgun approach, but a shotgun aimed with intelligence, experience, and accumulated wisdom. While far more thorough in scope, and more congenial to the scientific mind, the first method usually does require a significantly greater amount of protein. In those cases where the quantity of material is limiting, it may simply be impractical. The second technique provides much more opportunity for useful conditions to escape discovery, but in general requires less precious material.

The second approach also has, presently at least, one other major advantage, and that is convenient. There is currently on the commercial market, from numerous companies, a wide variety of crystallization screening kits. The availability and ease of use of these relatively modestly priced kits, which may be used in conjunction with a variety of crystallization methods (hanging and sitting drop vapor diffusion, dialysis, etc.) make them the first tool of choice in attacking a new crystallization problem. With these kits, nothing more is required than combining a series of potential crystallization solutions with one's protein of interest using a micropipette, sealing the samples, and waiting for success to smile. Often it does, but sometimes not, and this is when the crystal grower must begin using his own intelligence to diagnose the problem and devise a remedy.

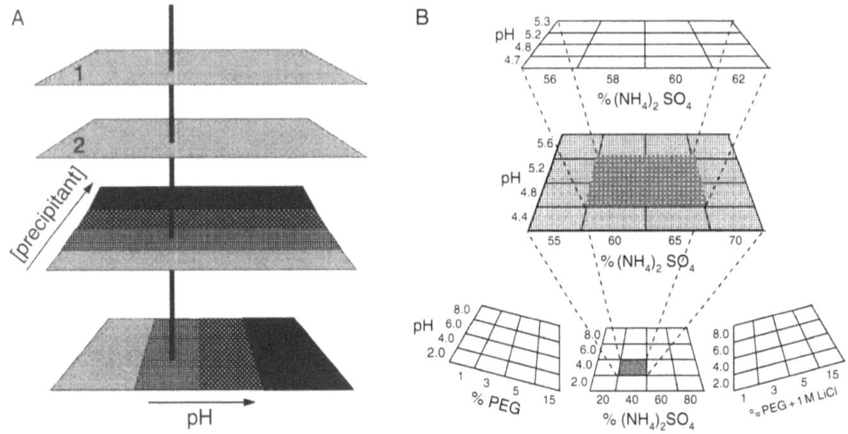

FIGURE 4 Diagram of the successive automated grid searches strategy for protein crystallization (Cox and Weber, 1988). In (A), components of the grid are displayed separately. The bottom square shows the variation in pH across the columns. The square above it shows the variation in precipitant concentration in the rows. The combination of these two layers produces the pH versus precipitant grid that serves as the basis for the two dimensional crystallization strategy. Fixed concentrations of other reagents can be added onto this grid as indicated by the upper squares labeled 1 and 2. The diagram in (B) illustrates how solution parameters are chosen using the approach for protein crystallization. Broad screen experiments (shown at the bottom) are set up using three different precipitating agents. Small ranges of pH and precipitant concentration are centered about droplets-containing crystals.

Once some crystals, even if only microcrystals, are observed and shown to be of protein origin (and one ardently hopes for this event) then optimization begins. Every component in the solution yielding crystals must be noted and considered (buffer, salt, ions, etc.) along with pH, temperature, and whatever other factors (see below) might have an impact on the quality of the results. Each of these parameters or factors is then carefully incremented in focused trial matrices encompassing a range spanning the conditions which gave the "hit." Because the problem is nonlinear, and one variable may be coupled to another, this process is often more complex and difficult than one might expect (McPherson, 1982, 1999). It is here that the amount of protein and the limits of the investigator's patience may prove a formidable constraint.

E. Methodologies for Producing Supersaturation

Practical techniques for creating supersaturation and crystallizing proteins, nucleic acids, and viruses, that have been used up to now, are presented in Table II. Currently, the most widely used method for bringing about

TABLE II

Methods for Attaining a Solubility Minimum

1.	Bulk crystallization
2.	Batch method in vials
3.	Evaporation
4.	Bulk dialysis
5.	Concentration dialysis
6.	Microdialysis
7.	Liquid bridge
8.	Free interface diffusion
9.	Vapor diffusion on plates (sitting drop)
10.	Vapor diffusion in hanging drops
11.	Sequential extraction
12.	pH-induced crystallization
13.	Temperature-induced crystallization
14.	Crystallization by effector addition

supersaturation in microdrops of protein mother liquid is vapor diffusion (Hampel et al., 1968; McPherson, 1976a,b, 1982, 1999). This approach may be divided into those procedures that use a "sitting drop" and those employing a "hanging drop." Whatever its form, the method relies on the transport of either water or some volatile agent between a microdrop of mother liquor, generally 1–10 μl volume, and a much larger reservoir solution of 0.75–25 ml volume. Through the vapor phase, the droplet and reservoir come to equilibrium. Because the reservoir is of much larger volume, the final equilibration conditions are essentially those of the initial reservoir state.

Through the vapor phase, then, water is removed slowly from the droplet of mother liquor, its pH may be changed, or volatile solvents such as ethanol may be gradually introduced. As with other methods, the procedure may be carried out at a number of different temperatures to gain advantage of that parameter as well.

In the popular "sitting drop" and "hanging drop" methods, illustrated in Figs. 5 and 6, respectively, a drop of protein-containing mother liquor, 1–5 μl in volume, is dispensed onto shallow depression in the chamber of a plastic plate. The chamber is contiguous with a reservoir-containing precipitant at higher concentration, or is at a different pH. Through the vapor phase, the concentration of salt or organic solvent in the reservoir equilibrates with that in the sample. In the case of salting-out, the droplet of mother liquor must initially contain a level of precipitant lower than the reservoir, and equilibration proceeds by distillation of water out of the droplet and into the reservoir. This holds true for nonvolatile organic solvents such as methylpentanediol

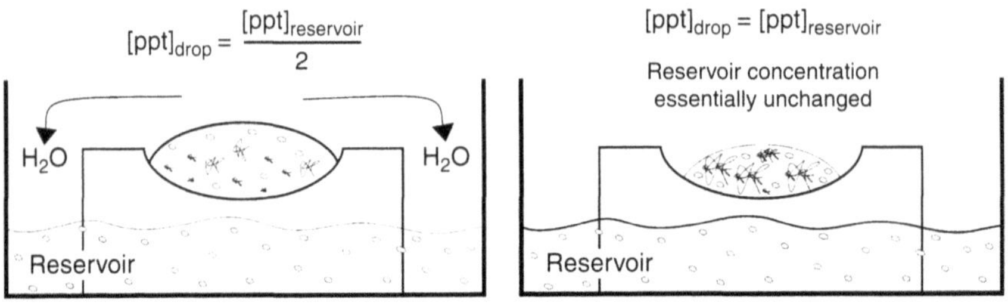

FIGURE 5 In sitting drop vapor diffusion, a small volume of protein sample, combined with, generally, but not necessarily, an equal volume of reservoir solution to make a drop of 1–10 μl. The drop is dispensed onto a small platform, which is contiguous through the vapor phase with a reservoir of much larger volume. Through the vapor phase, the microdroplet establishes equilibrium with the reservoir. By loss of water, as well as concentration of the protein, the droplet is brought to a state of supersaturation.

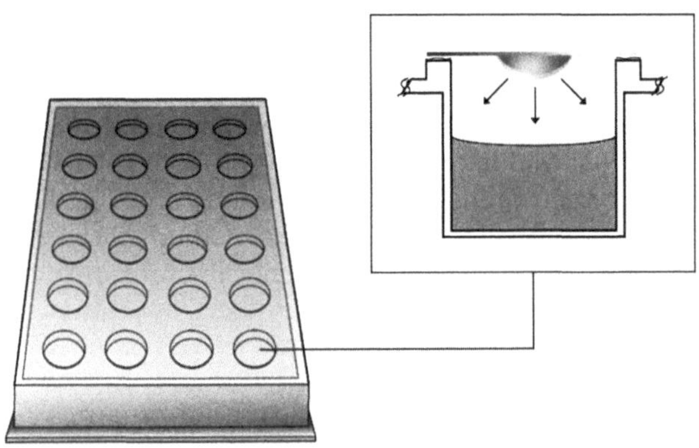

FIGURE 6 Vapor diffusion by hanging drop is essentially identical to that of using sitting drops, which is illustrated in Fig. 5. The major difference is that the protein-containing droplet hangs from the underside of a silicon coated coverslip, which may be of either glass or plastic. The reservoir chambers are provided by 24-well plastic plates.

and for poly(ethylene glycol) as well. In the case of volatile precipitants, none need to be added initially to the microdroplet, as distillation and equilibration proceed in the opposite direction.

In recent years there has been an increased emphasis on large scale crystallization screening experiments involving both large numbers of test conditions, and often large numbers of proteins. In addition, as the proteins addressed by X-ray crystallography have become increasingly more difficult to produce and purify, a premium is now placed on carrying out the crystallization trials with a small amount of material. In response to those pressures, efficient robotic systems have been designed, and are on the market that efficiently and accurately pipette droplets of mother liquor in the nanoliter range. These systems are generally accompanied by automated photoanalysis systems that also speed the examination and evaluation of trial conditions. With these systems, hundreds of trials per day can be deployed, observed, and recorded with virtually no human intervention. When clear plastic plates are used, large numbers of samples can be quickly inspected for crystals under a dissecting microscope and conveniently stored. The plastic ware and instructions for their use are now widely, commercially available. Some devices are shown in Fig. 7.

Initially, the parameters that one needs to establish as rapidly as possible are the appropriate concentrations for precipitants, optimal pH for solubilization and crystallization, and the effect of temperature. The precipitants that should be examined first are poly(ethylene glycol) 3350 and one or more representatives of salts. Probably ammonium sulfate or sodium malonate would be the best choice. They represent the two major classes of precipitants in use. If quantity of protein permits, then the additional two groups, organic solvents and short chain alcohols, should be investigated as well. The best representatives of the latter are ethanol and methylpentanediol, respectively.

A consideration in screening crystallization conditions is minimization of the number of trials that must be carried out. Even in those happy cases where the quantity of protein is not a limitation, reduction of trials means less time and effort. Thus, one seeks to avoid conditions that are certain to be unprofitable. For example, if the protein is observed to precipitate rapidly at salt concentrations greater than 50% saturation, or at pH below 5.0, or at 4 °C, then clearly the trials lying beyond those limits or at that temperature can be eliminated.

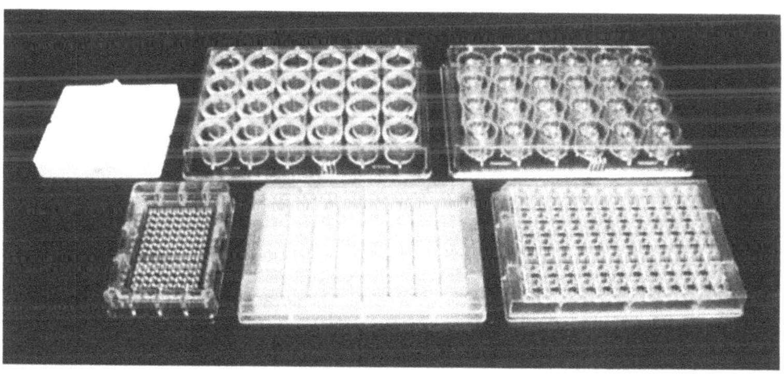

FIGURE 7 An array of commercially available and commonly used plastic plates for both sitting and hanging drop vapor diffusion crystallization. Also in the picture is a box of silicone coated cover slips for hanging drops. Courtesy of Hampton Research.

There are many variables that may be significant in protein crystallization, and a selection is shown in Table III. It is wise to remember, and reassuring as well, that for a specific protein, only a few of these variables may be meaningful. The objective is to determine those which are important, and those which are not. The entire strategy of crystallizing proteins is often a process of picking out those areas of variable space that have some chance of yielding success and intuiting those likely to produce failure. A major difficulty in this pursuit is that only a narrow range of conclusions are possible from each crystallization trial. The mother liquor (a) contains some amount of precipitate, (b) it is clear, (c) there is oiling out, or phase separation, (d) large crystals are present, or (e) microcrystals are present. It is always difficult to know how close a trial, or a set of conditions is to success unless crystals are actually present.

For proteins that are difficult to crystallize, it is essential to take all possible measures to purify the protein free of contaminants and to do whatever is necessary to engender a state of maximum structural and chemical homogeneity. Frequently, we are misled by our standard analytical approaches such as PAGE or IEF into believing that a specific protein preparation is completely homogeneous. This is frequently illustrated for us by distinctive differences in the crystallizability of several preparations, even when all analyses indicate they are identical. Imperceptible differences may be due to degrees of microheterogeneity within preparations that lie at the margin of our ability to detect them. Mild detergents may help achieve these objectives.

The utility of nonionic detergents is an important, if not crucial factor in the crystallization of membrane proteins and has been treated in detail elsewhere (Michel, 1990; Wiener, 2004). It is useful to point out, however, that detergents may be of value in the crystallization of otherwise soluble proteins as well (McPherson et al., 1986). Many protein molecules, particularly when they are highly concentrated and in the presence of precipitating agents such as poly(ethylene glycol) or methylpentanediol, tend to form transient and sometimes metastable, nonspecific aggregates. The existence of a spectrum of varying sizes, shapes, and charges presents problems not appreciably different from the crystallization of a protein from a heterogeneous mixture or an impure solution composed of dissimilar macromolecules. An objective in crystallizing proteins is to limit the formation of nonuniform states and reduce the population to a set of standard individuals that can form identical interactions with one another.

Because the key to successfully crystallizing a macromolecule often lies in the procedure, means, or solvent used to solubilize it, some careful consideration should be given to this initial step. This is particularly true of membrane, lipophilic, or other proteins which, for one reason or another, are only marginally soluble in water solutions. In addition to mild detergents there are

TABLE III

Factors Effecting Crystallization

Physical	Chemical	Biochemical
1. Temperature/temperature variation	1. pH	1. Purity of the macromolecule/impurities
2. Surfaces	2. Precipitant type	2. Ligands, inhibitors, effectors
3. Methodology/approach to equilibrium	3. Precipitant concentration	3. Aggregation state of the macromolecule
4. Gravity	4. Ionic strength	4. Posttranslational modifications
5. Pressure	5. Specific ions	5. Source of macromolecule
6. Time	6. Degree of supersaturation	6. Proteolysis/hydrolysis
7. Vibrations/sound/mechanical perturbations	7. Reductive/oxidative environment	7. Chemical modifications
8. Electrostatic/magnetic fields	8. Concentration of the macromolecules	8. Genetic modifications
9. Dielectric properties of the medium	9. Metal ions	9. Inherent symmetry of the macromolecule
10. Viscosity of the medium	10. Crosslinkers/polyions	10. Stability of the macromolecule
11. Rate of equilibration	11. Detergents/surfactants/amphophiles	11. Isoelectric point
12. Homogeneous or heterogeneous nucleants	12. Nonmacromolecular impurities	12. History of the sample

TABLE IV

Some Important Principles

1. *Homogeneity*: Begin with as pure and uniform as population of a molecular species as possible; purify
2. *Solubility*: Dissolve the macromolecule to a high concentration without the formation of aggregates, precipitate, or other phases
3. *Stability*: Do whatever is necessary to maintain the macromolecules as stable and unchanging as possible
4. *Supersaturation*: Alter the properties of the solution to obtain a system which is appropriately supersaturated with respect to the macromolecule
5. *Association*: Try to promote the orderly association of the macromolecules while avoiding precipitation, nonspecific aggregation, or phase separation
6. *Nucleation*: Try to promote the formation of a few critical nuclei in a controlled manner
7. *Variety*: Explore as many possibilities and opportunities as possible in terms of biochemical, chemical, and physical parameters
8. *Control*: Maintain the system at an optimal state, without fluctuations or perturbations, during the course of crystallization
9. *Impurities*: Discourage the presence of impurities in the mother liquor, and the incorporation of impurities and foreign materials into the lattice
10. *Preservation*: Once the crystals are grown, protect them from shock and disruption, maintain their stability

chaotropic agents that can also be employed for the solubilization of proteins. These include compounds such as urea, guanidinium hydrochloride, and relatively innocuous anions such as SCN^-, ClO_4, I^-, Br, and NO (Hatefi & Hanstein, 1969). These compounds, even at relatively low concentrations, may serve to increase dramatically the solubility of a protein under conditions where it would otherwise be insoluble.

Table IV presents 10 principles that have, over the years, been found to bear heavily on the success or failure of the crystallization enterprise. It is wise to keep them in mind, or refer to them occasionally as one embarks on a new crystallization venture.

References

Aschaffenburg, R., Green, D. W., & Simmons, R. M. (1965). Crystal Forms of Beta-lactoglobulin. *Journal of Molecular Biology, 13*, 194–201.

Bailey, K. (1940). A Crystalline Albumin Component of Skeletal Muscle. *Nature, 145*, 934–935.

Bailey, K. (1942). Some Methods for the Preparation of Large Protein Crystals. *Transactions of the Faraday Society, 38*, 186–191.

Cohn, E. J., & Ferry, J. D. (1950). *Proteins, amino acids and peptides.* New York: Reinhold.

Cox, M. J., & Weber, P. C. (1988). Efficient Optimization of Crystallization Conditions by Manipulation of Drop Volume Ratio and Temperature. *Journal of Crystal Growth, 90*, 318–324.

Czok, R., & Buecher, T. (1960). Crystallized enzymes from the myogen of rabbit skeletal muscle. *Advances in Protein Chemistry, 15*, 315–415.

Hampel, A., Labanauskas, M., Connors, P. G., Kirkegard, L., RajBharday, U. L., & Siglar, P. B. (1968). Single crystals of transfer RNA from formylmethionine and phenylalanine transfer RNA's. *Science, 162*(860), 1384–1387.

Hatefi, Y., & Hanstein, W. G. (1969). Solubilization of particulate proteins and nonelectrolytes by chaotropic agents. *Proceedings of the National Academy of Sciences of the United States of America, 62*(4), 1129–1136.

Matthews, B. W. (1968). Solvent content of protein crystals. *Journal of Molecular Biology, 33*(2), 491–497.

McPherson, A., Jr. (1976a). Crystallization of proteins from polyethylene glycol. *The Journal of Biological Chemistry, 251*(20), 6300–6303.

McPherson, A., Jr. (1976b). The growth and preliminary investigation of protein and nucleic acid crystals for X-ray diffraction analysis. *Methods of Biochemical Analysis, 23*(0), 249–345.

McPherson, A. (1982). *The preparation and analysis of protein cystals.* New York: Wiley.

McPherson, A. (1985). Crystallization of macromolecules: General principles. *Methods in Enzymology, 114*, 112–120.

McPherson, A. (1989). Macromolecular crystals. *Scientific American, 260*(3), 62–69.

McPherson, A. (1999). *Crystallization of biological macromolecules.* Cold Spring Harbor, NY: Cold Spring Harbor Press.

McPherson, A., Koszelak, S., Axelrod, H., Day, J., Williams, R., Robinson, L., et al. (1986). An experiment regarding crystallization of soluble proteins in the presence of beta-octyl glucoside. *The Journal of Biological Chemistry, 261*(4), 1969–1975.

Michel, H. (1990). *Crystallization of membrane proteins.* CRC Press, Boca Raton, FL.

Northrop, J. H., Kunitz, M., & Harriott, R. (1948). *Crystalline enzymes.* New York: Columbia University Press.

Wiener, M. C. (2004). A pedestrian guide to membrane protein crystallization. In A. McPherson, (Ed.), Academic Press, London, *Methods,* Vol. 34, #3, pp. 364–372.

CHAPTER 2

In Vitro Synthesis of Posttranslationally Modified Membrane Proteins

Stefan Kubick,* Michael Gerrits,* Helmut Merk,* Wolfgang Stiege,* and Volker A. Erdmann[†]

*RiNA GmbH, Takustrasse 3, D-14195 Berlin, Germany
[†]Institut für Chemie und Biochemie, Freie Universität Berlin, Thielallee 63, D-14195 Berlin, Germany

I. OVERVIEW

Membrane proteins have become an important focus of the current efforts in structural and functional genomics and the rapid progress of various genome sequencing projects has greatly accelerated the discovery of novel genes encoding membrane proteins. In contrast, the molecular analysis of

Current Topics in Membranes, Volume 63 1063-5823/09 $35.00

DOI: 10.1016/S1063-5823(09)63002-7

membrane proteins lags far behind that of cytosolic soluble proteins. Preparing high quality samples of functionally folded proteins represents a major bottleneck that restricts further structural and functional studies.

Cell-free protein expression systems, in particular those of eukaryotic origin, have recently been developed as promising tools for the rapid and efficient production of a wide variety of membrane proteins. A large number of these proteins, however, require posttranslational modifications for optimum function. Several membrane proteins have been expressed *in vivo* to date, most of them being functionally, antigenically, and immunogenically similar to their authentic counterparts. This is mainly due to the properties of cultured eukaryotic cells, which are able to carry out many types of posttranslational modifications such as the addition of N- and O-linked oligosaccharides, but also palmitoylation, myristylation, and phosphorylation.

Based on these versatile properties of cultured cell lines, we have developed a technique for the standardized production of translationally active eukaryotic lysates from insect cells. In contrast to other cell-free protein synthesis systems (e.g., rabbit reticulocyte lysates and wheat germ extracts) our homogenization procedure avoids any serious breakdown of membrane vesicles already existing in the cytoplasm of the prepared eukaryotic cells. We have demonstrated the functional integrity of these subcellular components by showing signal peptide cleavage as well as glycosylation of *in vitro* expressed membrane proteins.

The development of this novel eukaryotic *in vitro* translation system now expands the possibilities of cell-free protein synthesis, since posttranslational modifications significantly alter the physical and chemical properties of proteins, including their folding and conformational distribution and these modifications are frequently a fundamental prerequisite for functional activity.

II. INTRODUCTION

Membrane proteins represent approximately 30% of the total proteins from an organism and an increasing number of membrane protein sequences without attributed function are continuously discovered in various genome-sequencing projects. Integral membrane proteins are involved in essential biochemical processes as they perform critical roles in the cell cycle by regulating signaling, metabolism, transport, and recognition. Dysregulation of their biological activity in response to bacterial or viral infection, as well as in cancer often leads to severe diseases. Considering that these proteins play a central role in drug discovery as potential pharmaceutical targets, it is now imperative to go deeper into their structure to understand their molecular and physiological function. However, membrane proteins are highly underrepresented in structural data banks due to tremendous difficulties that occur upon approaching their structural and functional analysis.

One of the major difficulties in the study of membrane proteins is to recover suitable amounts of recombinant proteins in correctly folded structures, displaying proper functions and activities. Cotranslational integration of membrane proteins into an appropriate lipid bilayer constitutes a critical step often necessary for obtaining native-like and functionally active membrane proteins. Unfortunately, membrane integration of recombinant membrane proteins upon their production in conventional cell culture systems of bacterial, yeast, mammalian, or insect origin often affects the integrity of the cellular membranes resulting in growth retardation or even lysis of the host cells. Blocking cellular transport and posttranslational processing systems by overproduced heterologous membrane proteins may cause further toxic effects. Examples of successfully overproduced membrane proteins in preparative amounts are relatively rare and even frequently associated with the aggregation and inactivation of the recombinant membrane protein by inclusion body formation or proteolytic degradation.

Any attempt to develop a new method for expressing membrane proteins must consider these challenging problems. In this context, cell-free translation platforms are gaining importance in structural as well as functional genomics as these systems do not depend on cellular integrity and can therefore meet the increasing demands for the synthesis of "difficult-to-express" membrane proteins. In general, cell-free systems offer several advantages over traditional cell-based expression methods, including easy adaptation of reaction conditions to favor protein folding, decreased sensitivity to product toxicity, and suitability for high-throughput strategies. Significant improvements made to the configuration, energetics, and robustness of *in vitro* translation reactions in conjunction with the general advantages of these systems, have led to achievements such that cell-free systems have become powerful tools to synthesize any desired protein, including native proteins, proteins toxic to living cells, and artificially modified proteins.

Eukaryotic *in vitro* translation systems in particular, have generated increased interest in their use for tackling fundamental problems in biochemistry and pharmacology. Rabbit reticulocyte lysates and wheat germ extracts, for example, are widely used to characterize proteins and investigate mRNA translational mechanisms (Endo & Sawasaki, 2003; Erickson & Blobel, 1983; Jackson & Hunt, 1983; Madin, Sawasaki, Ogasawara, & Endo, 2000; Pelham & Jackson, 1976). Both systems are suitable to produce correctly folded eukaryotic proteins but have significant shortfalls. For expression of integral membrane proteins or to introduce posttranslational modifications, the reticulocyte lysate requires the addition of a heterologous membrane fraction.

Furthermore, the quality of rabbit reticulocyte lysates are subject to variation as they are based on cells collected from living animals. The wheat germ system can be scaled up but production of posttranslationally modified proteins is limited (Vinarov et al., 2004). As this system is prepared from different natural sources, it shows batch-to-batch variations in its translational activity. Additionally, lysates prepared from mammalian sources, such as Ehrlich ascite cells, human HeLa, or mouse L-cells, have also given us efficient tools to study the synthesis and cell-free assembly of multiple proteins, in particular the *in vitro* generation of virus particles from mRNA (Bergamini, Preiss, & Hentze, 2000; Molla, Paul, & Wimmer, 1991).

The development of cell-free translation systems from selected cell lines has recently reached a very important stage, that is, the large-scale production of eukaryotic lysates displaying properties which are optimized for the synthesis of a wide range of structurally and functionally divergent membrane proteins. In this chapter, we describe such a system and detail how we have taken a widely used *in vivo* expression system, based on baculovirus-infected insect cells, and developed a standardized method for production of lysates which are suitable for *in vitro* translation and posttranslational modification of membrane proteins.

Baculovirus expression systems are frequently used for large-scale *in vivo* production of recombinant membrane proteins required for research and clinical applications (Midgett & Madden, 2007). High levels of protein expression are achieved in these systems, based on the use of recombinant viruses which are generated by homologous recombination between baculovirus genomic DNA and a cotransfected plasmid that harbors the foreign gene in insect cells (Luckow & Summers, 1988; O'Reilly, Miller, & Luckow, 1992; Summers & Smith, 1987). During the late phase of infection, the inserted heterologous genes are placed under the transcriptional control of the strong viral polyhedrin promoter, and recombinant products are expressed in place of the naturally occurring polyhedrin protein. The major advantage of the baculovirus expression system is the impressive level of recombinant proteins obtained in virus-infected insect cells, which often exceeds 100 mg/l (Hill-Perkins & Possee, 1990; Smith, Summers, & Fraser, 1983; Vlak et al., 1988).

Several hundred genes have been expressed in this way to date, most of them being functionally, antigenically, and immunogenically similar to their authentic counterparts. This is mainly due to the properties of the insect cell's subcellular transport and posttranslational processing system. Cultured insect cells are able to carry out many types of posttranslational modifications such as addition of N- and O-linked oligosaccharides, but also palmytoylation, myristylation, and phosphorylation. Taken together, such insect cells represent a promising resource for the preparation of novel cell-free membrane protein synthesis systems.

III. MATERIALS AND METHODS

Kits for template generation and cell-free protein expression were codeveloped by RiNA GmbH and Qiagen GmbH. Amplification of templates, cloning of expression constructs, cell-free protein synthesis, and subsequent analysis of expressed proteins was performed at RiNA GmbH (http://www.rina-gmbh.de/).

A. Construction of DNA Templates

Proteins were expressed using either PCR products or plasmid DNA as template. Appropriate cDNAs were amplified by PCR and cloned into an expression vector (pIX4.0 Vector, Qiagen). Alternatively, linear expression constructs were generated by PCR using the EasyXpress Linear Template Kit PLUS (Qiagen), following the instructions given in the kit's manual. Replacement of the natural signal sequence with the *mellitin* signal sequence was achieved using the expression-PCR system (Linear Template Kit Signal Peptide, RiNA GmbH).

Genomic DNA was isolated from cultured HeLa cells using trizol reagent (Invitrogen).

B. Protein Expression

Cell-free protein expression was performed using lysates from *Spodoptera frugiperda* (*Sf*) cells. Preparation of lysates are described later in this chapter. The depicted insect lysates have been developed further for commercialization (EasyXpress Insect Kit II, Qiagen). Transcription and translation of membrane proteins were performed using the EasyXpress Insect Kit II. *In vitro* translation reactions were supplemented with 25% (v/v) *Sf* lysate, complete amino acids (200 μM), ^{14}C-labeled leucine, and energy-regenerating components, according to the manufacturer's instructions (high-yield protocol, Qiagen). Translation reactions were performed in a 50 μl volume in a thermomixer.

C. Protein Analysis

Protein yields were estimated by hot trichloroacetic acid (TCA)-precipitation of 5 μl aliquots of the translation reaction after 90 min of incubation at 27 °C (duplicate analysis). To determine homogeneity and size of *in vitro* translated proteins, 7.5 μl aliquots of radiolabeled cell-free synthesis reactions were subjected to 15% SDS-PAGE. Subsequently, the gel was dried and radioactively labeled proteins were visualized with a phosphoimager (Typhoon, Amersham).

D. Protein Deglycosylation

Protein glycosylation was analyzed by protein shift after enzymatic deglycosylation of the translated protein or by tunicamycin-mediated inhibition of glycosylation during translation (Duksin & Mahoney, 1982). Glycosylated and deglycosylated proteins were subjected to 15% SDS-PAGE. PNGase F and Endo H deglycosidases were from New England Biolabs and used according to the manufacturer's recommendations. Tunicamycin was from Sigma-Aldrich and applied prior to the protein synthesis reaction in a final concentration of 10 μg/ml reaction volume to inhibit protein glycosylation.

E. Luciferase Activity Assay

Luciferase activity was determined using the Luciferase Assay System (Promega) according to the manufacturer's recommendations.

IV. RESULTS

A. Extract Preparation

High-yield *in vivo* expression of biologically active recombinant protein is frequently achieved in cell lines derived from the fall army worm *Sf* or from the cabbage looper *Trichoplusia ni* (Vaughn, Goodwin, Tompkins, & McCawley, 1977). Recombinant baculoviruses containing the gene of interest are usually propagated in these insect-derived cell lines. Particular insect cell lines, for example, *Sf* cells, grow well in suspension cultures and these cells can be easily scaled up for the large scale production of recombinant proteins. Therefore, *Sf* cells were grown in well controlled fermenters at 27 °C in an animal component-free insect cell medium. During a period of exponential growth, at a density of approximately 4×10^6 cells/ml, *Sf* cells were collected by centrifugation and washed with a HEPES-based homogenization buffer consisting of 40 mM HEPES-KOH (pH 7.5), 100 mM KOAc, and 4 mM DTT. Finally, the *Sf* cell pellet was resuspended in an appropriate volume of homogenization buffer to achieve a cell density of approximately 2×10^8 cells/mm. Resuspended *Sf* cells were lysed mechanically and the homogenate was centrifuged at 10,000 × *g* for 10 min at 4 °C to spin out the nuclei and debris. The resulting supernatant was applied to a Sephadex G-25 column and fractions with the highest RNA/protein concentrations were pooled. Aliquots of the *Sf* lysate were immediately frozen in liquid nitrogen and then stored at −80 °C to preserve maximum activity. Due to this mild treatment, the obtained final extracts retain intact subcellular membranous structures derived from the endoplasmic reticulum (ER). These vesicular structures are an essential prerequisite for the expression of functionally active membrane proteins and their subsequent posttranslational modification.

B. In Vitro Translation

Efficient cell-free translation systems are mainly based on crude cell extracts, including ribosomes, tRNAs, soluble enzymes, and factors essential for initiation and elongation. This primary lysate must be supplemented with amino acids, ribonucleoside triphosphates (NTPs), and an NTP-regenerating system. Upon addition of exogenous messenger RNA and incubation of this *in vitro* translation reaction at the appropriate temperature, some initial translational activity might occur. However, protein synthesis is a complex multistep pathway that can be exerted at many levels, and careful optimization of reaction parameters is essential to obtain an effective *in vitro* translation system. Translation in eukaryotes is primarily regulated at the initiation step. Among the factors involved in translation initiation, eukaryotic translation initiation factors 4F (eIF4F) and eIF2 play pivotal roles in translational regulation (Holcik & Sonenberg, 2005). The activity of the protein synthesis machinery mainly depends on the phosphorylation status of the initiation factor eIF2 (Farrell, Balkow, Hunt, & Jackson, 1977; Singh, Aroor, & Wahba, 1994; Welsh, Miller, Loughlin, Price, & Proud, 1998). Changes in energy charge strongly influence initiation and elongation of protein synthesis. In particular, the rate of translation initiation is highly sensitive to changes in the ADP:ATP and GDP:GTP ratios (Rupniak & Quincey, 1975). Cell-free protein synthesis requires biochemical energy, supplied by the hydrolysis of nucleoside triphosphates. To ensure efficient translation, the ATP and GTP concentration is maintained by an energy-regeneration system. Eukaryotic *in vitro* translation systems therefore usually use the creatine phosphate-creatine phosphokinase ATP regenerating system. *In vitro* translation reactions performed in the fixed volume of a test tube (batch format) usually stop within 20-60 min as essential substrate is exhausted, or any product reaches inhibitory concentrations. In contrast, the eukaryotic system for cell-free protein synthesis described here, shows prolonged translational activity. The linked transcription-translation procedure can be performed in two modes. In the high-throughput mode, an aliquot of the transcription step is directly pipetted into the extract to initiate translation, thus enabling multiparallel and automated processing of various reactions. The high-yield mode is designed to maximize protein yield which is achieved by an intermediate gel filtration step to clean up the mRNA prior to addition to the cell-free extract. Coupled transcription-translation reactions combine the RNA polymerase with the insect cell lysate and the reaction starts upon addition of linear or circular DNA (Fig. 1). For the production of proteins in a larger scale, the reaction volume can be linearly scaled up into the milliliter range.

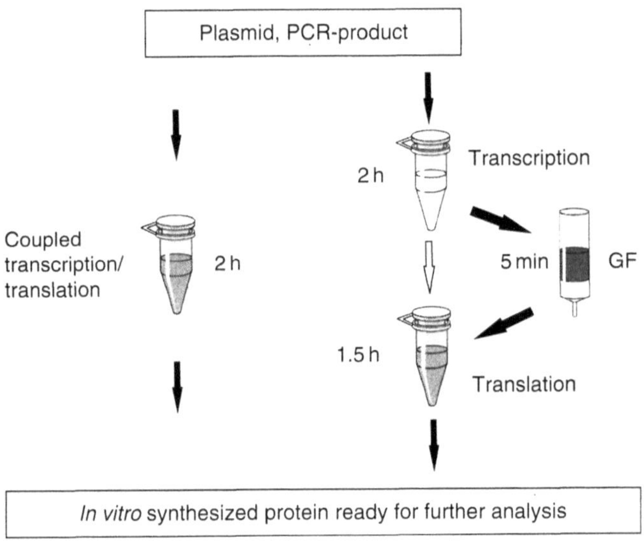

FIGURE 1 General principle of cell-free protein synthesis in insect cell extracts. The linked transcription-translation procedure is performed in the high-yield mode. An aliquot of the initial transcription step is purified by an intermediate gel filtration (GF) step (DyeEx spin columns, Qiagen) to clean up the mRNA prior to addition to the cell-free extract. Proteins can be expressed from a variety of DNA templates including circular and linearized plasmid DNA as well as PCR Products (generated using the EasyXpress Linear Template Kit Plus). For production of proteins in a larger scale the reaction volume can be linearly scaled up into the milliliter range. In the high-throughput mode, an aliquot of the transcription step is directly pipetted into the extract to start the translation, thus enabling multiparallel and automated processing of various reactions. Alternatively, the coupled transcription-translation reaction combines the RNA polymerase with the insect cell lysate to form a single master mix. The reaction starts upon addition of the linear or circular DNA-template.

A key goal for *in vitro* translation is to synthesize properly folded and biologically active proteins. At this point cell-free systems have a clear advantage over *in vivo* protein synthesis: Environmental conditions can be adjusted easily in these "open" systems and strategies to improve protein folding include the addition of a variety of reagents and folding catalysts to the reaction. To investigate the functionality of protein expression in the insect cell-free system, firefly luciferase enzymatic activities and total protein yields were analyzed (Fig. 2A). Full activity of luciferase is known to be dependent on the eukaryotic folding machinery as refolding of the denatured protein requires the Hsp90 chaperone (Schneider et al., 1996). Synthesis was performed under optimized conditions in batch formatted reactions. A maximum yield of novel synthesized and functionally active protein was reached after 90–180 min. Up to 50 μg active luciferase per ml reaction volume can be expressed as determined by a multiwell plate-based activity assay. The significant portion of 90% of the total expressed protein was determined to be functional and synthesized luciferase remained active in the lysate for more than 4 h. A slight decrease in functionally active protein can be detected after prolonged incubation of 6 h, which may be due to endogenous protease activity (Fig. 2B).

Further alternatives to optimize the conditions of the translation reaction have been investigated. In particular, the temperature dependence of the protein-synthesis yield and luciferase activity has been analyzed in the insect lysate (Fig. 3). Cell-free systems enable protein synthesis at a broad temperature range whereas *in vivo* expression systems are restricted to the requirements of the utilized organism. Luciferase expression in the insect lysate results in highest yields of functionally active protein at an incubation temperature of 30 °C which differs from the optimal growth temperature of 27 °C for cultured *Sf* cells. These experiments emphasize the potential of insect cell lysates as a system for cell-free expression of structurally divergent proteins to clarify the question how these proteins are assembled and how to think about the thermodynamics of the stability of protein complexes.

C. Template Generation

The design of an expression template has significant influence on the efficiency of heterologous protein production. This fact may be due to mRNA secondary structure, leading to inefficient initiation of translation. Otherwise, secondary structures may lead to increased mRNA stability and substantially higher protein yields. Tag sequences and positions of these tags may also influence the yield and solubility of the expressed protein (Zacharias et al., 2004). Furthermore, some full-length proteins cannot be expressed heterologously, while a truncated mutant of these proteins leads to successful

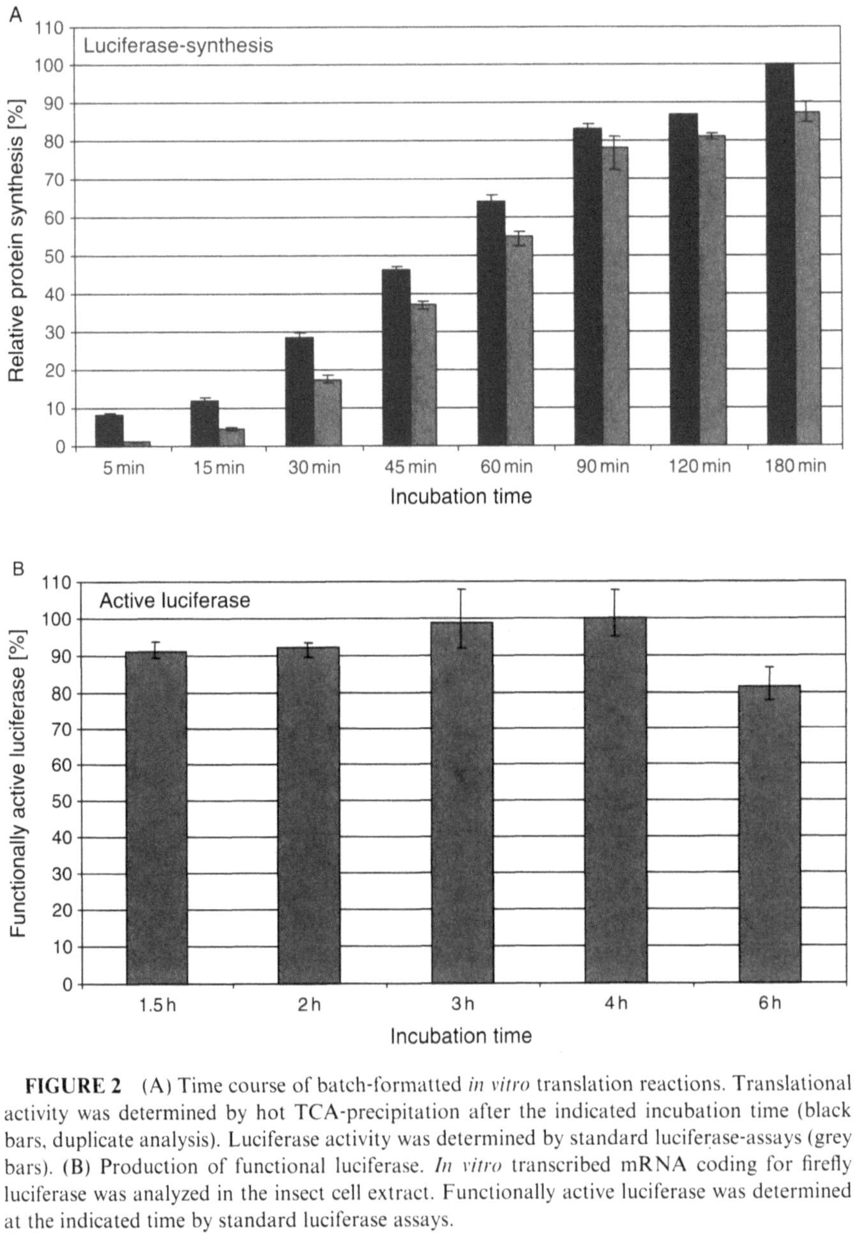

FIGURE 2 (A) Time course of batch-formatted *in vitro* translation reactions. Translational activity was determined by hot TCA-precipitation after the indicated incubation time (black bars, duplicate analysis). Luciferase activity was determined by standard luciferase-assays (grey bars). (B) Production of functional luciferase. *In vitro* transcribed mRNA coding for firefly luciferase was analyzed in the insect cell extract. Functionally active luciferase was determined at the indicated time by standard luciferase assays.

protein expression (Cornvik et al., 2005). These observations point out that evaluation of different constructs is required to successfully express a heterologous protein. Subsequent screening of several templates in cell-free

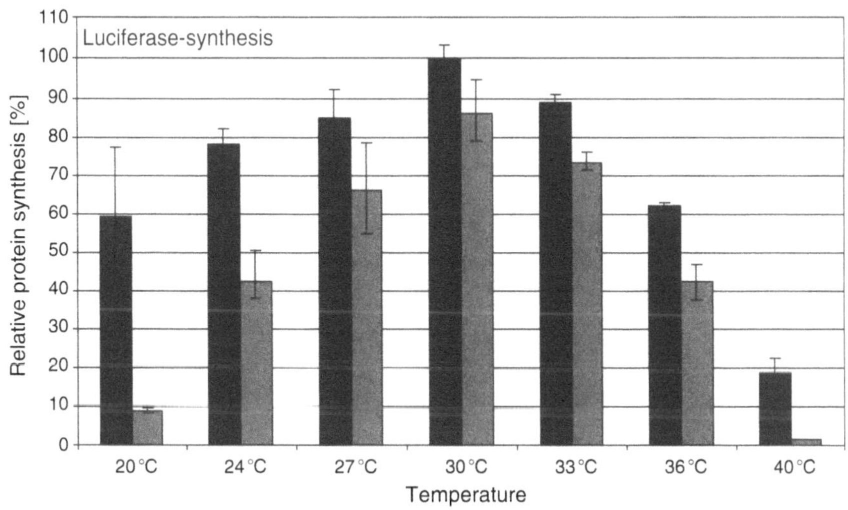

FIGURE 3 Temperature-dependent protein synthesis in insect cell lysates. *In vitro* translation reactions were composed of 25% (v/v) lysate, mRNA encoding firefly luciferase, complete amino acids (200 μM), including ^{14}C-labeled leucine and energy regenerating components. Translational activity was determined by hot TCA-precipitation of 5 μl aliquots after the incubation time of 90 min at the indicated temperature (black bars, duplicate analysis). Functionally active Luciferase was determined by standard luciferase-assays (grey bars).

systems is a powerful tool to speed up protein production projects significantly. The generation of linear templates by PCR is a fast and convenient methodology to determine the optimal expression construct. The so-called Expression-PCR (E-PCR) was originally developed for prokaryotic cell-free systems and this methodology is based on the amplification of a single gene within a two-step PCR, while simultaneously supplying the PCR product with the necessary regulatory elements for transcription and translation and optional elements for purification (Merk, Meschkat, & Stiege, 2003). The T7 promoter and an additional sequence stretch for unhindered transcription, a hairpin sequence protecting the 5′-end of the mRNA against degradation, the epsilon sequence, and the ribosome binding site from T7 gene10, as well as a spacer sequence to the translation start codon ATG, are introduced upstream of the translated gene. A further spacer sequence is introduced downstream of the translation stop codon which separates the following T7 transcription terminator from the translated sequence. Additional sequences coding for N- or C-terminal affinity tags can be introduced selectively into the template (Fig. 4). The E-PCR comprises a two step reaction. In the first PCR-step, two gene-specific primers are used which hybridize to the target gene as well as to long and universally applicable primers of the second PCR

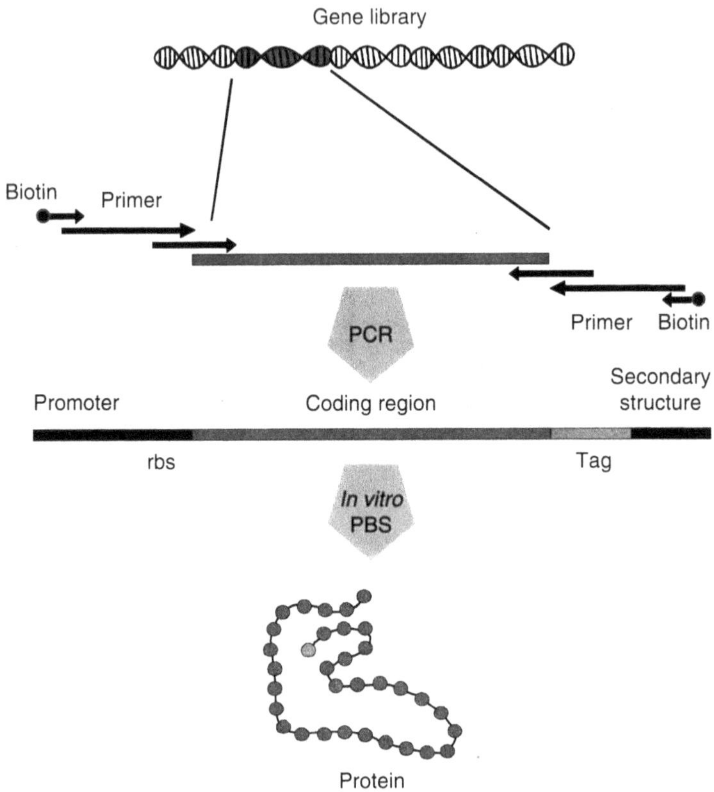

FIGURE 4 Generating PCR Products suitable for *in vitro* translation reactions. The Expression-PCR system (E-PCR) uses a two-step procedure to generate PCR products suitable for translation in prokaryotic as well as in eukaryotic cell-free systems (EasyXpress Linear Template Kit Plus, EasyXpress *E. coli* and Insect-based Kits, Qiagen). In the first PCR step, the target gene is amplified with gene-specific primers. The second PCR step completes the product of the first PCR step by adding essential regulatory sequences for transcription and translation as well as a sequence for an affinity tag. Long primers used in the second PCR step are universally applicable. A biotin-based stabilization strategy allows survival of the linear template in the lysate. E-PCR products can be used without any further purification in cell-free systems.

step. Each of these long PCR primers has a length of approximately 100 bases coding for the necessary regulatory elements for protein expression. Two additional primers complete the PCR product while introducing restriction endonuclease recognition sites enabling an optional cloning step. Linear templates generated by E-PCR are suitable templates for the expression in *Escherichia coli*-based cell-free systems as well as in insect cell lysate-based *in vitro* translation systems. Additionally, insect cell expression, for example, in baculovirus/*Sf*-systems, can now be accelerated by using cell-free screening approaches. Utilizing E-PCR products as templates in cell-free systems saves time by eliminating several rounds of tedious and time consuming transfections, virus constructions, and cell culture optimizations. The best construct determined in the *in vitro* screening approach can subsequently be used for efficient baculovirus construction and cloning into a vector suitable for upscaling of protein production both *in vitro* and *in vivo*.

D. Cell-Free Synthesis of Membrane Proteins

One of the greatest and potentially most far-reaching impacts of cell-free protein synthesis is predicted to be in the area of membrane protein production. We analyzed the cell-free synthesis of the type 1 transmembrane protein heparin-binding EGF-like growth factor (HB-EGF) in the insect cell-based lysate. The HB-EGF gene was fused N-terminally to enhanced yellow fluorescent protein (EYFP) to develop an efficient fluorescent probe for *in vitro* translation studies. In addition to the phosphorylation status of translational components, the concentrations of Mg^{2+} and K^+ ions are critical parameters for the high-yield expression of proteins in cell-free systems. The range of Mg^{2+} concentration for optimal translation in eukaryotic cell-free systems is very narrow, and therefore small changes in Mg^{2+} levels can dramatically affect the efficiency and fidelity of translation. In a series of translation reactions, we tested various concentrations of Mg^{2+} in *Sf* lysates. Highest translational activities were observed at a well-defined optimum of 2.5 mM $Mg(OAc)_2$ (Fig. 5). In contrast, translation in HeLa extracts occurs at higher Mg^{2+} concentrations (2.5–4 mM $Mg(OAc)_2$), whereas *Drosophila* extracts require a significantly lower Mg^{2+} optimum (0.2–0.6 mM Mg $(OAc)_2$) (Bergamini et al., 2000; Castagnetti, Hentze, Ephrussi, & Gebauer, 2000; Gebauer, Corona, Preiss, Becker, & Hentze, 1999; Lie & Macdonald, 2000).

In vitro expression of membrane proteins was further investigated in insect cell lysates applying the optimal conditions determined in previously described experiments. Therefore, E-PCR product-based constructs, encoding a broad range of structurally and functionally divergent transmembrane proteins, were used as expression templates. The sizes of the encoded integral membrane proteins ranged from 30 to 120 kDa. Cell-free expression efficiency was analyzed qualitatively and quantitatively by autoradiography and by TCA-precipitation of the radiolabeled protein. All membrane proteins could be produced successfully in the *Sf* lysate (Fig. 6). Pharmacologically relevant G-protein coupled receptors, for example, the $\beta2$-adrenergic receptor, the full-length cannabinoid receptor 1, and endothelin receptor were synthesized in the cell-free system. Furthermore, ion channels and carriers, for example, transient receptor potential cation channels (TRPV1, TRPV4)

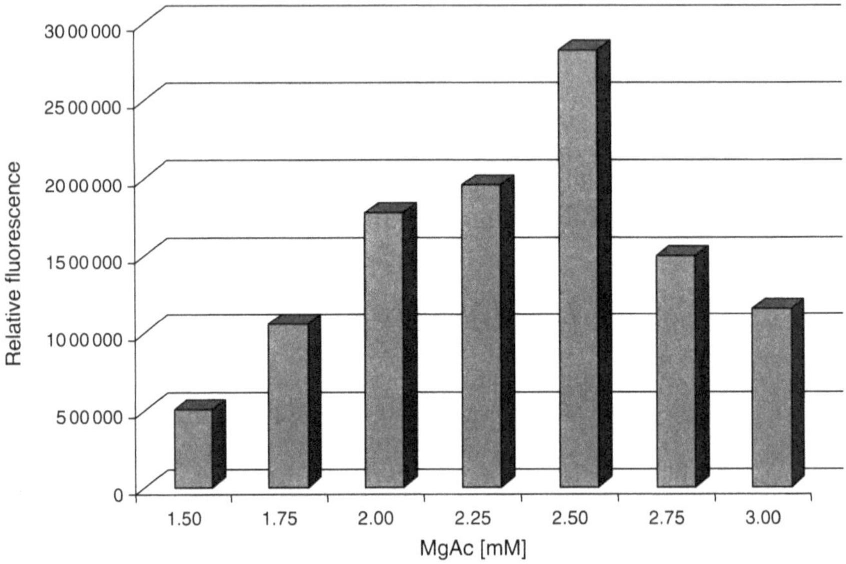

FIGURE 5 Mg^{2+}-dependent *in vitro* translation of a single pass transmembrane protein. The Heparin-binding EGF-like growth factor gene was N-terminally fused to EYFP. *In vitro* transcribed mRNA from this construct was analyzed in the insect cell-free system at the indicated Mg^{2+} concentrations. EYFP mediated fluorescence was quantified in the Amersham Typhoon Imager.

and even membrane proteins known as "difficult to express" proteins, such as aquaporin 6 (AQP6), were properly synthesized. Membrane proteins of prokaryotic origin, for example, the voltage dependent K^+ channel (KVAP) from *E. coli* and mitochondrial integral membrane proteins such as the 2-oxoglutarate/malate carrier (OGCP) and uncoupling proteins (UCP1, UCP4) could be produced in the insect cell lysate. The fast batch-formatted cell-free system enables the parallel synthesis of different membrane proteins within 90 min leading to protein yields in the range of 9.2–23.9 μg/ml (Fig. 7).

E. High-Throughput In Vitro Translation Systems

One of the most challenging applications of *in vitro* translation is the combination of fast template generation starting from total human DNA and the subsequent cell-free synthesis of individual proteins in a highly parallel manner. We approached this task by amplifying and expressing specific members of the odorant receptor gene superfamily. These membrane proteins are known to form the largest multigene family in the human genome. Approximately 350 human odorant receptor sequences have an

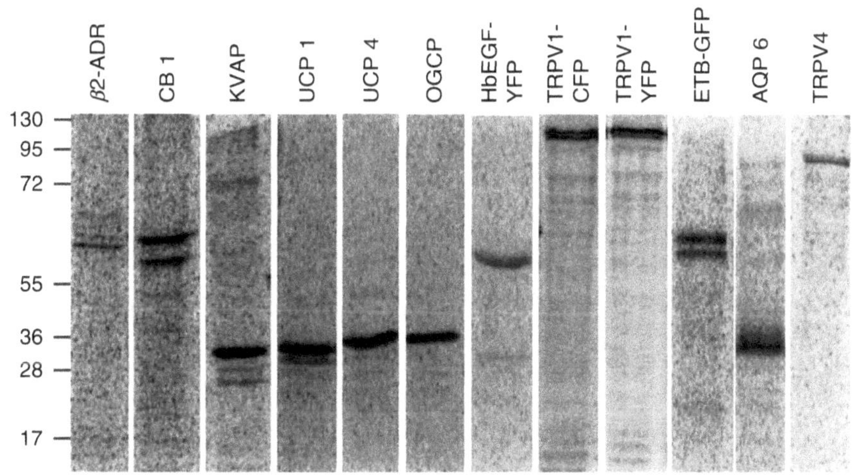

FIGURE 6 Cell-free synthesis of structurally divergent membrane proteins. *In vitro* transcribed mRNA coding for membrane proteins was analyzed in the insect cell extract (EasyXpress Insect Kit II, Qiagen). A 7.5 μl aliquot of the radiolabeled cell-free synthesis reaction was separated in a 15% SDS-PAGE and analyzed in the phosphoimager (Typhoon, Amersham). Expressed integral membrane proteins: Beta 2-adrenergic Receptor (β2-ADR), cannabinoid receptor 1 (CB1), voltage-dependent K^+ channel (KVAP), uncoupling protein 1 (UCP1), uncoupling protein 4 (UCP4), 2-oxoglutarate/malate carrier protein (OGCP), heparin-binding EGF-like growth factor (Hb-EGF-YFP), transient receptor potential cation channel, subfamily V, member 1 (TRPV1 CFP, TRPV1-YFP), endothelin receptor (ETB-GFP), aquaporin 6 (AQP6), transient receptor potential cation channel, subfamily V, member 4 (TRPV4). Individual genes were fused N-terminally to the indicated fluorescent reportergene YFP, CFP, or GFP.

intact open reading frame, and are potentially functional. A prerequisite for further functional analysis of these G-protein coupled receptors is the conversion of their coding sequences into efficient templates for protein synthesis. The structure of odorant receptor coding regions, devoid of introns, greatly facilitates this endeavor, as it is much easier to obtain genomic DNA from an individual than cDNA from the olfactory mucosa. We isolated genomic DNA from cultured HeLa cells and amplified defined olfactory receptor genes by Expression-PCR directly from this complex gene mixture. All E-PCR products were detected predominantly as homogenous bands with the expected size and restriction fragment length polymorphism (RFLP) analysis indicates amplification of individual odorant receptor encoding sequences (Fig. 8). This result demonstrates the applicability and specificity of E-PCR with human DNA. Selected E-PCR products were used directly without the need for further purification as templates for linked transcription-translation reactions in the insect cell-free system. Radio-labeled *in vitro* translation products were separated on an SDS gel

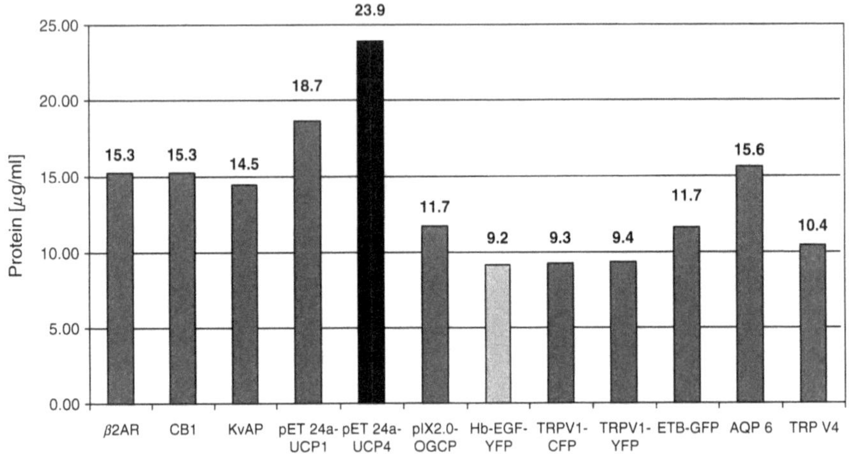

FIGURE 7 Productivity of the batch-formatted insect cell lysates. *In vitro* translation reactions were supplemented with mRNA encoding individual membrane proteins. Reactions were performed under identical conditions in a 50 μl volume in a thermo mixer. Expression levels were measured after an incubation time of 90 min in the batch-formatted system and translational activity was determined by hot TCA-precipitation of 5 μl aliquots, respectively.

and synthesized membrane proteins migrate at their expected positions (Fig. 9). In this way, membrane proteins were expressed within 1 day, starting from genomic human DNA.

The direct consequence of using linear templates generated by E-PCR is that cell-free expression systems are accessible methods for high-throughput protein expression because they prevent the time-consuming cloning and labor-intensive steps. Additionally, batch formatted cell-free protein synthesis offers tremendous flexibility for fast and parallel analysis of the products. The presented technology offers an economically feasible option for subsequent structural and functional analysis, for example, ligand screening technologies to deorphanize G-protein coupled receptors.

F. Glycosylation

Glycosylation is the most widespread and complex form of posttranslational modification in eukaryotes (Lowe & Marth, 2003). Proteins are translocated to the lumen of the ER where their leader peptide is cleaved and they acquire the oligosaccharide chain. During the synthesis of N-linked glycans in mammalian cells, a 14 saccharide core unit is assembled as a membrane-bound dolichylpyrophosphate precursor by enzymes located on both sides of the ER membrane (Burda & Aebi, 1999; Gahmberg & Tolvanen, 1996;

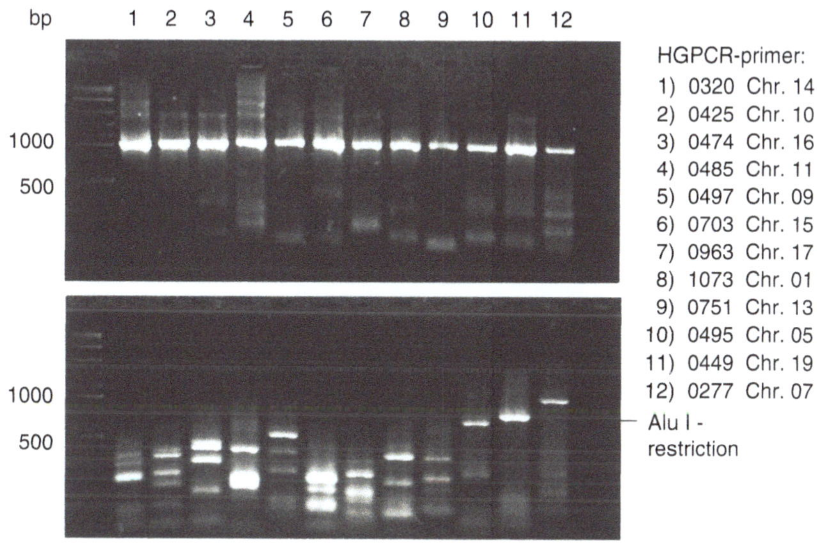

FIGURE 8 E-PCR-based amplification of individual representatives from a membrane protein multigene family. Genomic DNA was used as a template for the amplification of intronless olfactory receptors. Gene specific primers were designed against the coding sequence of individual olfactory receptors located on different chromosomes (HGPCR-primers). Amplification products were analyzed subsequent to the second E-PCR step. Four microliters of each E-PCR reaction was separated in an ethidium bromide stained 1% agarose gel. All PCR products are detected primarily as homogenous bands with the expected size (upper gel). Restriction fragment length polymorphism analysis shows that individual members of the multigene family of olfactory receptors were amplified (Alu I restriction, lower gel).

Kornfeld & Kornfeld, 1985). The completed core oligosaccharide is transferred from the dolichylpyrophosphate carrier to the growing, nascent polypeptide chain, and is coupled through an N-glycosidic bond to the side chain of an asparagine residue. The oligosaccharyltransferase responsible for this transfer is a membrane protein complex with its enzymatic active site in the ER lumen (Silberstein & Gilmore, 1996). Protein glycosylation is acknowledged as one of the major posttranslational modifications with significant effects on protein folding, structure, and functional activity.

In a set of experiments, we analyzed the potential of our cell-free system to synthesize glycoproteins. One of the most abundant glycoproteins in human plasma, erythropoietin (EPO), was analyzed in the linked transcription-translation procedure in the insect cell-free system. RNA encoding EPO was *in vitro* transcribed and purified by gel filtration. An aliquot of the mRNA was translated under standard conditions, whereas a second aliquot was translated in the presence of tunicamycin, an inhibitor of protein N-glycosylation (Fig. 10A). *In vitro* translation products were subsequently visualized by

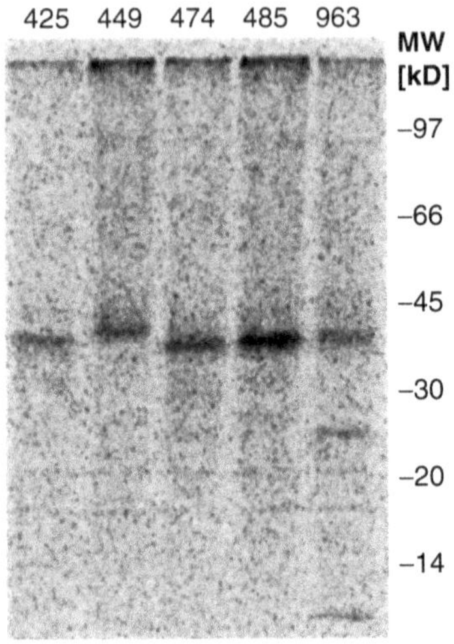

FIGURE 9 Cell-free synthesis of G-Protein coupled receptors. E-PCR products encoding olfactory receptors were used as a template for *in vitro* transcription and translation in the insect cell lysate (EasyXpress Insect Kit II, Qiagen). The G-protein coupled receptors were expressed in the presence of ^{14}C-Leucin, and 7.5 μl aliquots of the radiolabeled cell-free synthesis reactions were subjected to 15% SDS-PAGE and analyzed in the phosphoimager (Typhoon, Amersham). The cell-free synthesized membrane proteins migrate at the expected positions (human G-protein coupled receptor HGPCR 425, 449, 474, 485, and 963).

autoradiography (Fig. 10B). In the insect cell-free system, EPO migrates at two distinct positions, when synthesized under standard conditions. The molecular masses of the reaction products are similar to that of glycosylated EPO and an additional nonglycosylated EPO precursor protein, respectively. No glycosylated protein was detected in the presence of tunicamycin, but a signal peptide cleaved version of EPO was apparently visible. Tunicamycin inhibits N-glycosylation in eukaryotes by blocking the transfer of GlcNAc-1-P from UDP-GlcNAc to dolichyl-P, thereby decreasing dolichyl-PP-GlcNAc (Heifetz, Keenan, & Elbein, 1979). The reaction is catalyzed by GlcNAc phosphotransferase and synthesis starts on the cytosolic surface of the ER membrane by the addition of sugars, one by one, to dolicholphosphate. Most of the ER biosynthetic machinery faces the cytosol and can therefore directly use precursors such as sugar nucleotides provided by cytosolic enzymes.

In this context, effective signal peptide cleavage and synthesis of glycosylated EPO in our cell-free system indicates the presence of translocation competent ER compartments such as microsomes in the insect cell lysate. Furthermore, our lysate production methodology preserved full activity of core protein glycosylation enzymes and additionally provided a sufficient

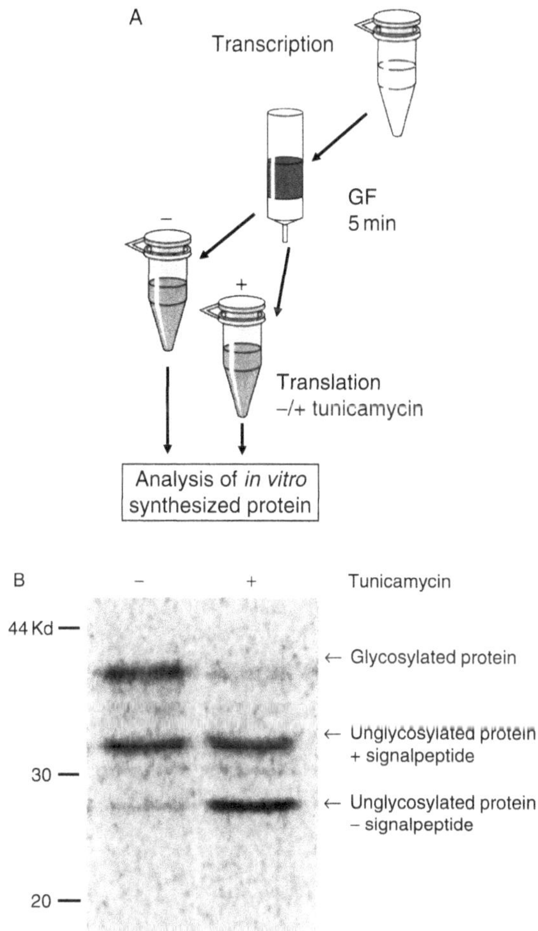

FIGURE 10 (A) Parallel synthesis of glycosylated and nonglycosylated protein. Glycoproteins are produced in a linked transcription-translation procedure. *In vitro* transcribed mRNA encoding a glycoprotein is purified by a gel filtration step (DyeEx spin columns included in the EasyXpress Insect Kit II, Qiagen). An aliquot of this mRNA is translated in the standard high-yield mode, a second aliquot is translated in the presence of 10 μg/ml tunicamycin. (B) *In vitro* Glycosylation on demand. *In vitro* transcribed mRNA coding for erythropoietin was analyzed in the insect cell extract in the presence of tunicamycin and without tunicamycin (f.c. 10 μg/ml, EasyXpress Insect Kit II, Qiagen). A 7.5 μl aliquot of each radiolabeled cell-free synthesis reaction was separated in a 15% SDS-PAGE and analyzed in the phosphoimager (Typhoon, Amersham).

amount of sugar nucleotide precursors for quantitative glycosylation. Cell-free production of glycoproteins is thus achieved within a homogenous system without the need for supplementing the reaction with additional membrane vesicles.

G. Optimization of Membrane Protein Expression

Any attempt to develop a new method for expressing membrane proteins must consider some challenging problems such as the efficient integration of the membrane protein into a lipid bilayer in a correctly folded structure. A decisive step in the biosynthesis of membrane proteins is their complete translocation across the eukaryotic ER membrane. Usually these proteins are translocated through a protein conducting channel formed by a conserved, heterotrimeric membrane-protein complex, the Sec61 complex (reviewed in Rapoport, 2007). The channel allows soluble polypeptides to cross the membrane and hydrophobic transmembrane segments of membrane proteins to exit laterally into the lipid phase. Secreted proteins and a huge number of membrane proteins are directed to the ER membrane by signal sequences, and these sequences widely differ in their ability to facilitate protein translocation. Moreover, foreign signal sequences, for example, mammalian sequences could be less efficient in insect cells. Therefore, substitution of the foreign protein's own signal sequence for a powerful insect signal sequence often results in more efficient protein translocation, a higher protein expression level, and improved biological activity. The honeybee mellitin signal sequence, for example, enables efficient translocation of proteins into the ER of *Sf* cells (Tessier, Thomas, Khouri, Laliberté, & Vernet, 1991).

To investigate whether signal peptides affect the synthesis of glycosylated membrane proteins in our cell-free insect system, we analyzed the expression of the inducible costimulatory (ICOS) receptor, a member of the CD28 family of costimulatory molecules. Various ICOS receptor expression templates were designed using the E-PCR methodology (Linear Template Kit Signal Peptide, RiNA GmbH). As a result, the native signal sequence of the ICOS receptor was either deleted or replaced by E-PCR using a synthetic oligonucleotide encoding the melittin signal sequence as sense primer (Fig. 11A). Resulting E-PCR products were applied directly to the linked transcription-translation procedure for cell-free protein synthesis. Radiolabeled *in vitro* translation products were separated on an SDS gel and ICOS receptor synthesis in the insect cell lysate was analyzed.

The results of this experiment demonstrated that expression of the ICOS receptor without a signal peptide at the N-terminus resulted in the accumulation of the nonglycosylated protein within the lysate as expected (Fig. 11B). Glycosylated ICOS receptor could be detected when expressing the membrane protein in its native form. However, only a small fraction of the synthesized protein was found to be glycosylated. In contrast, glycosylation of the expressed ICOS receptor was progressively increased by eliminating the native signal sequence and replacing it by the cleavable melittin signal

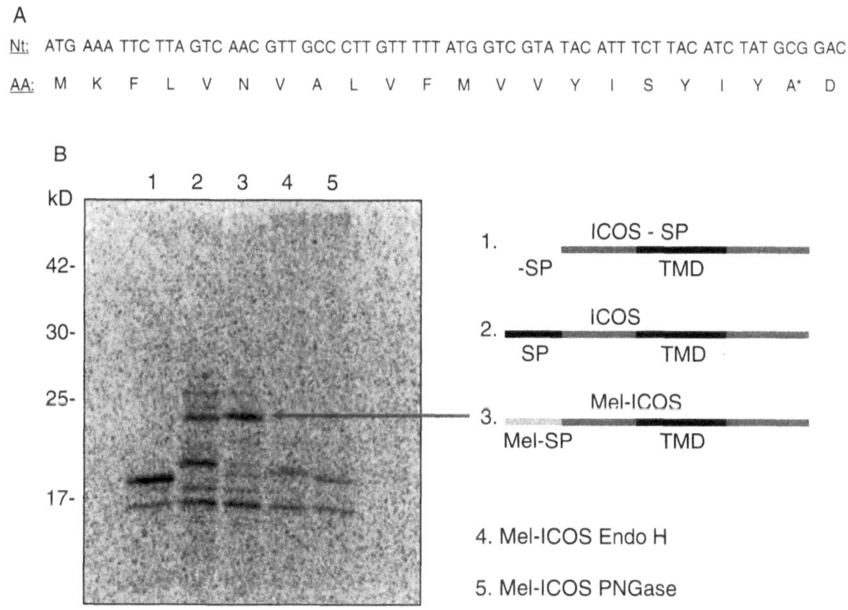

FIGURE 11 (A) The honeybee *melittin* signal sequence. Nucleotide sequence (Nt) and deduced amino acid composition (AA) of the honeybee *melittin* signal sequence. Replacement of the natural signal sequence with the *melittin* signal sequence can be achieved using the E-PCR system (Linear Template Kit Signal Peptide, RiNA GmbH). An asterisk indicates the position where the signal peptide is cleaved off. (B) Functionality of the *melittin* signal peptide in cell-free membrane protein synthesis. *In vitro* transcribed mRNA coding for the inducible costimulatory receptor (ICOS) was analyzed in the insect cell extract (EasyXpress Insect Kit II, Qiagen). A 7.5 μl aliquot of the radiolabeled cell-free synthesis reaction was separated in a 15% SDS-PAGE and analyzed in the phosphoimager (Typhoon, Amersham). Lane 1: cell-free expression of the ICOS receptor without signal peptide. Lane 2: *in vitro* translation of the native ICOS receptor. Lane 3: improved expression and posttranslational modification of the ICOS receptor with the melittin signal sequence substituted for the native signal sequence. Lanes 4 and 5: aliquot of the translation reaction analyzed in lane 3 after enzymatic deglycosylation by Endo H (lane 4) and PNGase F (lane 5), respectively.

peptide. Endo H treatment as well as PNGase F digestion reduced the apparent molecular weight of the glycosylated membrane protein indicating that the ICOS receptor was efficiently glycosylated. These results demonstrate that the signal sequence of the ICOS receptor is a key factor in determining its expression and posttranslational modification in the cell-free insect system. In this context, E-PCR-based template construction in conjunction with *in vitro* translation offers a unique opportunity to modify and analyze templates coding for integral membrane proteins in order to ensure their optimal cotranslational integration into the lipid bilayer.

V. CONCLUSIONS

In vitro synthesis of proteins has found a large variety of low- and high-throughput applications suitable for functional and structural proteomics. There is immense potential of cell-free expression systems to make a major contribution toward the synthesis of difficult-to-express proteins and in particular for the production of membrane proteins.

We have exploited and optimized the technique of lysate preparation and protein expression in our eukaryotic cell-free translation system (Kubick et al., 2003, 2005; von Groll, Kubick, Merk, Stiege, & Schäfer, 2007). The discussed development is based on the properties of versatile eukaryotic *in vivo* expression systems thereby expanding the possibilities of *in vitro* protein engineering. Besides the expression of soluble and functionally active proteins, cell-free translation enables the synthesis of toxic proteins as well as protein variants with unnatural amino acids (Merk, Stiege, Tsumoto, Kumagai, & Erdmann , 1999; Stiege & Erdmann, 1995). Moreover, obtaining proteins with improved properties depends on efficient translation and correct protein folding as a prerequisite for molecular evolution. Lysates from eukaryotic cell lines therefore provide several chaperone assisted protein folding mechanisms, enabling the correct folding of the *in vitro* expressed polypeptide into its three-dimensional structure. In the case of membrane proteins, insertion into a native lipid bilayer strongly facilitates correct assembly. We have adapted the production procedure of our insect lysate to this requirement. As a result, translocation of cell-free synthesized membrane proteins across the membrane of intact microsomal vesicles derived from fragmented ER represent feasible achievements. The formation of posttranslational modifications in particular expands the possibilities of this eukaryotic cell-free system, as these covalent modifications are often an essential prerequisite to obtain functionally active proteins. Our technology has been applied to a broad range of membrane proteins, demonstrating a versatile method for the production of structurally and functionally different membrane proteins. In the near future, eukaryotic *in vitro* translation systems may represent an attractive alternative for the coexpression of multi-subunit membrane protein complexes and cotranslational incorporation of supplied artificial cofactors. A clear application for cell-free protein expression involves miniaturization and protein micro arrays as *in vitro* translation reactions can easily be scaled down to submicroliter volumes. In the field of genetic diagnostics, cell-free expression has attracted considerable interest via use of the protein truncation test as a method for detecting mutations in marker genes (Gite et al., 2003). Finally, protein interaction validation studies are greatly facilitated in cell-free expression systems, for instance by combining *in vitro* cotranslational labeling with fluorescence microscopic analysis or fluorescence cross-correlation spectroscopy. In this context, cell-free protein synthesis systems offer the unique opportunity to use the highly efficient translational machinery while introducing novel functional groups at specific sites in proteins (Gerrits, Kubick, Merk, Strey, & Stiege, 2007a). A well-established method of protein modification is the site-directed cotranslational incorporation of unnatural amino acids (Gerrits et al., 2007b). In this way, *in vitro* translation systems are valuable tools for the production of protein conjugates. Such proteins with unique functional groups, for example affinity labels, offer a high-potential in many biotechnical or biomedical investigations. Various *in vitro* translation systems are now utilized for the synthesis of protein domains, the introduction of point mutations, the synthesis of glycosylated proteins, the site-specific incorporation of biotin, and the immobilization of proteins. The synthesis of stable isotope-labeled proteins in cell-free protein synthesis systems facilitates structural proteomics projects (Gourdon et al., 2008).

The rapidly accumulating information obtained from these high-throughput approaches and systematic reaction condition screens will provide a comprehensive and reliable database of knowledge for the preparative-scale cell-free synthesis of many membrane protein targets.

Acknowledgments

Plasmid DNA encoding membrane proteins was kindly provided by Prof. C. Harteneck (Universität Tübingen), Prof. R. Kroczek (Robert Koch Institut, Berlin), PD Dr. E. Pohl (Charite, Berlin), Prof. P. Pohl (Universität Linz, Austria), and Prof. M. Schäfer (Universität Leipzig).

References

Bergamini, G., Preiss, T., & Hentze, M. W. (2000). Picornavirus IRESes and the poly(A) tail jointly promote cap-independent translation in a mammalian cell-free system. *RNA, 6*, 1781–1790.

Burda, P., & Aebi, M. (1999). The dolichol pathway of *N*-linked glycosylation. *Biochimica et Biophysica Acta (BBA)—General Subjects, 1426*(2), 239–257.

Castagnetti, S., Hentze, M. W., Ephrussi, A., & Gebauer, F. (2000). Control of oskar mRNA translation by Bruno in a novel cell-free system from *Drosophila* ovaries. *Development, 127*, 1063–1068.

Cornvik, T., Dahlroth, S.-L., Magnusdottir, A., Herman, M. D., Knaust, R., Ekberg, M., Nordlund, P. (2005). Colony filtration blot: A new screening method for soluble protein expression in *Escherichia coli*. *Nature Methods 2*, 507–509.

Duksin, D., & Mahoney, W. C. (1982). Relationship of the structure and biological activity of the natural homologues of tunicamycin. *Journal of Biological Chemistry, 257*, 3105–3109.

Endo, Y., & Sawasaki, T. (2003). High-throughput, genome-scale protein production method based on the wheat germ cell-free expression system. *Biotechnology Advances, 21*, 695–713.

Erickson, A. H., & Blobel, G. (1983). Cell-free translation of messenger RNA in a wheat germ system. *Methods Enzymology, 96*, 38–50.

Farrell, P. J., Balkow, K., Hunt, T., & Jackson, R. J. (1977). Phosphorylation of initiation factor eIF-2 and the control of reticulocyte protein synthesis. *Cell, 11*, 187–200.

Gahmberg, C., & Tolvanen, M. (1996). Why mammalian cell surface proteins are glycoproteins. *Trends in Biochemical Sciences, 21*(8), 308–311.

Gebauer, F., Corona, D. F. V., Preiss, T., Becker, P. B., & Hentze, M. W. (1999). Translational control of dosage compensation in *Drosophila* by sex-lethal: Cooperative silencing via the 5 and 3 UTRs of msl-2 mRNA is independent of the poly(A) tail. *EMBO Journal, 18*, 6146–6154.

Gerrits, M., Kubick, S., Merk, H., Strey, J., & Stiege, W. (2007a). *In vitro* synthesis of modified proteins. In A. Kyriakopoulos, B. Michalke, A. Graebert, & B. Behne, (Eds.). *Proceedings of the 5th fall conference on metallobiolomics.* München: Herbert Utz Verlag.

Gerrits, M., Strey, J., Claußnitzer, I., von Groll, U., Schäfer, F., Rimmele, M., & Stiege, W. (2007b). Cell-free synthesis of defined protein conjugates by site-directed cotranslational labeling. In T. Kudlicki, F. Katzen, & R. Bennett, (Eds.). *Cell-free expression.* Austin: Landes Bioscience.

Gite, S., Lim, M., Carlson, R., Olejnik, J., Zehnbauer, B., & Rothschild, K. (2003). A high-throughput nonisotopic protein truncation test. *National Biotechnology, 21*, 194–197.

Gourdon, P., Alfredsson, A., Pedersen, A., Malmerberg, E., Nyblom, M., Widell, M., *et al.* (2008). Optimized *in vitro* and *in vivo* expression of proteorhodopsin: A seven-transmembrane proton pump. *Protein Expression and Purification, 58*, 103–113.

Heifetz, A., Keenan, R. W., & Elbein, A. D. (1979). Mechanism of action of tunicamycin on the UDP-GlcNAc: Dolichyl-phosphate GlcNAc-1-phosphate transferase. *Biochemistry, 18*, 2186–2192.

Hill-Perkins, M. S., & Possee, R. D. (1990). A baculovirus expression vector derived from the basic protein promoter of *Autographa californica* nuclear polyhedrosis virus. *Journal of General Virology, 71*, 971–976.

Holcik, M., & Sonenberg, N. (2005). Translational control in stress and apoptosis. Nature reviews. *Molecular Cell Biology, 6*, 319–327.

Jackson, R. J., & Hunt, T. (1983). Preparation and use of nuclease-treated rabbit reticulocyte lysates for the translation of eukaryotic messenger RNA. *Methods Enzymology, 96*, 50–74.

Kornfeld, R., & Kornfeld, S. (1985). Assembly of asparagine-linked oligosaccharides. *Annual Review of Biochemistry, 54*, 631–664.

Kubick, S., Merk, H., Stiege, W., von Groll, U., Drees, J., Schäfer, F. (2005). Insect-cell based *in vitro* synthesis of posttranslationally modified proteins. In Qiagen News, Issue 2, 41–43.

Kubick, S., Schacherl, J., Fleischer-Notter, H., Royall, E., Roberts, L. O., & Stiege, W. (2003). *In Vitro* translation in an insect-based cell-free system. In J. R. Swatz, (Ed.), *Cell-free protein expression* pp. 209–217. Berlin-Heidelberg-New York: Springer.

Lie, Y. S., & Macdonald, P. M. (2000). *In vitro* translation extracts prepared from *Drosophila* ovaries and embryos. *Biochemical Biophysical Research Communications, 270*, 473–481.

Lowe, J. B., & Marth, J. D. (2003). A genetic approach to mammalian glycan function. *Annual Review of Biochemistry, 72*, 643–691.

Luckow, V. L., & Summers, M. D. (1988). Trends in the development of baculovirus expression vectors. *Biotechnology, 6*, 47–55.

Madin, K., Sawasaki, T., Ogasawara, T., & Endo, Y. (2000). A highly efficient and robust cell-free protein synthesis system prepared from wheat embryos: Plants apparently contain a suicide system directed at ribosomes. *PNAS, 97*, 559–564.

Merk, H., Meschkat, D., & Stiege, W. (2003). Expression-PCR: From gene pools to purified proteins within 1 day. In J. R. Swartz, (Ed.), *Cell-free protein expression* (1st ed., pp. 209–217). Berlin: Springer-Verlag.

Merk, H., Stiege, W., Tsumoto, K., Kumagai, I., & Erdmann, V. A. (1999). Cell-free expression of two single-chain monoclonal antibodies against lysozyme: Effect of domain arrangement on the expression. *Journal of Biochemistry, 125*, 328–333.

Midgett, C. R., & Madden, D. R. (2007). Breaking the bottleneck: Eukaryotic membrane protein expression for high-resolution structural studies. *Journal of Structural Biology, 160*(3), 265–274.

Molla, A., Paul, A. V., & Wimmer, E. (1991). Cell-free, *de novo* synthesis of poliovirus. *Science, 254*, 1647–1651.

O'Reilly, D. R., Miller, L. K., & Luckow, V. A. (1992). *Baculovirus expression vectors: A laboratory manual.* New York: Freeman.

Pelham, H. R. B., & Jackson, R. J. (1976). An efficient mRNA-dependent translation system from reticulocyte lysates. *European Journal of Biochemistry, 67*, 247–256.

Rapoport, T. A. (2007). Protein translocation across the eukaryotic endoplasmic reticulum and bacterial plasma membranes. *Nature 450*, 663–668.

Rupniak, H. T., & Quincey, R. V. (1975). Small changes in energy charge affect protein synthesis in reticulocyte lysates. *FEBS Letters, 58*, 99–101.

Schneider, C., Sepp-Lorenzino, L., Nimmesgern, E., Ouerfelli, O., Danishefsky, S., Rosen, N., Hartl, F. U. (1996). Pharmacologic shifting of a banlance between protein refolding and degradation mediated by Hsp90. *Proceedings of the National Academy of Science, 93*, 14536–14541.

Silberstein, S., & Gilmore, R. (1996). Biochemistry, molecular biology, and genetics of the oligosaccharyltransferase. *FASEB Journal, 10*, 849–858.

Singh, L. P., Aroor, A. R., & Wahba, A. J. (1994). Phosphorylation of the guanine nucleotide exchange factor and eukaryotic initiation factor 2 by casein kinase II regulates guanine nucleotide binding and GDP/GTP exchange. *Biochemistry, 33*, 9152–9157.

Smith, G. E., Summers, M. D., & Fraser, M. J. (1983). Production of human beta interferon in insect cells infected with a baculovirus expression vector. *Molecular Cell Biology, 3*, 2156–2165.

Stiege, W., & Erdmann, V. A. (1995). The potentials of the *in vitro* protein biosynthesis system. *Journal of Biotechnology, 41*, 81–90.

Summers, M. D., Smith, G. E. (1987). A manual of methods for baculovirus vectors and insect cell culture procedures. *Texas Agricultural Experiment Station Bulletin,* 1555.

Tessier, D. C., Thomas, D. Y., Khouri, H. E., Laliberté, F., & Vernet, T. (1991). Enhanced secretion from insect cells of a foreign protein fused to the honeybee melittin signal peptide. *Gene, 98*(2), 177–183.

Vaughn, J. L., Goodwin, R. H., Tompkins, G. J., & McCawley, P. (1977). The establishment of two cell lines from the insect *Spodoptera frugiperda* (Lepidoptera; Noctuidae). *In Vitro, 13*, 213–217.

Vinarov, D. A., Lytle, B. L., Peterson, F. C., et al. (2004). Cell-free protein production and labeling protocol for NMR based structural proteomics. *Nature Methods, 1*(2), 1–5.

Vlak, J. M., Klinkenberg, F. A., Zaal, K. J. M., et al. (1988). Functional studies on the p10 gene of *Autographa californica* nuclear polyhedrosis virus using a recombinant expression of a p10-Bglactosidase fusion gene. *Journal of General Virology, 69*, 765–776.

von Groll, U., Kubick, S., Merk, H., Stiege, W., & Schäfer, F. (2007). Advances in insect-based cell-free protein expression. In T. Kudlicki, F. Katzen, & R. Bennett, (Eds.). *Cell-free expression.* Austin: Landes Bioscience.

Welsh, G. I., Miller, C. M., Loughlin, A. J., Price, N. T., & Proud, C. G. (1998). Regulation of eukaryotic initiation factor eIF2B: Glycogen synthase kinase-3 phosphorylates a conserved serine which undergoes dephosphorylation in response to insulin. *FEBS Letters, 421*, 125–130.

Zacharias, A., Schäfer, F., Müller, S., von Groll, U. (2004). Recombinant-protein solubility screening using the EasyXpress *in vitro* translation system. QIAGEN, Qiagen News, 2004. e6.

CHAPTER 3

Harnessing Photosynthetic Bacteria for Membrane Protein Production

Deborah K. Hanson, Donna L. Mielke, and Philip D. Laible
Biosciences Division, Argonne National Laboratory, Argonne, Illinois 60439

Current Topics in Membranes, Volume 63

1063-5823/09 $35.00
DOI: 10.1016/S1063-5823(09)63003-9

I. OVERVIEW

Membrane proteins present unparalleled challenges for structural biology initiatives, nascent functional genomics efforts, and drug discovery experiments. To obtain sufficient quantities of membrane proteins for these studies, an expression system based upon the *Rhodobacter* species of photosynthetic bacteria is being employed. This system exploits this organism's unique physiology whereby strongly induced heterologous expression of target membrane proteins can be coordinated with synthesis of new membranes, thereby favoring membrane insertion of natively folded polypeptides. A series of vectors utilizing promoters responding to oxygen and/or light has been constructed. A variety of engineered host strains is available that carry deletions of one or more native transmembrane complexes, thereby increasing the membrane volume available for accommodation of foreign protein. Using this approach, data suggest that many target membrane proteins from a variety of organisms can be produced and purified at levels that equal or exceed those of native membrane protein complexes (10 mg/l). Standardized strategies have been developed for semiautomation of cloning and purification. Analysis of nearly 500 expression strains (including representatives encompassing an entire membrane proteome) has shown that ~60% of the membrane proteins are expressed in *Rhodobacter* at levels that exceed 1 mg/l of cell culture (many at levels of 10-20 mg/l), are localized within intracytoplasmic membranes (ICMs), and—most importantly—display structural and functional integrity. The *Rhodobacter* system represents an advance towards the development of an integrated strategy for obtaining structures of this important class of proteins at a more rapid pace.

II. INTRODUCTION

The functions performed by membrane proteins are extremely important for all organisms and have an overwhelming impact on human health. Membrane proteins represent approximately 30% of every genome and comprise the majority of all drug targets. Despite this prominent role, only around 150 unique, unrelated structures have been determined to date (http://blanco.biomol.uci.edu/Membrane_Proteins_xtal.html) in contrast with unique structures representing more than 9800 soluble protein families (http://www.rcsb.org/pdb). Of the experimental approaches used to solve these structures, X-ray diffraction—requiring high-quality, three-dimensional crystals—resulted in approximately 85% of them. Other strategies—including NMR, cryo-electron microscopy, and electron diffraction—have been used far less frequently (http://www.mpdb.ul.ie).

A. The Scarcity of Membrane Protein Structures

A major factor influencing the paucity of membrane protein structures is that the expression levels of membrane proteins in native tissue are generally low relative to the expression levels of many soluble proteins. Thus, it is rare that adequate amounts of starting material are available for purification of a membrane protein from its natural source organism. Discovery of purification strategies in these scenarios is protein-specific, time-consuming, labor-intensive, and expensive. The parameter space that must be searched is quite large and no precedent exists for a single "magic bullet" set of generic conditions (detergents, temperatures, incubation time, protein/surfactant ratios, etc.) under which membrane proteins can be extracted from native hosts and purified in the high yield and functional form needed for extensive biochemical studies and crystallization trials. It is no coincidence, then, that most of the few membrane protein structures obtained prior to the year 2000 were for proteins which were abundant in their native-host organisms.

B. Recombinant Protein Expression as a Solution

To make membrane proteins more readily available, recombinant systems are often employed today for heterologous expression. *Escherichia coli*-based strategies offer many advantages such as simplicity, low cost, and rapid growth. Alternatively, many eukaryotic protein expression systems are also available and have been employed. However, the biggest drawbacks of the latter are that they are cumbersome and expensive for the preparation of the quantities of membrane proteins that are necessary for structure determination experiments. Other limitations of both prokaryotic and eukaryotic host organisms include inadequate membrane volume for accommodation of heterologously expressed membrane proteins (Fig. 1A; Arechaga et al., 2000; Miroux & Walker, 1996) and saturation of the secretory machinery for integration of the heterologous protein into the membrane (Essen, 2002). Thus, overexpression strategies often result in cell death or precipitation of aggregates of the heterologously expressed membrane protein as inclusion bodies. Finding conditions that yield a functionally active and structurally relevant membrane protein from solubilization and "refolding" of recombinant protein contained within inclusion bodies is not straightforward. These techniques can sometimes produce adequate amounts of material which is native enough for functional or immunological studies (see, e.g., Kiefer, Maier, & Vogel, 1999; reviewed in Grisshammer & Tate, 1995) but rarely can they provide the quantity of homogenous, natively folded protein that is necessary for exhaustive functional characterization or crystallization trials.

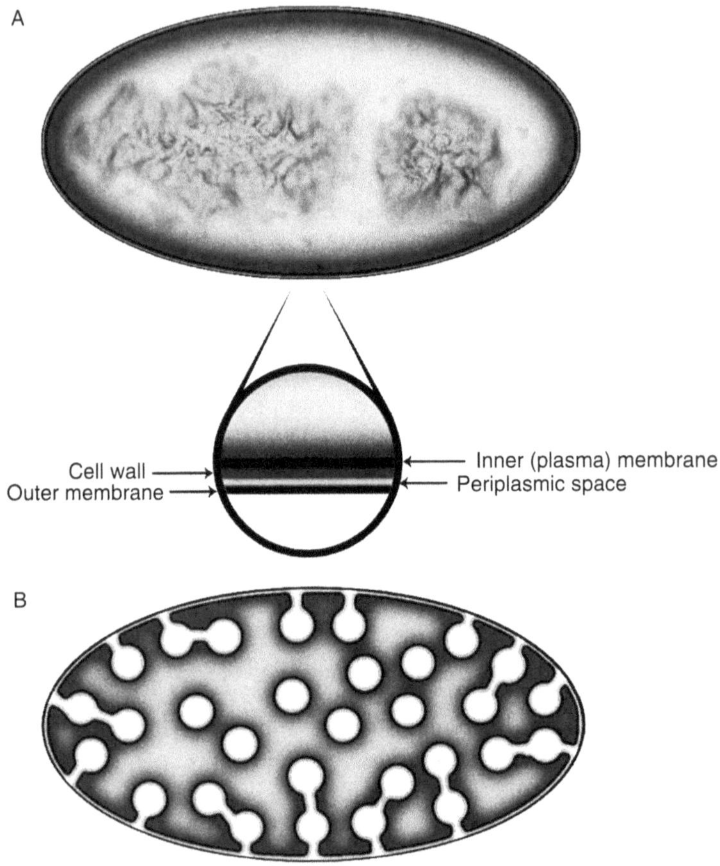

FIGURE 1 Membrane morphologies of prokaryotic cells. The volume of the inner membrane in *E. coli* is relatively small (~3%; panel A), posing a serious limitation for the incorporation of heterologously expressed membrane proteins. Organisms, like *Rhodobacter*, which increase this membrane fraction (e.g., panel B) are better suited for expression and sequestration of greater amounts of membrane proteins in native form.

The examples illustrated above pertain to systems that were designed for overexpression of soluble proteins that are being employed for membrane proteins. They fail with membrane proteins because the space in the cell's membranes is already occupied. To gain a sense of the limited space available in a typical *E. coli* cell, consider that approximately 5% of the cell's volume is the inner membrane (Fig. 1A). Assuming a lipid:protein ratio of 60:40 in the inner membrane, this leaves only 3% of the total volume of the cell to accommodate both the cell's native inner membrane proteins as well as any heterologously expressed membrane proteins. In contrast, 84% of the volume of this typical prokaryotic cell is cytoplasm, explaining why *E. coli* is so useful as an expression host for soluble proteins.

Alternatively, several other expression systems that are tailored to the expression of membrane proteins are in use or in development. Expression strategies utilize a variety of promoters and induction conditions, and they employ different means of targeting expressed proteins to cellular compartments, providing samples of utility for various studies of structure and function. A theme common to systems that are being developed is a proliferation of membranes (Fig. 1B) that can address the problem of compartment space for the incorporation of heterologously expressed membrane proteins. Proliferating membranes are a property of the C41 and C43 strains of *E. coli* (Arechaga et al., 2000; Miroux & Walker, 1996). *Tetrahymena thermophila* (Gaertig, Gao, Tishgarten, Clark, & Dickerson, 1999) and cell wall-deficient L-forms of *Proteus mirabilis*, *E. coli*, *Bacillus subtilis*, and *Streptomyces hygroscopicus* (Gumpert and Hoischen, 1998; Hoischen et al., 2002) have been used for surface display of expressed membrane proteins. *Halobacterium salinarum* and its inducible purple membrane have been employed for the heterologous expression of membrane proteins (Turner, Reusch, Winter-Vann, Martinez, & Betlach, 1999), while the yeast *Hansenula polymorpha* (van Dijk, Faber, Kiel, Veenhuis, & van der Klei, 2000) and strains of methylotrophic bacteria (e.g., Nguyen, Elliott, Yip, & Chan, 1998) and photosynthetic bacteria (Collins & Cheng, 2004; De Smet, Kostanjevecki, Guisez, & Van Beeumen, 2001; Graichen et al., 1999; Laible & Hanson, 2002; Laible, Scott, Henry, & Hanson, 2004) feature an inducible ICM that has been exploited for the incorporation of newly synthesized foreign membrane proteins.

C. Coordinating Membrane and Protein Synthesis

In regard to the last case, photosynthetic bacteria are particularly intriguing for addressing the membrane protein expression problem since they produce extremely large amounts of ICM under certain growth conditions in response to changes in light intensity and/or oxygen tension (as reviewed in Drews & Golecki, 1995; Verméglio, Joliot, & Joliot, 1995). Members of the *Rhodobacter* genus are facultative photoheterotrophs characterized by a metabolic diversity that allows them to adapt readily to a wide variety of environmental conditions. They are known to reduce nitrogen compounds, fix carbon dioxide, utilize carbon sources in an aerobic environment, grow photosynthetically under anaerobic conditions, or grow anaerobically in the dark in the presence of exogenous electron acceptors (reviewed in Imhoff, 1995). In particular, under conditions of light and/or lowered oxygen tension, the membrane surface in the organism increases manyfold as an ICM is elaborated as invaginations of the cytoplasmic membrane (Fig. 2A; reviewed in Drews and Golecki, 1995). Concomitantly, the

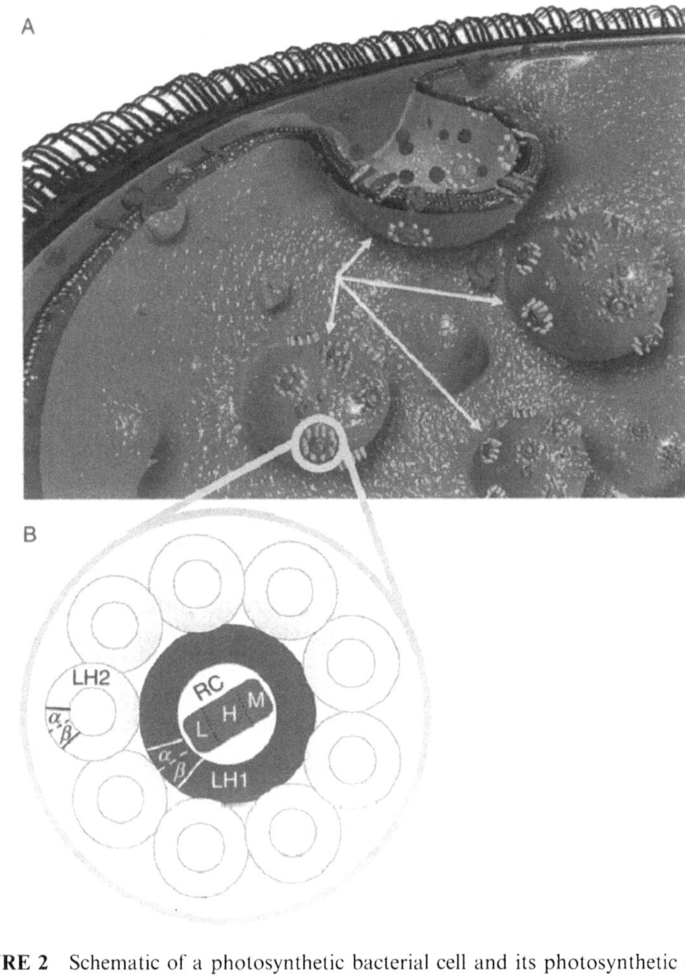

FIGURE 2 Schematic of a photosynthetic bacterial cell and its photosynthetic apparatus. (A) A model of a *Rhodobacter* cell, underscoring key features of its physiology which are vital to its use in a novel expression system for membrane proteins. Synthesis of the specialized ICM (arrows) is induced by light and/or lowered oxygen tension; the ICM houses the transmembrane complexes of the cell's photosynthetic apparatus. (B) The protein components of the native photosynthetic unit include the reaction center (RC; consisting of subunits L, M, and H), the core light-harvesting antenna complex (LH1, consisting of multiple pairs of $\alpha\beta$ heterodimers) and the peripheral light-harvesting antennae (LH2, also consisting of multiple $\alpha\beta$ heterodimers).

same environmental cues induce synthesis of the photosynthetic apparatus—the peripheral (LH2) and core (LH1) light-harvesting assemblies and the reaction center (RC). The new ICM sequesters these complexes that are composed of transmembrane polypeptides and their associated hydrophobic redox and energy transfer cofactors (Fig. 2B; reviewed in Kiley & Kaplan, 1988).

III. FOREIGN GENE EXPRESSION IN *RHODOBACTER SPHAEROIDES*

Members of the *Rhodobacter* (*R.*) genus are versatile organisms that may be cultured to high-cell densities in the presence or absence of light (without or with oxygen, respectively) on a variety of rich or defined minimal media. This inducible ICM of *Rhodobacter* has been exploited for the expression of foreign membrane proteins. A suite of plasmids is available for insertion of foreign genes behind promoters that direct synthesis of proteins of the photosynthetic apparatus. The expression system also employs host strains which carry deletions of genes encoding some of these native proteins. With these engineered tools, the synthesis of a membrane protein of interest is induced by the same environmental cues which induce the ICM concomitantly. The expressed heterologous membrane protein thus has a membrane destination—the ICM—into which it can insert and assemble. This coupled synthesis gives *Rhodobacter* an advantage over other host organisms where the discontinuity between protein synthesis and membrane synthesis often leads to the degradation of the expressed membrane proteins or the formation of insoluble aggregates or inclusion bodies (Columbus et al., 2006; Kiefer et al., 1999; Korepanova et al., 2005). The color of the cell culture reports that conditions leading to induction of both ICM synthesis and expression of the foreign membrane protein have been achieved. ICM vesicles enriched in heterologous membrane protein can be isolated easily after cell lysis, enabling a straightforward purification of the desired membrane protein in native form for subsequent structural or functional studies. Components of this membrane protein expression system and results from implementation of it are described in the following sections.

A. Vector Design

The well-established fact that the transmembrane protein-cofactor complexes that constitute the photosynthetic apparatus differ substantially in their relative abundances (Fig. 2B) figured prominently into the design of expression plasmids. In the native ICM, a single three-subunit RC assembly is encompassed by the LH1 core antenna complex that consists of 14–18 $\alpha\beta$ heterodimers (Jamieson et al., 2002; Karrasch, Bullough, & Ghosh, 1995; Qian, Hunter, & Bullough, 2005; Qian et al., 2003; Roszak et al., 2003). Each ring of the peripheral LH2 antennae complexes is made of eight to nine $\alpha\beta$ heterodimers (Koepke, Hu, Muenke, Schulten, & Michel, 1996; McDermott et al., 1995; Walz & Grigorieff, 1998). The relative amount of the LH2 complex varies, increasing dramatically in low light to a level of 10–20 LH2 per LH1/RC complex (Bahatyrova, Frese, Siebert, et al., 2004; Bahatyrova, Frese, van der Werf, et al., 2004; Clayton, 1980; Drews, 1985; van Grondelle, Hunter, Bakker, & Kramer, 1983). If genes encoding these three complexes were replaced with foreign genes, expectations follow that replacement of the LH2 structural genes with foreign genes would lead to the highest expression level, replacement of LH1 structural genes would result in an intermediate level of expression, and replacement of RC structural genes would lead to relatively low expression of a foreign protein. The upper limits of these expectations are based on the amounts of protein that can be purified from wild-type strains expressing the native photosynthetic apparatus; typically, RCs can be purified from chemoheterotrophic or photosynthetic cultures in yields of 10 mg/l, and yields of $\geq$100 mg/l of LH1 and LH2 complexes can be obtained. Taking advantage of these differences in natural abundance in vector design allows for high-level expression or modulation of the extent of expression of foreign proteins in cases where there is concern about toxicity, saturation of the translocon, and/or propensity of the expressed protein to aggregate.

In platform vectors pRKPLHT1 and pRKPLHT4, the oxygen-regulated *puf* promoter drives expression of foreign genes. In pRKPLHT4 (Fig. 3A), the *puf* L and *puf* M structural genes that encode the L and M subunits of the RC were replaced by a cassette carrying cloning sites, a C-terminal heptahistidine tag and two stop codons. A similar cassette replaced the *pufB* and *pufA* structural genes that encode the LH1 subunits in pRKPLHT1 (Fig. 3B). These two locations for foreign genes in the *puf* operon exploit the relative stoichiometry of the LH1 and RC polypeptides. This ratio is determined largely by differential transcript stabilities that are related to a region of RNA secondary structure elements located between the *pufA* and *pufL* genes that protects the *pufBA* segment of the transcript from exonuclease digestion (Klug, 1995). In pRKPLHT7, the *pucB* and *pucA* structural genes for the LH2 complex were replaced by an analogous cassette, yielding a platform vector in which expression of foreign genes is driven by the *puc* promoter that is responsive to both light and oxygen (Fig. 3c).

In practice, PCR-based cloning of foreign genes into these expression vectors is directed by N-terminal oligonucleotides that incorporate the consensus ribosome binding site of the photosynthetic gene cluster

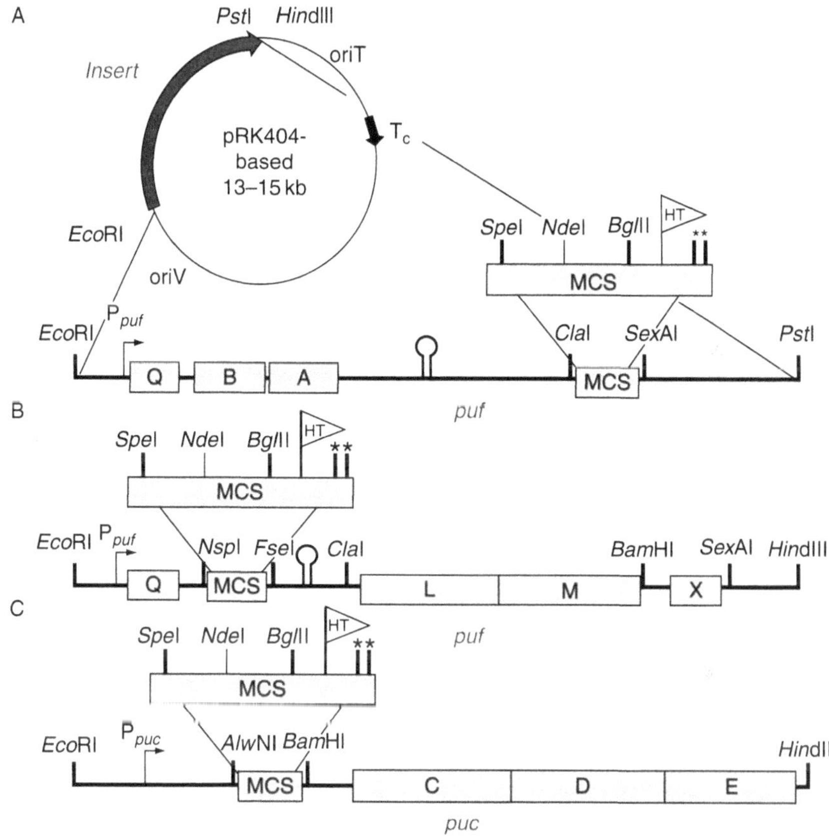

FIGURE 3 Vectors for expression of proteins in *Rhodobacter*. Platform expression vectors place foreign genes under control of the oxygen- and/or light-regulated *puf* (P*puf*) or *puc* (P*puc*) promoters. These vectors are derivatives of broad-host-range plasmid pRK404. Vectors based on the *puf* operon feature cloning cassettes replacing genes encoding either the RC subunits L and M (panel A) or LH1 subunits β and α (panel B). The third vector (panel C) is based on the *puc* operon and a cassette was used to replace the structural genes for the LH2 complex. Cloning sites are followed by a C-terminal heptahistidine tag fused in frame with two stop codons. A region of RNA secondary structure (hairpin) that dictates transcript stability is indicated in panels (A) and (B).

(Naylor, Addlesee, Gibson, & Hunter, 1999) and by C-terminal oligonucleotides that fuse the coding sequence in frame with the tag and stop codons. The same amplicon can be cloned into all three platform vectors. All vectors utilize transcription terminators present in the native operons. After cloning steps are completed from DNA propagated in *E. coli* hosts, expression plasmids are transferred to *R. sphaeroides* via conjugation from *E. coli* donor strain S17-1 (Simon, Priefer, & Puhler, 1983).

Expression plasmids are derivatives of broad-host-range plasmid pRK404 (Ditta et al., 1985), which is maintained stably *in trans* in *R. sphaeroides* by selection for tetracycline resistance (1 mg/l). The inventory of expression vectors also includes plasmids for either restriction enzyme-based or ligation-independent cloning that feature both cleavable and uncleavable N-terminal affinity and epitope tags, extended N- and C-terminal affinity tags (10× His, 13× His), N-terminal signal and membrane anchor peptides, and a C-terminal epitope tags (e.g., rhodopsin 1D4; Molday & MacKenzie, 1983).

B. Host Design

The inducible ICM was targeted for localization of the heterologously expressed membrane proteins, thus the effects of engineered deletions of native ICM proteins were investigated. To test the hypothesis that a partially depleted ICM could accommodate more foreign membrane protein, four strains of *R. sphaeroides* were evaluated as hosts for heterologous expression. These strains differ in both the nature and number of native complexes of the photosynthetic apparatus present in the ICM (Fig. 4A). They range from a true wild type to a strain that is deleted for the three transmembrane protein complexes of the photosynthetic apparatus: wild-type ATCC17023 (RC^+ $LH1^+LH2^+$; PS^+), PUC705-BA ($RC^+LH1^+LH2^-$; PS^+; Lee, Kiley, & Kaplan, 1989), PUFΔLMX21 ($RC^-LH1^+LH2^+$; PS^-; Farchaus & Oesterhelt, 1989), and ΔΔ11 ($RC^-LH1^-LH2^-$; PS^-; Pokkuluri et al., 2002). Strains ATCC17023 and PUC705-BA are photocompetent (PS^+), while deletion of the RC in strains PUFΔLMX21 and ΔΔ11 renders them incapable of photosynthetic growth (PS^-).

1. ICM Morphology

The effect of deletion of native transmembrane complexes on the morphology and volume of the ICM was examined by transmission electron microscopy. In the wild-type organism, the ICM appeared as vesicles (ATCC17023; Fig. 4B). Deletion of the LH2 complex in the PUC705-BA strain of *R. sphaeroides* yielded a strain characterized by tubular membranes (Drews & Golecki, 1995; Fowler et al., 1995; Golecki & Heinrich, 1991; Golecki, Ventura, & Oelze, 1991; Hunter, Pennoyer, Sturgis, Farrelly, & Niederman, 1988; Kiley, Varga, & Kaplan, 1988; Fig. 4E). The PUFΔLMX21 strain that synthesizes the LH1 and LH2 complexes, but carries a deletion of the RC, looked much like the native strain (Fig. 4D), and a strain lacking all three complexes of the photosynthetic apparatus (ΔΔ11) was characterized by a less structured ICM (Fig. 4C). Complementation of these deletions with wild-type LH1/RC or LH2 complexes expressed recombinantly *in trans* restored

A Host name	Native complexes LH2	LH1	RC
ΔΔ11	–	–	–
PUC705-BA	–	+	+
PUFΔLMX21	+	+	–
ATCC17023	+	+	+

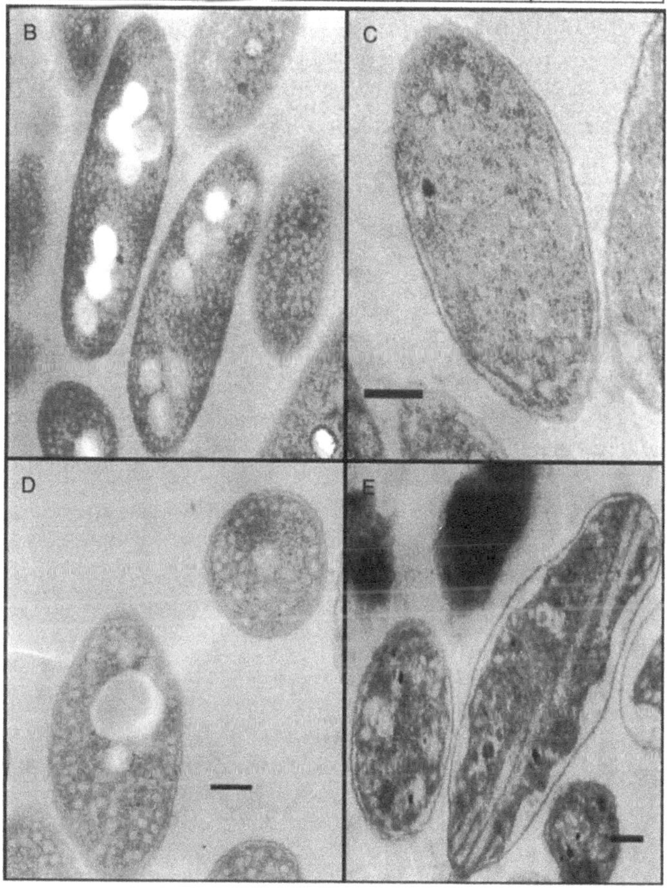

FIGURE 4 Host strains for foreign membrane protein expression in *Rhodobacter*. (A) Host strains of *R. sphaeroides* used in this study differ in both the nature and number of native complexes of the photosynthetic apparatus present in the ICM (Fig. 1). (B–E) Transmission

the volume and many of the morphological features of the native ICM (not shown). Similarly, spherical vesicles were restored in ΔΔ11 carrying an expression plasmid producing foreign membrane proteins, and the size of the membrane fraction in cell lysates is comparable in semiaerobic cultures of all ΔΔ11 expression strains.

2. Space to Accommodate Foreign Proteins

The yield of heterologous expression in semiaerobic cultures of each of these host strains was measured to determine whether deletion of native complexes favored ICM incorporation of expressed foreign membrane protein. Membrane-bound cytochrome c_y of *Rhodobacter capsulatus* was used as a reporter protein to indicate the level of membrane insertion in a particular host strain. Data revealed that yields were highest in two deletion strains—ΔΔ11 (fully deleted) and PUC705-BA (LH2 deleted). The wild-type host and the PUFΔLMX21 strain (retaining both light-harvesting complexes) incorporated lesser amounts of expressed protein under these conditions.

C. Autoinduction of Heterologous Expression

In all of these strains, protein expression is controlled easily. By manipulating the culture conditions of this versatile organism with regard to the induction cues (i.e., oxygen and/or light), expression of the desired protein is coordinated with the synthesis of the ICM. Depending upon the host strain employed, cultures can be grown either chemoheterotrophically or photosynthetically. In photosynthetic cultures, concomitant synthesis of ICM and heterologous protein is autoinduced by anaerobiosis and/or light. In semiaerobic chemoheterotrophic cultures, coupled synthesis is autoinduced when the oxygen tension lowers as the cell density increases. Either type of growth strategy used to overexpress heterologous protein can employ either rich or minimal defined media [e.g., YCC (Taguchi et al., 1992) or MR26 (Laible, Hata, Crawford, & Hanson, 2005), respectively].

Even in semiaerobic chemoheterotrophic growth modes, *Rhodobacter* is a highly pigmented organism, and pigment biosynthesis (reviewed in Marrs, Young, Bauer, & Williams, 1990) responds to the same induction cues as those employed for concomitant induction of ICM and protein synthesis.

electron micrographs of thin sections of cells from induced (semiaerobic, dark) cultures of the four host strains described herein. Effects of deletion of native complexes of the photosynthetic apparatus (panel A) on the ultrastructure of the ICM are illustrated: (B) ATCC17023, (C) ΔΔ11, (D) PUFΔLMX21, (E) PUC705-BA. Scale bars represent 200 nm.

In the early stages of semiaerobic growth, the culture is unpigmented because the cell density is low and the oxygen levels are relatively high. As the cell density increases, the oxygen concentration becomes limiting and the culture becomes increasingly pigmented. Although culture color does not imply success in accumulation of heterologously expressed protein, the natural pigmentation of semiaerobic *Rhodobacter* cultures can be exploited to indicate that the induction conditions which result in the concomitant synthesis of target membrane protein and new membrane have been achieved. The culture shown in Fig. 5B is expressing a foreign membrane protein whose synthesis has been autoinduced in semiaerobic growth conditions (dark, 125 rpm) and is highly pigmented. Doubling times for these conditions range from 2.5 to 6 h, depending upon the richness of the medium. Repression of the *puf* and *puc* promoters occurs under aerobic conditions; these colorless cultures are achieved by sparging the culture with oxygen or by rapid shaking (350 rpm) of flasks carrying a relatively small amount of medium (Fig. 5C). This strategy can be employed when induction of ICM and heterologous protein synthesis need to be controlled tightly, as in the case of expression of a toxic protein or incorporation of selenomethionine (or other labeled compounds) into induced proteins (Laible et al., 2005).

When a photocompetent strain (e.g., ATCC17023 or PUC705-BA; Fig. 5A) is employed as the expression host, the ICM is induced maximally, culture densities reach higher levels, and anaerobiosis and/or light can be used to induce expression of genes under the control of the *puf* and *puc* promoters. From semiaerobic chemoheterotrophic culture, it is easy to

FIGURE 5 *Rhodobacter* cultures. (A) Photosynthetic culture of a *R. sphaeroides* expression strain; (B) semiaerobic, chemoheterotrophic culture of a *R. sphaeroides* expression strain in which production of the target membrane protein is autoinduced; (C) aerobic culture of a *R. sphaeroides* expression strain in which production of the target membrane protein is repressed.

obtain >10 mg of many purified proteins per liter of cell culture, yet others are expressed at lower levels. It is possible that photosynthetic growth could increase the amounts of protein expressed in all strains, and this culture regime is of particular interest for those proteins that fall into low- or borderline-expressing categories. Under these conditions, the amount of protein that is produced could reach useful levels, thus possibly rescuing some strains that did not appear promising initially. Recently, extremely high cell densities have been achieved under modified (high light) photosynthetic conditions and doubling times have been reduced to 30 min (B. Curtis and W. R. Curtis, personal communication).

D. Host/Vector Combinations for Production of Membrane Proteins

The ability of a combination of vector and host to produce quantities of foreign membrane proteins for structural and functional studies depends upon a number of factors. These include the ability to grow the culture to high cell density under conditions where foreign gene expression is coupled to ICM expression, leading to the incorporation of the expressed membrane protein into the ICM. To identify host/vector combinations where these criteria can be met adequately to produce useful quantities of foreign membrane proteins, two reporter proteins—the soluble green fluorescent protein (GFP) from *Aequoria victoria* (Yang, Moss, & Phillips, 1996) and the membrane-bound cytochrome c_y—were cloned into the pRKPLHT1, pRKPLHT4, and pRKPLHT7 vectors and their expression levels in the four different host strains were quantified by UV-vis spectroscopy following purification of the polyhistidine-tagged proteins by immobilized metal affinity chromatography (IMAC). Expression of the soluble GFP served to indicate whether chemoheterotrophic growth conditions were optimal for induction of the *puf* and *puc* promoters, while expression of the membrane-bound cytochrome c_y indicated the level of membrane insertion of the expressed protein in a particular host strain. Synthesis of GFP in semiaerobic cultures was greatest from the pRKPLHT1 vector in each host strain, bearing out the expectation that replacement of LH1 genes by the foreign gene would result in higher yields than replacement of RC genes (pRKPLHT4) in the *puf* operon-based vectors. The yield of protein from the *puc*-based vector (pRKPLHT7) was expected to be the highest of the three platform vectors, and that expectation was borne out when the yield of protein from semiaerobic cultures was quantified on a per cell basis. However, oxygenation conditions that induce this promoter and produce respectable cell densities are more difficult to achieve in shake flasks, thus affecting the absolute yield of heterologous expression on a per volume basis. When all of the expression results from both cytochrome c_y and GFP were considered, the highest yields of foreign protein in chemoheterotrophic culture were obtained with the host/vector combination of ΔΔ11/pRKPLHT1. This expression system has now been used to synthesize hundreds of proteins, examining the versatility of *Rhodobacter* for expressing a wide variety of target membrane proteins.

E. Detection and Quantification of Expressed Proteins

To analyze a large number of expression constructs, generic and efficient screening methods have been developed to determine whether a target protein is expressed successfully and incorporated into the ICM. In brief, the expression strains were grown in 80-ml chemoheterotrophic cultures. Proteins from whole cell lysates and membrane fractions of aliquots of these cultures were prepared for SDS-polyacrylamide gel electrophoresis (SDS-PAGE; Fig. 6A, upper panel). Immunoblotting techniques were applied routinely for detection and rough quantitation of expression levels in *Rhodobacter* of polyhistidine-tagged foreign membrane proteins. These immunoblots (Fig. 6A, lower panel) are conclusive, general, and help identify membrane proteins that are often characterized by anomalous mobility on SDS gels.

Success in expressing a target membrane protein was measured by comparing the immunoblot signal from the target protein with that of a positive control protein that is known to be expressed at a level of 1 mg/l culture (+ control; Fig. 6B). A recombinant strain harboring an "empty" platform vector served as the negative control (− control; Fig. 6B). If the signal from a target protein in the immunoblot was equivalent to or greater than that of the positive control, it was scored as a "hit" and was a candidate for purification and characterization. Some target membrane proteins were expressed at levels below 1 mg/l culture, and, while purification of these expressed proteins would be extremely cumbersome, it could be pursued, depending upon the value of the target.

F. Cellular Localization of Heterologously Expressed Proteins

The screening process routinely incorporates the determination of whether the foreign membrane proteins were incorporated into the ICM. This specialized membrane is contiguous with the cytoplasmic membrane, but differs from the latter in its chemical and protein composition, its morphological and physical properties, and in its kinetics of biogenesis (Drews and Golecki, 1995; Verméglio et al., 1995). Upon cell lysis via mechanical

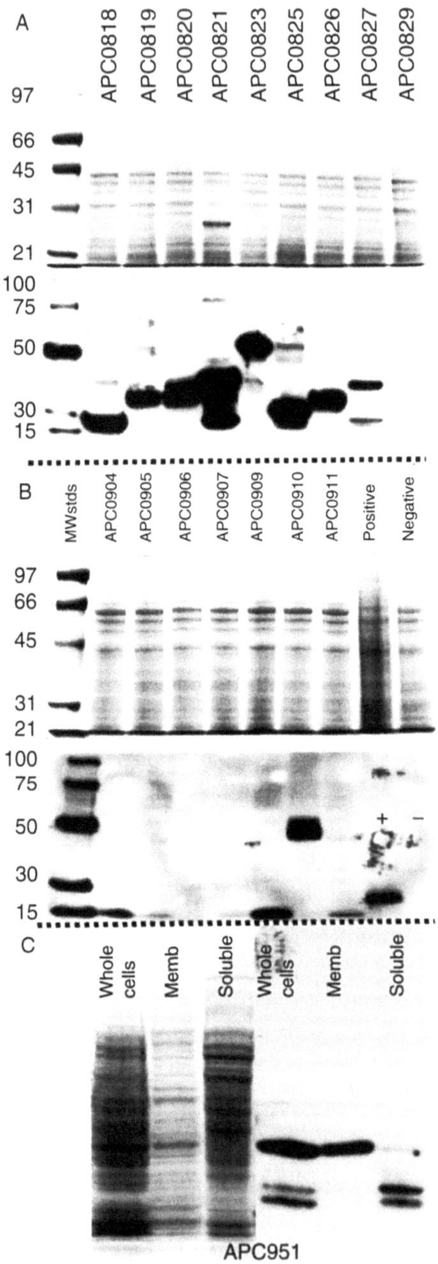

FIGURE 6 Screening for protein expression and cellular localization. (A,B) Results of screening for successful *Rhodobacter* expression and ICM insertion from small-scale cultures. Membrane fractions are extracted and probed using immunoblot analysis (e.g., anti-His, Novagen; bottom panels). Overexpressed bands are not always clearly visible in Coomassie-stained gels (top panels). Signals from experimental lanes are compared with signals from positive (+)

breakage (e.g., French press, microfluidizer), the ICM invaginations break apart from the cytoplasmic membrane, becoming sealed inside-out vesicles that can be recovered easily using differential centrifugation. In the wild-type *Rhodobacter* strains, this fraction is rich in the integral membrane proteins that constitute the photosynthetic apparatus. In engineered expression strains, this fraction contains the heterologous membrane protein.

The cellular localization of the target protein was tracked using appended tags. The immunoblot methods employed also reported whether any tagged proteins were present in inclusion bodies or had been cleaved by proteases. Typical results are shown in Fig. 6C. Here, target membrane protein APC00951 (28 kDa) was found almost exclusively in the membranes. The small amount of target protein that was found in the soluble fraction resulted from small membrane fragments that were not harvested during ultracentrifugation. These small ICM vesicles were pelleted quantitatively by centrifugation of greater duration or force (not shown). Cross-reacting host proteins (or his-tagged degradation products) resided exclusively in the soluble fraction. ICM localization of the expressed foreign protein was taken as an indicator that the protein possessed at least some degree of structural integrity that directed membrane insertion. Examination of various cell fractions for hundreds of expressed proteins has failed to produce clear evidence for the formation of inclusion bodies in any *R. sphaeroides* strain expressing foreign proteins.

G. A Case Study

Genes representing the entire *E. coli* membrane proteome were selected for cloning and expression studies by bioinformatics analysis of the annotated *E. coli* genome. From the 1030 hydrophobic proteins identified, 444 were selected as representatives of unique membrane proteins encoded by this genome whose structure is unknown (i.e., they shared $<30\%$ homology with any structures deposited in the Protein Data Bank). The majority of these targets were unannotated or "hypothetical" membrane proteins. This set of genes was cloned into the ΔΔ11[pRKPLHT1] host-vector combination and expression was autoinduced in chemoheterotrophic cultures.

Analysis of more than 400 of the expression strains has shown that approximately 60% of the *E. coli* membrane proteins were expressed in *R. sphaeroides* at levels that exceed 1 mg/l. Proteins that were expressed at

and negative (−) control strains. (C) Membrane proteins expressed heterologously in *Rhodobacter* are localized to the ICM. Here, target protein APC951 (28 kDa) is found almost exclusively in the membranes.

or above this threshold level encompassed a wide range of size (14–86 kDa), isoelectric points (5.9–11.9), and transmembrane spans (1–14). Subcellular fractionation revealed that these expressed *E. coli* membrane proteins were localized within the *Rhodobacter* ICM, and many could be purified to yields of >10 mg/l of culture, rivaling the expression levels of native ICM proteins.

H. Application to Eukaryotic Target Proteins

After this initial case study, targets of increasing complexity have been introduced into the *Rhodobacter* expression system. This expanded target set includes membrane proteins from a variety of functional classes and organisms—including eukaryotes. Among these targets are G-protein-coupled receptors (GPCRs) and ion channels that are of enormous interest in the development of pharmaceuticals and biomimetic devices. Success has been achieved in expressing GPCRs in *Rhodobacter* (Fig. 7), including human HIV coreceptors CCR5 and CXCR4. *Rhodobacter* also excels in expression of multisubunit complexes, including those requiring protein maturation and attachment of complex cofactors (De Smet et al., 2001; Graichen et al., 1999; Kappler and McEwan, 2002; Kirmaier et al., 2002; Laible et al., 2004).

IV. MEMBRANE PROTEIN PREPARATIONS FOR STRUCTURAL AND FUNCTIONAL STUDIES

As the overall objective using the *Rhodobacter* membrane protein expression system is to produce membrane protein samples in native form, the exclusive localization of target membrane proteins to the *Rhodobacter* ICM in concert with semiautomated chromatographic protocols allows for overexpression and efficient purification of target membrane proteins in the desired state(s).

A. Production Protocols

Rapid, generic, and reproducible methods have been developed for solubilizing and purifying heterologously expressed proteins from the membranes of *R. sphaeroides* (e.g., Laible et al., 2004). Steps in the production of milligram quantities of purified membrane proteins from *Rhodobacter* cells follow schemes typical for soluble protein production in more typical prokaryotic hosts (e.g., *E. coli* cells) with important key distinctions (indicated by outlines in Fig. 8). Membrane protein production requires the

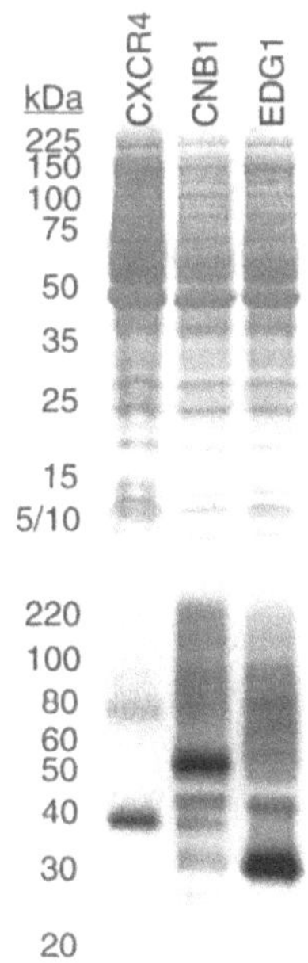

FIGURE 7 Eukaryotic protein expression in *Rhodobacter*. Expression of human GPCRs in the *Rhodobacter* ICM. (Upper panel) Coomassie-stained gel of membrane samples; (Lower panel) corresponding immunoblot (anti-1D4; Flintbox, Vancouver, BC) showing expression of HIV coreceptor CXCR4, cannabinoid receptor CNB1, and endothelial differentiation sphingolipid receptor EDG1.

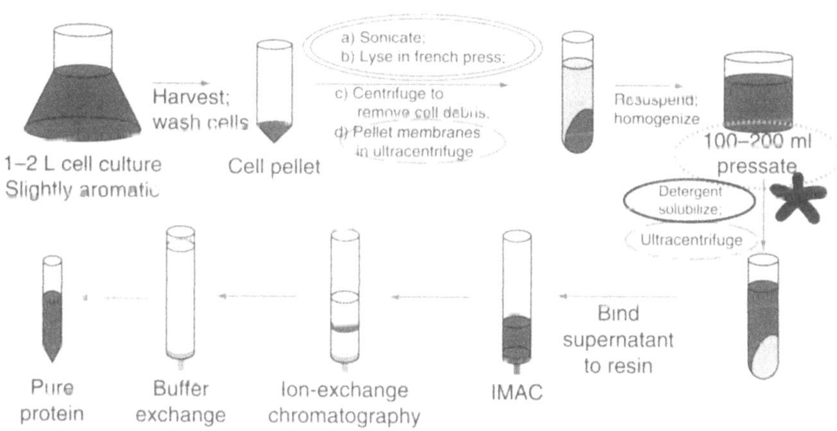

FIGURE 8 Production scheme. Steps in the production of milligram quantities of purified membrane protein samples from *Rhodobacter* cells.

ultracritical step of detergent solubilization (* outline) that removes them from the lipid bilayer and prepares them for chromatographic separation using aqueous phases. The addition of detergents requires careful matching of it to both protein and lipid levels and dictates that the volume of starting material for chromatography is rather large (dotted outline). This scheme also requires two critical ultracentrifugation steps (dashed outlines): (i) a step after cell breakage used to separate membranes from the soluble fraction before detergent solubilization and (ii) a step to remove membrane debris, preparing the solubilized membrane protein suspension for column chromatography. The last difference to note is that efficient recovery of ICM vesicles requires mechanical breakage (double outline) of *Rhodobacter* cells using a French press or microfluidizer at 12,000–18,000 psi.

B. Exploiting Affinity Tags in Purification

Utilizing peptide tags engineered into the expression vectors, detergent-solubilized target membrane proteins can be purified readily by affinity chromatography (e.g., immobilized metal or antibody columns). These methods are specific, their rapidity can facilitate purification of the target protein in its native state, and their general utility eliminates the need to determine *de novo* the type of chromatography which will be successful for each protein. By combining affinity chromatography sequentially with gel-filtration and/or ion-exchange steps, highly purified heterologously expressed target proteins can be recovered rapidly from the ICM. These protocols have been used successfully with a wide variety of detergents (zwitterionic, charged, nonionic, etc.). The uniformity and yield of the purified samples is increased significantly if the cells are harvested and are lysed in the presence of protease inhibitors; it is important to use EDTA-free inhibitor cocktails if pursuing divalent metal affinity strategies in any of the downstream steps.

Some examples of final purity levels reached for heterologously expressed target proteins encompassing a molecular weight range of 17–48 kDa are shown in Fig. 9. These samples were purified in two dimensions, that is, they were subjected to affinity chromatography followed by a gel filtration step. Purities of >90% were achieved with the two-step process, yielding both quantities (~1–10 mg) and purity levels that were sufficient for studies of structure and function. The ease of purification of a particular membrane protein is correlated strongly with expression and solubilization yields. In this vein, purification of proteins expressed at low levels is more difficult but can be successful and worthwhile, depending upon the value of the target. For vigilant assessment of purity, samples of purified target membrane proteins were routinely concentrated and overloaded (at least 25 μg protein per lane) on SDS gels. Following purification, target proteins were stored by

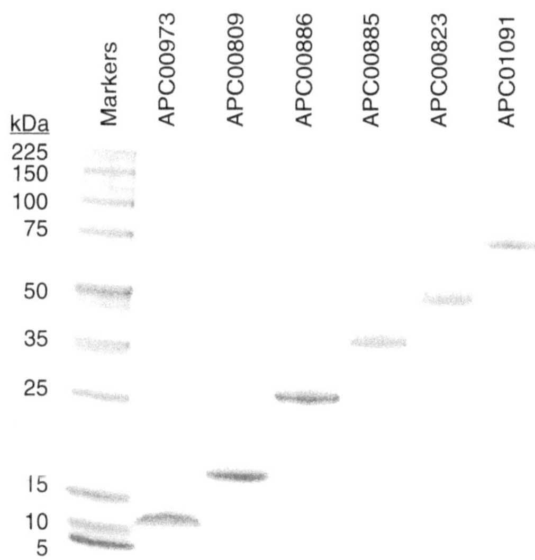

FIGURE 9 Pure membrane proteins of varied size. Molecular weight range of target membrane proteins expressed heterologously in *Rhodobacter*, analyzed by SDS-PAGE after purification. Protein concentrations range from 1 to 5 μg.

suspension in stabilizing detergent micelles or by reconstitution into a lipid environment (e.g., liposomes, lipidic-cubic phases, etc.). Structural and functional integrity of foreign membrane proteins expressed in *Rhodobacter* was confirmed by activity assays, when available and practical. For those proteins lacking specific activity assays, structural and functional integrity was suggested by the presence of strong signals from α-helices and β-sheets in circular dichroism spectra (not shown).

C. A Higher Throughput Approach Towards Purification

The protocols for purification of membrane proteins from the *Rhodobacter* ICM are applicable to both native and SeMet-labeled target membrane protein. The size-exclusion and ion-exchange "polishing" steps became more important for SeMet-substituted proteins since they were commonly obtained with reduced yields and purities. In addition, the efficiency by which these protocols can be carried out benefited from their adaptation to automation using state-of-the-art, fast-protein-liquid-chromatography (FPLC™) workstations (model ÄKTA™) designed by GE Healthcare (Uppsala, Sweden). Following two manual steps—isolation of the ICMs and extraction of the target proteins from this *Rhodobacter* lipid bilayer by the use of detergents—the chromatography could be carried out serially and automatically by these specialized FPLCs (Laible et al., 2004). These systems allowed for flexibility to be introduced into the purification schemes. For example, binding of polyhistidine-tagged target membrane proteins to Ni-charged columns can be rather slow—possibly due to restricted accessibility of the peptide tag because the target membrane protein is suspended in detergent micelles—and the automated routines can be set up to allow the protein to pass over the IMAC column several times to increase binding efficiency.

Protocols utilized with a single ÄKTA-FPLC can purify up to three different proteins—most readily overnight—using two chromatographic steps each, for example, IMAC followed by buffer exchange. In this simplest of chromatographic combinations, the latter step removes the eluant imidazole from the IMAC step and prepares the sample for concentration and crystallization trials. The purity of the target proteins following IMAC and buffer exchange was usually greater than 80%, as judged by SDS-PAGE; however, the purity of target membrane proteins overexpressed extremely well (>10 mg/l culture) typically exceeds 90%. In more complex purification schemes, buffer exchange can be replaced by either ion-exchange or gel-filtration chromatographies, but no more than two steps can be run in an automated fashion for each of the three proteins due to hardware limitations.

Purification cycles are highly reproducible—both in terms of sample yields, sample fractionation, and resultant sample purity. The largest factor in any variability that is seen is due to the cell culture and expression induction phases. If using IMAC in the initial steps of target membrane protein preparation, final sample purities can be influenced significantly by the affinity resins employed. So, if purity is of utmost concern, exploration of variations in the metal employed and the organic linker used to attach it to agarose beads has proven to be beneficial in decreasing the numbers and types of impurities observed after multiple column steps, and can allow for more stringent washing and sharper elution profiles. Yields have reached up to 10-20 mg of pure protein/l of culture. Greater than 100 mg of a single heterologously expressed target protein have been purified in a given FPLC run. Thus, the capacity of an ÄKTA-FPLC is not limiting, as the two membrane isolation steps (dashed outlines highlighted in Fig. 8) remain the slowest steps in the global process.

D. Matching Membrane Proteins with Detergents

Following purification—or, routinely, during purification—it is often necessary to replace the detergent used for solubilization and purification of a membrane protein with a secondary detergent that is more suitable for its stabilization, characterization, and crystallization. This idea stems from the fact that not all detergents serve as good solubilizing agents, and not all detergents safeguard the native, functional state of a membrane protein (Caffrey, 2003; le Maire, Champeil, & Moller, 2000; Nollert, 2005; Prive, 2007; Seddon, Curnow, & Booth, 2004; Wiener, 2004). The functional half-lives of membrane proteins are usually short once they are outside the lipid bilayer, necessitating the need for rapid methods for both the detergent-exchange and quantification processes to allow timely utilization of samples in structural and functional studies before significant degradation occurs. Passing large amounts of buffer by protein while it is bound to ion-exchange or affinity columns has proven to be an effective means to replace a detergent used for purification with a detergent of utility for further studies (Seddon et al.). In contrast, detergent exchange using dialysis proceeds relatively slowly and has been proven to be efficient only when replacing a detergent which is characterized by a low molecular weight and a high critical micelle concentration (CMC); very poor results have been obtained when trying to replace detergents with low CMCs (Seddon et al.; Wiener). The success of any given detergent-exchange process can be assessed by thin-layer chromatography methods (Eriks, Mayor, & Kaplan, 2003) utilizing iodine staining for detergent visualization (Kates, 1986). These methods have the added benefits that they are not only inexpensive and fast, but also allow for the observation, identification, and quantification of lipids that copurify with the target membrane protein at the same time that the detergent content of the protein sample is being probed (Laible & Kors, in preparation). Thus, automated chromatographic workstations permit the most efficient type of detergent exchange, as on-column detergent exchange is effective and quantitative during affinity chromatography and ion-exchange chromatography steps providing that wash volumes exceed 10 times the bed volume of the column. When performing the exchange using ion-exchange columns, care must be taken to ensure that both the initial and final detergents are compatible with the chromatographic resin. Nonionic and zwitterionic detergents are the best types of surfactants to use with ion-exchange chromatographies.

V. Practical Aspects of Heterologous Protein Expression in *Rhodobacter*

A. Potential Limitations

Strategies for heterologous protein expression employing *R. sphaeroides* can be designed to take full advantage of this organism's metabolic diversity to control the timing and rate of protein synthesis, increasing the likelihood of success. *Rhodobacter* differs from a typical *E. coli* expression host in that it has not been as highly engineered to afford it as many "user-friendly" features. The G + C content of its genome is 68%, thus its codon preference is skewed accordingly. Some rare codons are common between *E. coli* and *Rhodobacter*, but the latter organism is characterized by its own unique set of underutilized codons. In practice, this feature has not been found to be a limitation in the expression of genes encoded by distantly related organisms. Since *R. sphaeroides* is a prokaryote, the varieties of posttranslational modifications that are characteristic of eukaryotic hosts do not occur. Because posttranslational modifications are often incomplete or heterogeneous in eukaryotic cells, they often introduce a level of variability in purified protein samples that thwarts their crystallization. In this regard, production of such target proteins in *Rhodobacter* may yield a protein product that is much more amenable to the growth of well-diffracting crystals.

B. Diversity of Rhodobacter Membrane Fractions

The strength of the *Rhodobacter* membrane protein expression system is the temporal coupling of the induction of the synthesis of ICMs with the synthesis of foreign membrane protein. Many types of membrane preparations may be isolated from *Rhodobacter* cells, thus the design of particular structural or functional experiments can benefit from selection of different cell lysis and membrane fractionation procedures in order to take full advantage of the characteristics of the ICM bilayer that houses the heterologously expressed membrane proteins. A factor affecting the nature of membrane samples (size, morphology, abundance, protein complement) that are produced is the identity of the host strain from which they were derived (Fig. 4A).

The types of *Rhodobacter* membrane subpopulations that can be obtained are outlined in Fig. 10A. A notable difference between these subpopulations is their "sidedness," and this feature can be exploited in the design of, for example, functional assays, strategies for attachment of oriented membrane fragments to a surface, or in the selection of affinity reagents that recognize a particular exposed domain of an embedded target membrane protein. Protocols for producing each class of membrane subpopulation are well documented, as are methods for characterizing their protein complement and orientation/sidedness, that is, their "purity" (Dierstein, Schumacher, & Drews, 1981; Jungas, Ranck, Rigaud, Joliot, & Vermeglio, 1999; Lommen & Takemoto, 1978; Reilly & Niederman, 1986; Takemoto & Bachmann, 1979). The three major classes are as follows.

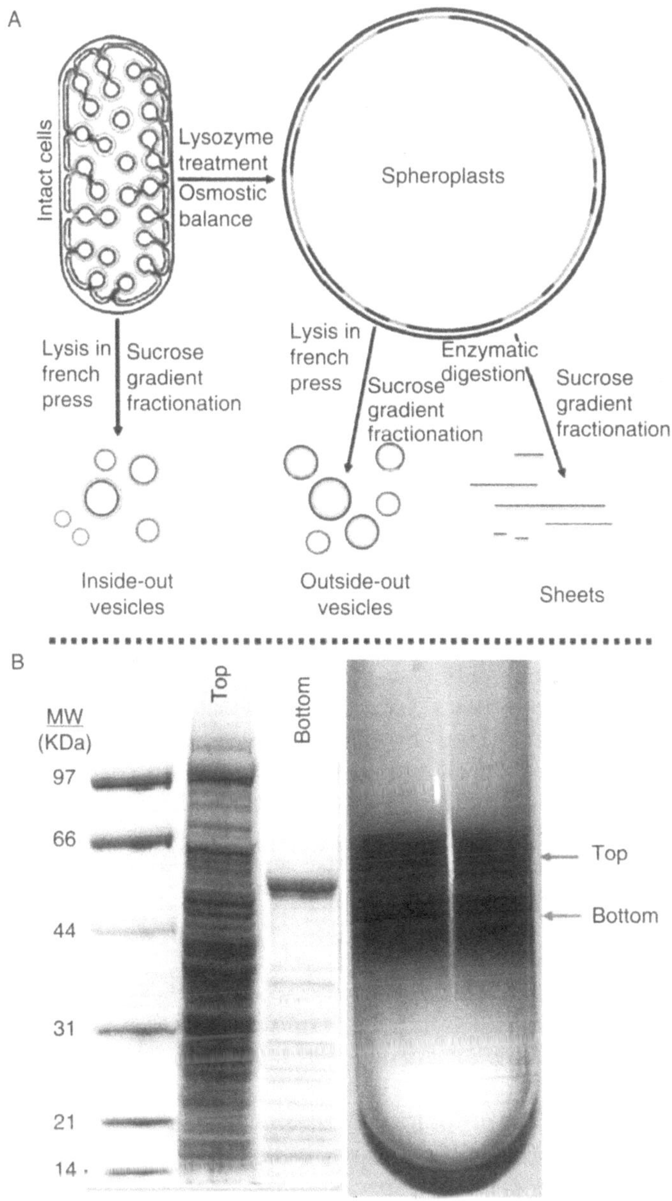

FIGURE 10 *Rhodobacter*-specific membrane preparations. (A) Protocols exist for the preparation of different types of vesicles and fragments from ICMs in *Rhodobacter* cells. In each of these protocols, ICM-specific fractions can be isolated easily from other cellular components via density separation. (B) *Rhodobacter* ICM fractions in which the target membrane proteins are predominant can be separated on sucrose gradients (right, arrows). Purity assessment of these "bands" on SDS-PAGE (left) suggests that the target protein constitutes ~60% of the total protein in the bottom band.

1. Inside-Out Vesicles

These vesicles are obtained easily by mechanical disruption of *Rhodobacter* cells. The outer surface of the particle is the cytoplasmic face of the ICM, and the inner face constitutes the periplasmic face of the membrane. The interior of the particle is a hydrophilic environment that encloses soluble components normally localized to the cell's periplasmic space. This is a reducing environment and includes enzymes that assist with protein maturation (e.g., disulfide bond formation and heme maturation and attachment). The outside surface of the membrane contains membrane-bound proteins that direct the folding and insertion of membrane proteins (Troschel, Eckhardt, Hoffschulte, & Muller, 1992). These vesicles are often used in functional assays since the expressed membrane protein resides in a "native" environment that is undisturbed by detergents. They also serve as the starting material in protein purification procedures.

2. Outside-Out Vesicles

This type of sample is produced via generation of spheroplasts by treatment of *Rhodobacter* cells with lysozyme and EDTA in an osmotically stabilized solution (Fig. 10A). Upon disruption of cells by osmotic shock or mechanical means, outside-out vesicles of varied size are produced (Lommen & Takemoto, 1978). The ones most enriched in lipids and proteins of the ICM can be isolated by density separation methods (vide infra; Fig. 10B). In this case, the membrane surfaces have the opposite orientation as the inside-out vesicles described in Section V.B.1; soluble cytoplasmic components are trapped on the inside while soluble periplasmic components are released into the medium and are removed when the vesicles are harvested by centrifugation.

3. Planar Preparations

Either mechanical or enzymatic lysis of spheroplasts can also produce membrane sheets as a subpopulation. This fraction often takes the form of ordered two-dimensional arrays of protein and lipid and can be separated from the other vesicle types with density gradient centrifugation (Fig. 10B; Siebert et al., 2004). The proportion of sheets can be increased by manipulating the membrane protein complement of the host strain (Fig. 4A; Siebert et al., 2004). These fragments present both membrane surfaces to the surrounding medium.

Separation of membrane subpopulations via sucrose density gradient centrifugation is a technique that has been used extensively in preparation of membrane fragments that are enriched in native or foreign membrane proteins or membrane protein complexes expressed in *Rhodobacter*. It is very clear that there are several distinct membrane fractions that can be separated effectively (Fig. 10B), and, in practice, the target protein has been found to constitute >60% of the total protein in a particular fraction. When "cleaner" membrane preparations are necessary, a two-step isolation procedure can be adopted that separates subpopulations by both size and density. Likewise, treatment of either inside-out or outside-out vesicles with sub-CMC concentrations of detergent (e.g., 0.03% LDAO) leaves the vesicles intact but permeabilizes them such that trapped soluble proteins can diffuse out.

C. *Localization to a Membrane of Defined Chemical Composition*

Post-expression steps benefit from the reality that a particular target membrane protein expressed in the *Rhodobacter* ICM resides in a lipid bilayer of well-defined chemical composition (Benning, 1998). Large-scale tests—that incorporate activity assays for target proteins—have identified a small number of detergents which optimize the first step of solubilization (dismantling the bilayer). The same set of detergents should work well for any membrane protein that is localized to the ICM and can be used in a generic approach. Subsequent testing for functionality will determine the robustness of micelles of this "dismantling detergent set" in stabilizing membrane proteins. Preliminary evidence suggests that they might be quite good, as the *Rhodobacter* ICM can be broken down with detergents that are considered to be fairly "gentle" (e.g., Deriphat 160; Kirmaier, Laible, Hindin, Hanson, & Holten, 2003). The development of these generic approaches is aided by decades of work on photosynthetic complexes and it implements detergents that have worked well in extracting complexes of the photosynthetic apparatus that are characterized by markedly different stabilities outside the lipid bilayer.

VI. CONCLUDING STATEMENTS

Production of membrane proteins for structure determination remains a very challenging endeavor. New, emerging expression systems tailored for the overexpression of membrane proteins are exploiting unique features of several unconventional organisms as hosts in order to overcome many bottlenecks commonly encountered in membrane protein production. In the case of *Rhodobacter*, this membrane protein expression system puts to task the organism's photosynthetic machinery in a special way to produce the milligram quantities of membrane proteins that are needed for structural studies. This integrated approach towards membrane protein expression, solubilization, and purification can dovetail with other technologies described in this series to overcome the myriad, unique problems associated with the generation of well-diffracting crystals of membrane proteins for use in structure determination experiments.

Acknowledgments

The authors have benefited greatly from the contributions of laboratory members (past and present) and colleagues at Argonne National Laboratory for various aspects of the development and evaluation of the *Rhodobacter* expression system. They also thank David Mets and Hewson Swift for expertise in and facilities for electron microscopy, Marc Wander and Kelsey Wander for assistance with graphic design and preparation of figures, and GE Healthcare for suggestions regarding adaptation of ÄKTA-FPLCs for semiautomated purification of membrane proteins. Funding for these efforts was provided by the National Institutes of Health (R01 GM61887, P50 GM62414, R01 GM71318, and P01 GM75913) and the United States Department of Energy, under contract no. W-31-109-ENG-38.

References

Arechaga, I., Miroux, B., Karrasch, S., Huijbregts, R., de Kruijff, B., Runswick, M. J., et al. (2000). Characterisation of new intracellular membranes in *Escherichia coli* accompanying large scale over-production of the b subunit of F(1)F(o) ATP synthase. *FEBS Letters, 482*(3), 215–219.

Bahatyrova, S., Frese, R. N., Siebert, C. A., Olsen, J. D., Van Der Werf, K. O., Van Grondelle, R., et al. (2004). The native architecture of a photosynthetic membrane. *Nature, 430*(7003), 1058–1062.

Bahatyrova, S., Frese, R. N., van der Werf, K. O., Otto, C., Hunter, C. N., & Olsen, J. D. (2004). Flexibility and size heterogeneity of the LH1 light harvesting complex revealed by atomic force microscopy: Functional significance for bacterial photosynthesis. *The Journal of Biological Chemistry, 279*(20), 21327–21333.

Benning, C. (1998). Membrane lipids in anoxygenic photosynthetic bacteria. In P. A. Siegenthaler & N. Murata, (Eds.), *Lipids in photosynthesis: Structure, function, and genetics*, pp. 83–101. Dordrecht: Kluwer.

Caffrey, M. (2003). Membrane protein crystallization. *Journal of Structural Biology, 142*, 108–132.

Clayton, R. K. (1980). *Photosynthesis: Physical mechanisms and chemical patterns.* Cambridge: Cambridge University Press.

Collins, M. L. P., & Cheng, Y. (2004). In U.S. Patent Office. *Host/vector system for expression of membrane proteins.* USA. Patent number: 6,680,179.

Columbus, L., Lipfert, J., Klock, H., Millett, I., Doniach, S., & Lesley, S. A. (2006). Expression, purification, and characterization of *Thermotoga maritima* membrane proteins for structure determination. *Protein Science, 15*(5), 961–975.

De Smet, L., Kostanjevecki, V., Guisez, Y., & Van Beeumen, J. (2001). A novel system for heterologous expression of flavocytochrome *c* in phototrophic bacteria using the *Allochromatium vinosum* rbca promoter. *Archives of Microbiology, 176*(1–2), 19–28.

Dierstein, R., Schumacher, A., & Drews, G. (1981). On insertion of pigment-associated polypeptides during membrane biogenesis in *Rhodopseudomonas capsulata. Archives of Microbiology, 128*, 376–383.

Ditta, G., Schmidhauser, T., Yakobsen, E., Lu, P., Liang, X. W., Finlay, D. R., et al. (1985). Plasmids related to the broad host range vector, prk290, useful for gene cloning and for monitoring gene expression. *Plasmid, 13*, 149–153.

Drews, G. (1985). Structure and functional organization of light-harvesting complexes and photochemical reaction centers in membranes of phototrophic bacteria. *Microbiological Review, 49*(1), 59–70.

Drews, G., & Golecki, J. R. (1995). Structure, molecular organization, and biosynthesis of membranes of purple bacteria. In R. E. Blankenship, M. T. Madigan, & C. E. Bauer, (Eds.), *Anoxygenic photosynthetic bacteria*, pp. 231–257. Dordrecht: Kluwer.

Eriks, L. R., Mayor, J. A., & Kaplan, R. S. (2003). A strategy for identification and quantification of detergents frequently used in the purification of membrane proteins. *Analytical Biochemistry, 323*(2), 234–241.

Essen, L. O. (2002). Structural genomics of "non-standard" proteins: A chance for membrane proteins? *Gene Function and Disease, 3*, 39–48.

Farchaus, J. W., & Oesterhelt, D. (1989). A *Rhodobacter sphaeroides puf* L, M and X deletion mutant and its complementation *in trans* with a 5.3 kb *puf* operon shuttle fragment. *EMBO Journal, 8*, 47–54.

Fowler, G. J., Gardiner, A. T., Mackenzie, R. C., Barratt, S. J., Simmons, A. E., Westerhuis, W. H., et al. (1995). Heterologous expression of genes encoding bacterial light-harvesting complexes in *Rhodobacter sphaeroides. The Journal of Biological Chemistry, 270*(40), 23875–23882.

Gaertig, J., Gao, Y., Tishgarten, T., Clark, T. G., & Dickerson, H. W. (1999). Surface display of a parasite antigen in the ciliate *Tetrahymena thermophila. Nature Biotechnology, 17*(5), 462–465.

Golecki, J. R., & Heinrich, U. R. (1991). Ultrastructural and electron spectroscopic analyses of cyanobacteria and bacteria. *Journal of Microscopy, 162*(Pt. 1), 147–154.

Golecki, J. R., Ventura, S., & Oelze, J. (1991). The architecture of unusual membrane tubes in the B800-B850 light-harvesting bacteriochlorophyll-deficient mutant 19 of *Rhodobacter sphaeroides. FEMS Microbiology Letters, 77*, 335–340.

Graichen, M. E., Jones, L. H., Sharma, B. V., van Spanning, R. J., Hosler, J. P., & Davidson, L. (1999). Heterologous expression of correctly assembled methylamine dehydrogenase in *Rhodobacter sphaeroides. Journal of Bacteriology, 181*(14), 4216–4222.

Grisshammer, R., & Tate, C. (1995). Overexpression of integral membrane proteins for structural studies. *Quarterly Reviews of Biophysics, 28*(3), 315–422.

Gumpert, J., & Hoischen, C. (1998). Use of cell wall-less bacteria (L-forms) for efficient expression and secretion of heterologous gene products. *Current Opinion in Biotechnology, 9*(5), 506–509.

Hoischen, C., Fritsche, C., Gumpert, J., Westermann, M., Gura, K., & Fahnert, B. (2002). Novel bacterial membrane surface display system using cell wall-less L-forms of *Proteus mirabilis* and *Escherichia coli. Applied and Environmental Microbiology, 68*(2), 525–531.

Hunter, C. N., Pennoyer, J. D., Sturgis, J. N., Farrelly, D., & Niederman, R. A. (1988). Oligomerization states and associations of light-harvesting pigment protein complexes of *Rhodobacter sphaeroides* as analyzed by lithium dodecylsulfate polyacrylamide-gel electrophoresis. *Biochemistry, 27*, 3459–3467.

Imhoff, J. F. (Ed). (1995). Taxonomy and physiology of phototrophic purple bacteria and green sulfur bacteria. In R. E. Blankenship, M. T. Madigan, & C. E. Bauer, (Eds.), *Anoxygenic photosynthetic bacteria*. Dordrecht: Kluwer.

Jamieson, S. J., Wang, P., Qian, P., Kirkland, J. Y., Conroy, M. J., Hunter, C. N., et al. (2002). Projection structure of the photosynthetic reaction centre-antenna complex of *Rhodospirillum rubrum* at 8.5 Angstrom resolution. *EMBO Journal, 21*(15), 3927–3935.

Jungas, C., Ranck, J. L., Rigaud, J. L., Joliot, P., & Vermeglio, A. (1999). Supramolecular organization of the photosynthetic apparatus of *Rhodobacter sphaeroides*. *EMBO Journal, 18*(3), 534–542.

Kappler, U., & McEwan, A. G. (2002). A system for the heterologous expression of complex redox proteins in *Rhodobacter capsulatus*: Characterisation of recombinant sulphite:cytochrome *c* oxidoreductase from *Starkeya novella*. *FEBS Letters, 529*(2–3), 208–214.

Karrasch, S., Bullough, P. A., & Ghosh, R. (1995). The 8.5 angstrom projection map of the light-harvesting complex I from *Rhodospirillum rubrum* reveals a ring composed of 16 subunits. *EMBO Journal, 14*, 631–638.

Kates, M. (1986). Techniques of lipidology: Isolation, analysis, and identification of lipids. In *Laboratory techniques in biochemistry and molecular biology* (2nd revised Ed., Vol. 3, Pt. 2). New York, NY: Elsevier.

Kiefer, H., Maier, K., & Vogel, R. (1999). Refolding of G-protein-coupled receptors from inclusion bodies produced in *Escherichia coli*. *Biochemical Society Transactions, 27*, 908–912.

Kiley, P. J., & Kaplan, S. (1988). Molecular genetics of photosynthetic membrane biosynthesis in *Rhodobacter sphaeroides*. *Microbiological Reviews, 52*, 50–69.

Kiley, P. J., Varga, A., & Kaplan, S. (1988). Physiological and structural analysis of light-harvesting mutants of *Rhodobacter sphaeroides*. *Journal of Bacteriology, 170*(3), 1103–1115.

Kirmaier, C., Laible, P. D., Czarnecki, K., Hata, A. N., Hanson, D. K., Bocian, D. F., et al. (2002). Comparison of M-side electron transfer in *R. sphaeroides* and *R. capsulatus* reaction centers. *Journal of Physical Chemistry B, 106*(7), 1799–1808.

Kirmaier, C., Laible, P. D., Hindin, E., Hanson, D. K., & Holten, D. (2003). Detergent effects on primary charge separation in wild-type and mutant *Rhodobacter capsulatus* reaction centers. *Chemical Physics, 294*, 305–318.

Klug, G. (1995). Post-transcriptional control of photosynthesis gene expression. In R. E. Blankenship, M. T. Madigan, & C. E. Bauer, (Eds.), *Anoxygenic photosynthetic bacteria*, pp. 1235–1244. Dordrecht: Kluwer.

Koepke, J., Hu, X., Muenke, C., Schulten, K., & Michel, H. (1996). The crystal structure of the light-harvesting complex II (B800-850) from *Rhodospirillum molischianum*. *Structure, 4*, 581–597.

Korepanova, A., Gao, F. P., Hua, Y., Qin, H., Nakamoto, R. K., & Cross, T. A. (2005). Cloning and expression of multiple integral membrane proteins from *Mycobacterium tuberculosis* in *Escherichia coli*. *Protein Science, 14*(1), 148–158.

Laible, P. D., & Hanson, D. K. (2002). In U. P. a. T. Office, (Ed.), *Methods and constructs for expression of foreign proteins in photosynthetic organisms*. The University of Chicago, USA.

Laible, P. D., Hata, A. N., Crawford, A. E., & Hanson, D. K. (2005). Incorporation of selenomethionine into induced intracytoplasmic membrane proteins of *Rhodobacter* species. *Journal of Structural and Functional Genomics, 6*(2–3), 95–102.

Laible, P. D., Scott, H. N., Henry, L., & Hanson, D. K. (2004). Towards higher-throughput membrane protein production for structure genomics initiatives. *Journal of Structural and Functional Genomics, 5*, 167–172.

Lee, J. K., Kiley, P. J., & Kaplan, S. (1989). Post-transcriptional control of *puc* operon expression of B800-850 light-harvesting complex formation in *Rhodobacter sphaeroides*. *Journal of Bacteriology, 171*, 3391–3405.

le Maire, M., Champeil, P., & Moller, J. V. (2000). Interaction of membrane proteins and lipids with solubilizing detergents. *Biochimica et Biophysica Acta, 1508*(1–2), 86–111.

Lommen, M. A., & Takemoto, J. (1978). Comparison, by freeze-fracture electron microscopy, of chromatophores, spheroplast-derived membrane vesicles, and whole cells of *Rhodopseudomonas sphaeroides*. *Journal of Bacteriology, 136*(2), 730–741.

Marrs, B. L., Young, D. A., Bauer, C. E., & Williams, J. C. (1990). The regulation of bacteriochlorophyll synthesis in *Rhodobacter capsulatus*. In G. Drews, & E. A. Dawes, (Eds.), *Molecular biology of membrane-bound complexes in phototrophic bacteria*. New York: Plenum.

McDermott, G., Prince, S. M., Freer, A. A., Hawthornthwaite-Lawless, A. M., Papiz, M. Z., Cogdell, R. J., et al. (1995). Crystal structure of an integral membrane light-harvesting complex from photosynthetic bacteria. *Nature, 374*, 517–521.

Miroux, B., & Walker, J. E. (1996). Over-production of proteins in *Escherichia coli*: Mutant hosts that allow synthesis of some membrane proteins and globular proteins at high levels. *Journal of Molecular Biology, 260*, 289–298.

Molday, R. S., & MacKenzie, D. (1983). Monoclonal antibodies to rhodopsin: Characterization, cross-reactivity, and application as structural probes. *Biochemistry, 22*(3), 653–660.

Naylor, G., Addlesee, H., Gibson, L., & Hunter, C. N. (1999). The photosynthesis gene cluster of *Rhodobacter sphaeroides*. *Photosynthesis Research, 62*, 121–139.

Nguyen, H. H., Elliott, S. J., Yip, J. H., & Chan, S. I. (1998). The particulate methane monooxygenase from *Methylococcus capsulatus* (Bath) is a novel copper-containing three-subunit enzyme. Isolation and characterization. *The Journal of Biological Chemistry, 273*(14), 7957–7966.

Nollert, P. (2005). Membrane protein crystallization in amphiphile phases: Practical and theoretical considerations. *Progress in Biophysics and Molecular Biology, 88*(3), 339–357.

Pokkuluri, P. R., Laible, P. D., Deng, Y. L., Wong, T. N., Hanson, D. K., & Schiffer, M. (2002). The structure of a mutant photosynthetic reaction center shows unexpected changes in main chain orientations and quinone position. *Biochemistry, 41*(19), 5998–6007.

Prive, G. G. (2007). Detergents for the stabilization and crystallization of membrane proteins. *Methods, 41*(4), 388–397.

Qian, P., Addlesee, H. A., Ruban, A. V., Wang, P., Bullough, P. A., & Hunter, C. N. (2003). A reaction center-light-harvesting 1 complex (RC-LH1) from a *Rhodospirillum rubrum* mutant with altered esterifying pigments: Characterization by optical spectroscopy and cryo-electron microscopy. *The Journal of Biological Chemistry, 278*(26), 23678–23685.

Qian, P., Hunter, C. N., & Bullough, P. A. (2005). The 8.5A projection structure of the core RC-LH1-pufx dimer of *Rhodobacter sphaeroides*. *Journal of Molecular Biology, 349*(5), 948–960.

Reilly, P. A., & Niederman, R. A. (1986). Role of apparent membrane growth initiation sites during photosynthetic membrane development in synchronously dividing *Rhodopseudomonas sphaeroides*. *Journal of Bacteriology, 167*(1), 153–159.

Roszak, A. W., Howard, T. D., Southall, J., Gardiner, A. T., Law, C. J., Isaacs, N. W., et al. (2003). Crystal structure of the RC-LH1 core complex from *Rhodopseudomonas palustris*. *Science, 302*, 1969–1972.

Seddon, A. M., Curnow, P., & Booth, P. J. (2004). Membrane proteins, lipids and detergents: Not just a soap opera. *Biochimica et Biophysica Acta, 1666*(1–2), 105–117.

Siebert, C. A., Qian, P., Fotiadis, D., Engel, A., Hunter, C. N., & Bullough, P. A. (2004). Molecular architecture of photosynthetic membranes in *Rhodobacter sphaeroides*: The role of pufX. *EMBO Journal, 23*(4), 690–700.

Simon, R., Priefer, U., & Puhler, A. (1983). A broad host range mobilization system for *in vivo* genetic engineering: Transposon mutagenesis in gram negative bacteria. *Bio/Technology, 1*, 37–45.

Taguchi, A. K. W., Stocker, J. W., Alden, R. G., Causgrove, T. P., Peloquin, J. M., Boxer, S. G., et al. (1992). Biochemical characterization and electron-transfer reactions of *sym1*, a *Rhodobacter capsulatus* symmetry mutant which affects the initial electron donor. *Biochemistry, 31*, 10345–10355.

Takemoto, J., & Bachmann, R. C. (1979). Orientation of chromatophores and spheroplast-derived membrane vesicles of *Rhodopseudomonas sphaeroides*: Analysis by localization of enzyme activities. *Archives of Biochemistry and Biophysics, 195*(2), 526–534.

Troschel, D., Eckhardt, S., Hoffschulte, H. K., & Muller, M. (1992). Cell-free synthesis and membrane integration of the reaction center subunit H from *Rhodobacter capsulatus. FEMS Microbiology Letters, 91*, 129–133.

Turner, G. J., Reusch, R., Winter-Vann, A. M., Martinez, L., & Betlach, M. C. (1999). Heterologous gene expression in an membrane-protein-specific system. *Protein Expression and Purification, 17*, 312–323.

van Dijk, R., Faber, K. N., Kiel, J. A., Veenhuis, M., & van der Klei, I. (2000). The methylotrophic yeast *Hansenula polymorpha*: A versatile cell factory. *Enzyme and Microbial Technology, 26*(9-10), 793–800.

van Grondelle, R., Hunter, C. N., Bakker, J. G. C., & Kramer, H. J. M. (1983). Size and structure of antenna complexes of photosynthetic bacteria as studied by singlet-singlet quenching of the bacteriochlrophyll fluorescence yield. *Biochimica et Biophysica Acta, 723*, 30–36.

Verméglio, A., Joliot, P., & Joliot, A. (1995). Organization of electron transfer components and supercomplexes. In R. E. Blankenship, M. T. Madigan, & C. E. Bauer, (Eds.), *Anoxygenic photosynthetic bacteria*, pp. 279–295. Dordrecht: Kluwer.

Walz, T., & Grigorieff, N. (1998). Electron crystallography of two dimensional crystals of membrane proteins. *Journal of Structural Biology, 121*, 142–161.

Wiener, M. C. (2004). A pedestrian guide to membrane protein crystallization. *Methods, 34*(3), 364–372.

Yang, F., Moss, L. G., & Phillips, G. N., Jr. (1996). The molecular structure of green fluorescent protein. *Nature Biotechnology, 14*(10), 1246–1251.

CHAPTER 4

Monoacylglycerols: The Workhorse Lipids for Crystallizing Membrane Proteins in Mesophases

M. Caffrey,* J. Lyons,* T. Smyth,* and D. J. Hart†

*Department of Chemical and Environmental Sciences, University of Limerick, Limerick, Ireland

†Department of Chemistry, The Ohio State University, Columbus, Ohio 43210

I. OVERVIEW

The *in meso* method for crystallizing membrane proteins has been shown to work with an array of different protein types. The method involves reconstituting the target protein into the bilayer of a bicontinuous lipid mesophase followed by an induced phase separation brought on by the

Current Topics in Membranes, Volume 63

1063-5823/09 $35.00
DOI: 10.1016/S1063-5823(09)63004-0

addition of a precipitant. A mechanism has been proposed for how *in meso* crystallogenesis happens at the molecular level and the assorted and critical roles that hosting and additive lipids play, highlights the need for a wider palette of molecular species to choose from. In support of this, a monoacylglycerol synthesis and purification program has been implemented, the details of which are outlined in this chapter.

II. INTRODUCTION

It is over a decade since Landau and Rosenbusch (1996) reported that a lipidic cubic mesophase could support the crystallization of a membrane protein. The observation triggered considerable interest and hope that it would accelerate the production of crystals for use in establishing how structure dictated function and interaction in membrane proteins. Some success with the method was achieved in the period immediately following its introduction (Raman, Cherezov, & Caffrey, 2006, http://www.mpdb.ul.ie/) but uptake by the community was slow. The stumbling block was the highly viscous and sticky nature of the cubic phase itself which made it difficult to handle. Thus, preparing crystallization trials proved technically challenging and expensive in the amount of valuable protein and lipid required. The tide began to turn with the introduction of tools for manually and robotically handling the cubic phase in amounts that required miniscule quantities of protein (Caffrey, 2008a; Cherezov & Caffrey, 2006; Cherezov, Peddi, Muthusubramaniam, Zheng, & Caffrey, 2004). These were used to expand the range of the method from α-helical cytoplasmic to β-barrel outer membrane bacterial proteins, and more recently to two human G protein-coupled receptors (GPCR) (Cherezov, Clogston, Papiz, & Caffrey, 2006; Cherezov et al., 2006, 2007, 2008; Raman et al., 2006; Jaakola et al., 2008).

A proposal for how the method works at a molecular level has been advanced (Caffrey, 2000, 2003, 2009); several aspects of it have been demonstrated experimentally. A report that more fully explores elements of the proposal, especially those relating to nucleation and crystal growth, has been published (Caffrey, 2008b). The system in which *in meso* crystallogenesis takes place is complex involving a bicontinuous lipid bilayer separating two interpenetrating but noncontacting aqueous channels, reconstituted protein, and a variety of precipitant solution components (Fig. 1).

The focus of this chapter is on the lipid that forms the bicontinuous mesophase in which crystallization occurs. We begin the chapter with an overview of the proposed mechanism by which *in meso* crystallogenesis of membrane proteins comes about. The underlying lipid phase behavior that is central to the method is then introduced. This is followed by an examination of the many roles played by lipid in *in meso* crystallogenesis. The rest of the

FIGURE 1 *In meso* crystallization of membrane proteins. The figure shows a cartoon representation of the events proposed to occur during the crystallization of an integral membrane protein from the lipidic cubic mesophase. The process begins with the protein reconstituted into the curved bilayers of the bicontinuous cubic phase (bottom left hand quadrant of the figure). Added precipitants shift the equilibrium away from stability in the cubic membrane. This leads to phase separation wherein protein molecules diffuse from the continuous bilayered reservoir of the mesophase by way of a sheet like or lamellar conduit (left upper quadrant of figure) to lock into the lattice of the advancing crystal face (right upper quadrant of figure). Salt (positive and negative signs) facilitates crystallization, in part, by charge screening. Co-crystallization of the protein with native lipid (cholesterol) is shown in this illustration. As much as possible, the dimensions of the lipid (light yellow oval with tail), detergent (pink oval with tail), native membrane lipid (purple), protein (blue; b2AR-T4L; PDB code 2RH1), bilayer and aqueous channels (dark blue) have been drawn to scale. The lipid bilayer is approximately 40 A° thick. For interpretation of the references to color in this figure legend, the reader is referred to the Science Direct version of this chapter available online. Redrawn from (Cherezov & Caffrey, 2007).

chapter is devoted to describing the synthesis and purification of the monoacylglycerol (MAG) lipids for use with the *in meso* method. In Section VII, an indication is given as to where the field of *in meso* crystallogenesis is headed and how further advances will be facilitated by the availability of novel lipids through rational design and synthetic organic chemistry.

Before launching into a description of the *in meso* method and how it works it is appropriate to introduce the reader to the N.T notation for naming MAGs. As noted, the bulk of the *in meso* crystallogenesis reported on in this chapter makes use of MAGs containing *cis*-monounsaturated fatty acids. A shorthand system for describing the chemical constitution of these

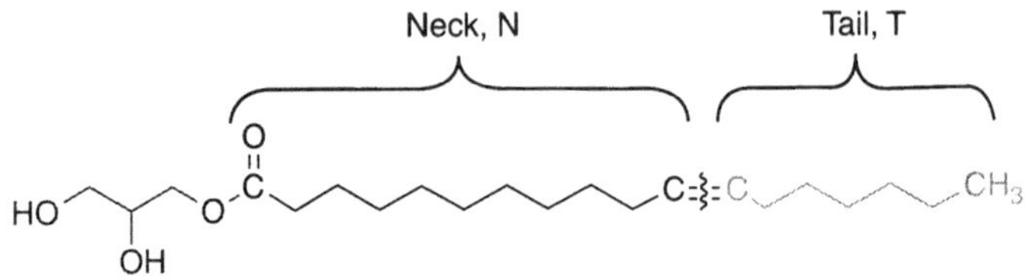

FIGURE 2 N.T MAG nomenclature. N, for Neck (blue), and T, for Tail (brown), represent the number of carbon atoms on either side of the double bond in the acyl chain, as indicated. The glycerol head group is shown in red. For interpretation of the references to color in this figure legend, the reader is referred to the Science Direct version of this chapter available online.

lipids is illustrated in Fig. 2 and is referred to as the N.T MAG notation (Misquitta & Caffrey, 2001). This is based on a rather simplistic view of the MAG molecule as an object consisting of a head, a neck, and a tail with the latter two joined by a trunk. Here the head is the glycerol headgroup. It is in ester linkage to the neck corresponding to that part of the acyl chain extending from its carboxyl carbon to the first carbon of the olefin. The trunk is the *cis*-double bond. The tail extends from the second carbon of the olefin to the chain's methyl terminus. In the N.T MAG notation, N and T correspond to the number of carbon atoms in the neck and tail, respectively. The total number of carbon atoms in the chain is the sum of N and T. Thus, 11.7 MAG represents monovaccenin, a MAG with a fatty acyl chain 18 carbon atoms long where the *cis*-double bond resides between carbon atoms 11 and 12. It is an olefinic isomer of 9.9 MAG also known as monoolein.

III. *IN MESO* CRYSTALLOGENESIS IN MOLECULAR DETAIL

A proposal has been advanced for how *in meso* crystallogenesis comes about at the molecular level (Caffrey, 2003) (Fig. 1). It begins with an isolated biological membrane which is treated with detergent to solubilize the target protein. The protein-detergent complex is purified by standard wet biochemical methods that usually involve a number of chromatographic steps. Homogenizing with a MAG effects reconstitution of the purified protein into the bilayer of the cubic phase. The latter is bicontinuous in the sense that both the aqueous and bilayer compartments are continuous in three-dimensional space (Fig. 1). The protein retains its native conformation and activity and is free to move within the plane of the cubic phase bilayer. A precipitant is added to the mesophase which triggers a phase separation. Under conditions leading to crystallization one of the separated phases is enriched in protein and it eventually grows into a bulk membrane protein crystal. The hypothesis includes a local lamellar phase that acts as a medium in which nucleation and crystal growth occurs. It also serves as a conduit or portal for proteins on their way from the cubic phase reservoir to the growing face of the crystal.

Experimental evidence in support of the hypothesis includes the following: (1) a nanometer-sized synchrotron beam has been used to probe the interface of an *in meso* crystal and has provided small-angle X-ray scattering (SAXS) evidence for the lamellar conduit (Cherezov & Caffrey, 2007), (2) all membrane proteins that have yielded structures from crystals grown *in meso* exhibit layered or Type I crystal packing (Michel, 1983) which is consistent with the proposed lamellar conduit (Cherezov et al., 2006; Raman et al., 2006), (3) atomic force (Qutub et al., 2004) and electron microscopic (Paas et al., 2003) evidence for the lamellar conduit next to the crystal has been observed, (4) electronic absorption and fluorescence and circular dichroic spectroscopic measurements on membrane proteins (BtuB, OpcA, bacteriorhodopsin) (Cherezov & Caffrey, 2006; Cherezov et al., 2006, 2008) reconstituted in the lipidic cubic phase support the view that the conformation *in meso* is native-like, (5) fluorescence quenching studies have demonstrated unequivocally that the protein is reconstituted into the bilayer of the cubic phase as a preliminary to crystallogenesis (Cherezov et al., 2006, 2008; Liu & Caffrey, 2005), (6) fluorescence measurements have been used to demonstrate that proteins reconstituted *in meso* are active (Cherezov et al., 2006, 2008), (7) failure to observe crystallization in the pure lamellar or H_{II} phases is consistent with the requirement for a bicontinuous mesophase (Misquitta et al., 2004a), and (8) visual observations with colored protein such as bacteriorhodopsin (Misquitta et al., 2004b) and fluorescence recovery after photobleaching (Cherezov et al., 2008) shows clear evidence for mobility in the cubic phase which is integral to crystallogenesis.

IV. MESOPHASE BEHAVIOR: FROM LAMELLAR TO CUBIC AND SPONGE PHASES

A. Phase Diagram

In meso crystallogenesis takes place in a hydrated, bicontinuous lipidic liquid crystal also referred to as a lyotropic mesophase. The many components present and conditions that prevail during crystallogenesis can impact phase behavior. Accordingly, it is important to be mindful of the phase properties of the relevant lipid/water system that forms the basis of the *in meso* trial. Monoolein is the lipid most commonly used and its temperature aqueous composition phase diagram (Qiu & Caffrey, 2000) (Fig. 3) will be described briefly to set the stage for the discussion that follows.

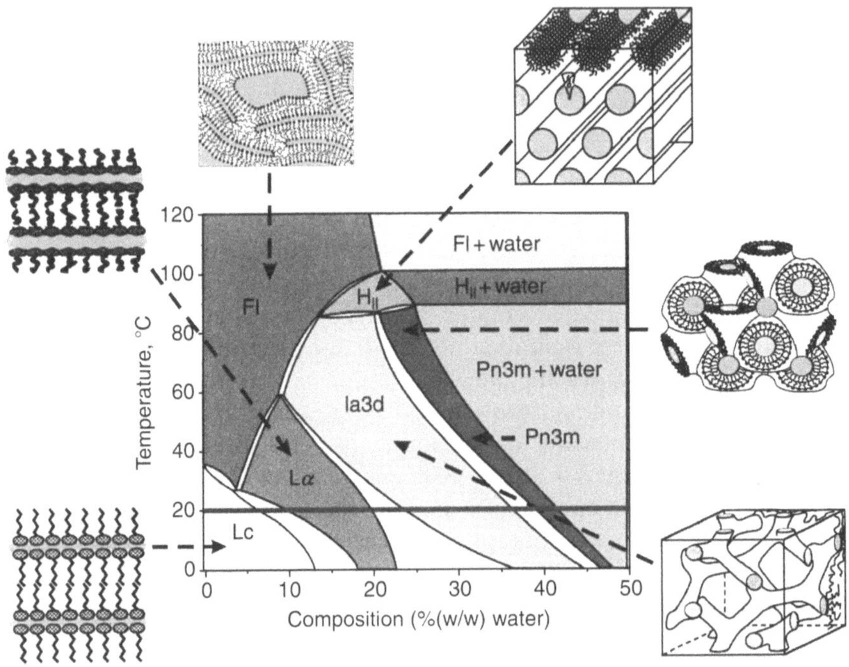

FIGURE 3 Temperature-composition phase diagram of the monoolein/water system. The phase diagram was determined under "conditions of use" in the heating and cooling directions from 20 °C. Redrawn from (Cherezov et al., 2006). A cartoon representation of the various phase states is included in which colored zones represent water. The 20 °C isotherm is shown as a horizontal blue line. The liquid crystalline phases below ~17 °C are metastable (Qiu & Caffrey, 2000). For interpretation of the references to color in this figure legend, the reader is referred to the Science Direct version of this chapter available online.

In the dry state, monoolein undergoes a melting transition from a solid lamellar crystal (Lc) to a liquid fluid isotropic (FI) phase at about 37 °C. Recooling the melt often leads to an undercooled liquid that can persist for extended periods at room temperature (~20 °C). The addition of water to the system gives rise to a number of lyotropic or water-induced mesophases, the identity of which depends on temperature. Thus, at high temperatures the inverted hexagonal (H_{II}) phase forms. This gives way to two different cubic phases upon lowering temperature; the cubic-Ia3d phase forms at lower hydration levels than the cubic-Pn3m phase. The latter can exist in equilibrium with excess water as a two-phase system over a wide temperature range. At intermediate hydration levels and temperatures the lamellar liquid crystalline (L_{α}) phase forms. The equilibrium phase diagram for the monoolein/water system shows that the mesophases are no longer stable below about

17 °C (Qiu & Caffrey, 2000). However, as was observed with the dry lipid, undercooling of the lamellar and cubic mesophases is possible and is commonly observed (Qiu & Caffrey, 2000).

B. Molecular Shape

Phase propensity reflects what has been referred to as the dynamically averaged molecular shape of the relevant amphiphiles. This concept is useful when explaining, in relatively simple geometrical terms, how phase behavior is determined by lipid or detergent molecular form or structure (Israelachvili, Mitchell, & Ninham, 1977). Thus, the planar lamellar phases come about as a result of the close packing of molecules with a dynamically averaged molecular shape that is cylindrical. Spherical, normal micelles form when cone shaped molecules spontaneously aggregate. The H_{II} phase is stabilized by lipids with a wedge shape where the relatively small polar headgroup is at the narrow end of the wedge. Finally, the inverted cubic phases, which form the basis of the *in meso* crystallogenesis methodology, incorporate lipid molecules with a dynamically averaged shape that can best be described as a splayed double-ended wedge (Briggs, Chung, & Caffrey, 1996; Hyde et al., 1997). The phase-shape relationship is significant in the context of the current chapter because MAGs with varying chain lengths and sites of unsaturation will have different dynamically averaged shapes and thus different tendencies to stabilize a crystal-growing mesophase.

C. Sponge Phase

In the presence of certain additives, the aqueous channels of the cubic phase enlarge and its lattice parameter, as monitored by SAXS, rises. These additives are substances commonly used to facilitate crystallogenesis and include pentaerythritol propoxylate (PPO), 2-methyl-2,4-pentanediol (MPD), jeffamine, *t*-butanol, 1,4-butanediol, polyethylene glycol (PEG) 400, and potassium thiocyanate (Cherezov et al., 2006). These so-called spongifying agents are proposed to act by interacting with the lipid headgroup and to increase the cross-sectional area per molecule at the aqueous/apolar interface of the mesophase. This causes the highly curved bilayer to "unbend" and for the aqueous compartment of the phase to enlarge. In parallel, the bilayer of the emerging so-called sponge phase becomes more flexible and the regular periodicity of the original cubic phase is lost. This is evidenced by a replacement of sharp X-ray diffraction by a more diffuse scatter at low-angles. Despite the dramatic microstructural changes undergone, the mesophase remains bicontinuous, which is key to the ability of the sponge phase to support *in meso* crystallogenesis. With the enlarged aqueous channels and the less highly curved lipid bilayer the prospect is that the sponge phase will offer advantages in the crystallization of proteins with large cross-sectional areas in the membrane plane and/or with large extramembranal domains. The sponge phase has been shown to support the crystallogenesis of several membrane proteins including LHII (Cherezov et al., 2006), BtuB (Cherezov et al., 2006), GPCRs (Cherezov et al., 2007; Jaakola et al., 2008), the photosynthetic reaction center from *Rhodobacter sphaeroides* (Wadsen et al., 2006), and a cytochrome oxidase from *Thermus thermophilus* (Slattery, Caffrey, & Soulimane, 2008). As might be expected the tendency to form the sponge phase depends on the acyl chain identity of the hosting MAG.

D. Detergents and Their Effects on Mesophase Properties

The solutions used to spontaneously form the protein-enriched cubic phase usually contain significant amounts of detergents that were employed initially to purify and to solubilize the membrane protein. By virtue of their amphiphilic and surface active natures, detergents have the potential to impact on the phase properties of the *in meso* system and, by extension, the outcome of the crystallization process. Accordingly, studies have been performed to quantify the effects that commonly used detergents have on the phase behavior of hydrated monoolein (Ai & Caffrey, 2000; Liu & Caffrey, 2005; Misquitta & Caffrey, 2003). Phase identity and microstructure were characterized by SAXS measurements on samples prepared to mimic *in meso* crystallization conditions. The results show that the cubic phase is relatively insensitive to small amounts of alkyl (hexyl, octyl, nonyl, decyl) glucosides, dodecyl maltoside, alkyl (dodecyl, hexadecyl) fos-cholines, lauryldimethylamine-oxide (LDAO), sodium dodecyl sulfate (SDS), and Cymal-6. However, at higher levels these detergents trigger a transition to the lamellar phase and, where studied, do so in a temperature- and a lipid and a salt concentration-dependent manner (Misquitta & Caffrey, 2003).

These data have important implications for *in meso* crystallization (Caffrey, 2008b). Firstly, a small amount of detergent may facilitate crystallogenesis by favoring formation of lamellar domains in which nucleation and crystal growth is proposed to take place. Secondly, proteins with a high concentration of detergent can give rise to the bulk lamellar phase upon homogenization with lipid as a preliminary to crystallization. If the precipitant solutions used have a high concentration of salt, the cubic-to-lamellar phase transition can be reversed for successful crystal growth, as has been demonstrated (Misquitta & Caffrey, 2003). Thirdly, the capacity of the bicontinuous mesophase to tolerate detergent, and thus to support crystallogenesis, will depend on the N.T character of the hosting MAG.

V. A CENTRAL ROLE FOR LIPID

A. Host Lipid

1. Making Lipids to Order

The default lipid for the bulk of the *in meso* crystallogenesis studies performed to date is monoolein or 9.9 MAG. However, there is no good reason why monoolein should be the preferred lipid for all membrane proteins. The latter come from an array of biomembrane types with varying properties that include hydrophobic thickness, intrinsic curvature, lipid makeup, and compositional asymmetry. Thus, it seems reasonable that screening for crystallizability based on the identity of the lipid creating the hosting mesophase would be worthwhile. For this, MAGs with differing acyl chain characteristics such as length and olefinic bond position must be available. A lipid rational design, synthesis, and purification program is in place in the author's laboratory to serve this need (Section VI).

The MAGs that have been used in successful structure determination studies based on *in meso*-grown crystals have had chains 16 and 18 carbon atoms long. A proposal was advanced that a shorter chained lipid producing a thinner bilayer would enhance crystallization. A 14 carbon MAG was chosen as the lipid with which to test the proposal. To be compatible with the *in meso* method, a *cis*-olefinic bond had to be placed in the acyl chain. Its position was arrived at by applying rational design principles to a set of temperature-composition phase diagrams for homologous MAGs. The relevant lipid, 2,3-dihydroxypropyl-(7*Z*)-tetradec-7-enoate (7.7 MAG), was identified, synthesized, and its phase properties characterized by SAXS. As designed, this short-chain lipid formed the requisite cubic mesophase at room temperature. Further, the overall change in cubic phase microstructure, when compared to monoolein, resulted in a decrease in bilayer thickness of 6.5 Å with the water channel radius increasing by 11 Å (Fig. 4) (Misquitta et al., 2004a). Interestingly, such an increase in the water channel radius of a bicontinuous phase formed by monoolein requires a spongifying agent, in addition to lipid and water (Cherezov et al., 2006; Wadsen et al., 2006). Consistent with the hypothesis, 7.7 MAG produced crystals of three different integral membrane proteins by the *in meso* method. These included bacteriorhodopsin, cytochrome caa_3 oxidase, and the bacterial outer membrane cobalamin (vitamin B_{12}) transporter, BtuB. The latter is notable in that it was the first β-barrel protein to be crystallized by the *in meso* method. Other short chained MAGs have been produced that are proving successful in the crystallogenesis of membrane proteins in *Pseudomonas aeruginosa*. The means by which these lipids facilitate crystallogenesis has not been established but it surely reflects a preferential partitioning of the protein between

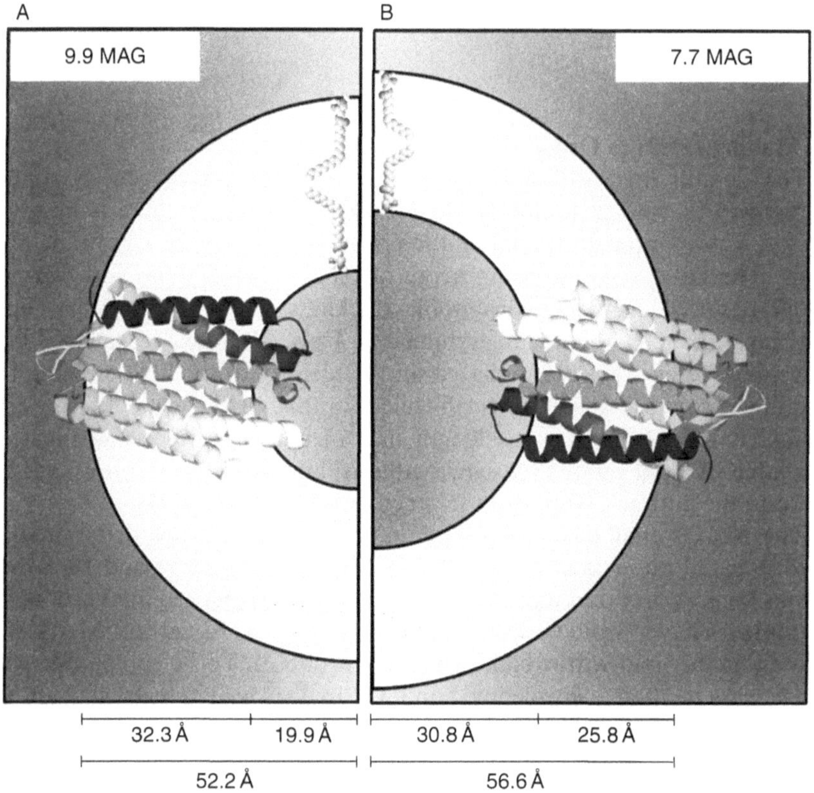

FIGURE 4 Comparison of the fully hydrated cubic phases of 9.9 and 7.7 MAG. Bilayer thickness and water channel radius calculations of (A) 9.9 MAG and (B) 7.7 MAG are based of SAXS data from fully hydrated samples at 40 °C. Figures are drawn to scale. A cartoon ribbon representation of bacteriorhodopsin (bR) (PDB ID: 1c3w) is added to highlight the location of the reconstituted protein. Figure is adapted from (Misquitta et al., 2004a).

the crystal and the hosting mesophase. Thus, bilayer thickness, or some other mesophase property, must not match the requirements of the protein such that it chooses to exist in the more ordered environment of a crystal.

2. Mesophase Microstructure

Reference has been made to the sensitivity of phase microstructure to lipid identity. Support for this statement is based on SAXS measurements performed on the cubic phase prepared with a homologous series of MAGs (Briggs & Caffrey, 1994a,b; Briggs et al., 1996; Caffrey, 2003; Qiu & Caffrey, 2000). The data show expected behavior in that as chain length decreases so too does the thickness of the lipid layer that creates the apolar fabric of the cubic phase, when evaluated at a single temperature. Less intuitive perhaps is the finding that the aqueous channel diameter drops as chain length increases. This is consistent with a "flattening" and an attenuating curvature at the polar/apolar interface with the shorter chained lipids.

While lipid identity can be used to tailor phase microstructure (Misquitta et al., 2004a,b), it is possible that the desired microstructure might not be accessible with a single lipid species in the temperature range of interest. In this case, it is possible to fine tune by using mixtures of MAGs with different acyl chain lengths where the mole ratio is adjusted to set microstructure at the desired intermediate value.

As noted, the microstructure of the mesophase can be engineered over relatively wide limits by manipulating temperature and/or lipid identity and composition. However, it is also important to note that the two metrics of the cubic phase—the polar and apolar compartment dimensions—are not independently adjustable and indeed are tightly coupled (Caffrey, 2003). Nonetheless, this feature of tunability is a valuable tool available to the crystal grower in search of a suitable lipid matrix in which to grow crystals. Thus, proteins with extramembranal domains that come in a variety of sizes can be accommodated as can those that originate from native membranes with different hydrophobic thicknesses (Munro, 1998).

3. Crystallization at Low Temperatures

The original *in meso* method does not work reliably at low temperatures, where proteins generally are more stable, because the hosting lipid, 9.9 MAG, is a solid (Qiu & Caffrey, 2000). The need exists therefore for a lipid that forms the cubic phase and that supports crystal growth at reduced temperatures. As with the 7.7 MAG in the example above (Section V.A.1), a database of phase diagrams was mined and used to design such a lipid. SAXS showed that the new lipid, 2,3-dihydroxypropyl-(7*Z*)-hexadec-7-enoate (7.9 MAG), exhibited expected phase behavior (Misquitta et al., 2004b). Further, it produced membrane protein crystals of diffraction grade by the *in meso* method at 6 °C. These results demonstrate that like their protein counterparts, lipids are amenable to rational design. The same approach, as used in these design studies, should find application in extending the range of membrane proteins amenable to *in meso* crystallization.

4. Changes in Lipid Profile During Crystallogenesis

MAGs form the basis of the *in meso* method. However, these lipids are intrinsically unstable in that the ester linkage between the glycerol headgroup and the fatty acid (Fig. 2) can hydrolyze and undergo transesterification (Clogston, Rathman, Tomasko, Walker, & Caffrey, 2000; Coleman et al.,

2004; Murgia, Monduzzi, Ljusberg-Wahren, & Nylander, 2002). The products of such reactions include glycerol, free fatty acids, the 2-isomer of the parent 1-MAG, and diacyl- and triacylglycerols (Clogston et al., 2000). Under extremes of pH the hydrolysis reaction can lead to rapid and extensive fatty acid production. However, most crystallogenesis trials are performed under mild conditions where the rate and extent of hydrolysis and transesterification of the hosting lipid are low. Nonetheless, they both do occur and in the course of a long trial it is possible for fatty acids and other products to accumulate to a point where they affect phase behavior (Clogston, 2005; Clogston, Graciun, Hart, & Caffrey, 2005). In this way the slow hydrolysis and transesterification reactions of the hosting MAG will alter the course of membrane protein nucleation and crystal growth. This might explain why the growth of crystals can be slow in particular systems. We are currently exploring the utility of stable, nonhydrolyzable lipids for *in meso* crystallogenesis.

B. Lipids as Additives

Hydrated *cis*-monounsaturated MAGs form the cubic mesophase that has been used for *in meso* crystallization of membrane proteins. To date, monoolein (9.9 MAG), monopalmitolein (9.7 MAG), monovaccenin (11.7 MAG) (Gordeliy et al., 2002), 7.7 MAG (Lyons, Soulimane, & Caffrey, 2008; Misquitta et al., 2004a; Slattery et al., 2008), and 7.9 MAG (Misquitta et al., 2004b) have served in this capacity. It is possible that the hosting cubic phase created by the MAG alone, which itself is a most uncommon membrane component, will limit the range of membrane proteins crystallizable by the *in meso* method. With a view to expanding the range of applicability of the method and to making the hosting cubic phase more "familiar" to its guest protein, the degree to which the reference, cubic-Pn3m phase formed by hydrated monoolein can be modified by other lipid types was examined by X-ray diffraction (Cherezov, Clogston, Misquitta, Abdel Gawad, & Caffrey, 2002). These included phosphatidylcholine (PC), phosphatidylethanolamine (PE), phosphatidylserine (PS), cardiolipin, lyso-PC, a polyethylene glycol-lipid (PEG-lipid), 2-monoolein, oleamide, and cholesterol. The study showed that all lipids investigated were accommodated in the cubic phase of 1-monoolein to some extent without altering phase identity. The positional isomer, 2-monoolein, was tolerated to the highest level. The least well tolerated were the negatively charged lipids, followed by lyso-PC. The others were accommodated to the extent of 20–25 mol%. These results should prove useful in rationally designing cubic phase crystallization matrices with lipid profiles that better match the needs of a greater range of membrane proteins. A case in point is the recent structure determination of the GPCRs where diffraction quality crystals were obtained only when the system contained up to 12 mol% cholesterol (Cherezov et al.,

2007; Hanson et al., 2008; Jaakola et al., 2008). The benefits of these added lipids range from altering mesophase microstructure, to favoring lamellar phase formation for nucleation and growth, all the way to stabilizing the protein as does a ligand or substrate. Undoubtedly, the effects seen will depend on the identity of the hosting lipid. It is important to realize too that while a lipid additive may destabilize the cubic phase, the latter mesophase can be recovered upon incubation with certain precipitant solutions, as noted.

VI. LIPID SYNTHESIS AND PURIFICATION

A. Synthesis

The MAGs used for *in meso* crystallization were provided initially by a separate project in the author's laboratory that focused on the relationship between the chemical constitution of lipids and their physical and biological properties. For this, the *cis*-monounsaturated MAGs were chosen as a subset of lipids with a relatively simple chemical constitution that exhibited diverse mesophase behavior as a function of temperature and aqueous composition. In order to decipher the relationship in question, it was necessary to have available a series of MAGs in which the acyl chain length and position of unsaturation along the chain varied in a systematic way. Initially, the needs of the project were met using commercially available MAGs (Briggs, 1994; Briggs and Caffrey, 1994a,b; Briggs et al., 1996; Misquitta & Caffrey, 2001; Qiu & Caffrey, 1999, 2000). Shortly thereafter it was realized that for the project to advance a much greater range of MAGs was needed and that a program of MAG synthesis and purification was required. The author enlisted the collaboration of a colleague, Professor D. J. Hart (The Ohio State University), to provide the needed support in synthetic organic chemistry. In short order, a modular strategy was devised for MAG synthesis which meant that the requisite N.T MAGs could be produced using parts or modules that were available commercially (Coleman et al., 2004). These MAGs have therefore served double duty in supporting structure-function studies as applied to both lipids and membrane proteins.

The modular strategy toward the synthesis of a generic N.T MAG can be described as follows. The neck (N) portion of the N.T MAG begins as a primary alcohol, N-2 carbon atoms long, with a terminal double bond. The alkyl chain is extended by one carbon atom, at the hydroxyl end, upon formation of a cyano intermediate which, in turn, is hydrolyzed to the corresponding carboxylic acid. For certain N.T MAGs, the latter fatty acid is available commercially in which case this initial synthesis can be bypassed. The monounsaturated N-1 acid is subsequently coupled to the protected glycerol by Steglich esterification yielding the corresponding so-called N-1 acetonide. Selective esterification is made possible by using the commercially available racemic solketal, a glycerol molecule with adjacent hydroxyls cross linked by acetone. This completes the synthesis of coupling partner 1 in the modular synthesis process.

The tail (T) portion of the N.T MAG begins as a 1-iodoalkane which is T-1 carbon atoms long. This is chain extended by two carbon atoms upon reaction with lithium acetylide to produce a 1-alkyne that is T + 1 carbon atoms long. For certain N.T MAGs, the alkyne is commercially available in which case the chain extension is not necessary. The alkyne is converted to a 1-iodoalkyne before subsequent conversion to *cis*-1-iodoalkene.[1] This completes the preparation of coupling partner 2 in the modular synthesis. Note that it is the double bond in this *cis*-vinyl iodide module that ends up as the olefinic bond in the final N.T MAG.

Suzuki-Miyaura coupling is then used to combine coupling partner 1, the N-1 acetonide, with coupling partner 2, the T + 1 *cis*-vinyl iodide, to produce the N.T acetonide with retention of the olefin stereochemistry.[2] Unblocking the acetonide via acid hydrolysis yields the desired N.T MAG. This completes the synthesis.

The reactions involved in the synthesis of N.T MAGs are detailed in Fig. 5. The particular synthesis chosen for illustration is that of 7.7 MAG. For illustrative purposes, we show the proposed synthesis of the starting N-1 acid and T + 1 alkyne. In practice, 5-hexenoic acid and 1-octyne are commercially available and are used as received. The synthesis consists of a series of discrete steps and after each the product formed must be verified and purified from other products, residual reactants, catalysts, and solvents. The principle at play here is given the acronym GIGO which stands for "garbage in, garbage out." The idea is that if impure reactants are fed into a reaction mix then the product will be difficult to identify and to purify subsequently. Thus, it is important to isolate in high purity the product of key reactions before using it in the next step.

The bulk of the purification following each step in the procedure is done using flash column chromatography with silica gel (230-400 mesh) as the solid support. Organic solvents ranging from pure hexanes to mixtures of hexanes and ethyl acetate are used as the mobile phase. Solvent quality used

[1]The published MAG protocol (Coleman et al., 2004) prepared the *cis*-1-iodoalkene by diimide reduction. In the current procedure this step was changed to a dicyclohexyl-borane reduction, which allows for a higher yield and a less cumbersome purification step by ruling out the concomitant over reduction to the 1-iodoalkane.

[2]It is important at this stage to ensure complete conversion of the olefin on the N-1 acetonide to the borate to maximize the yield of N.T acetonide.

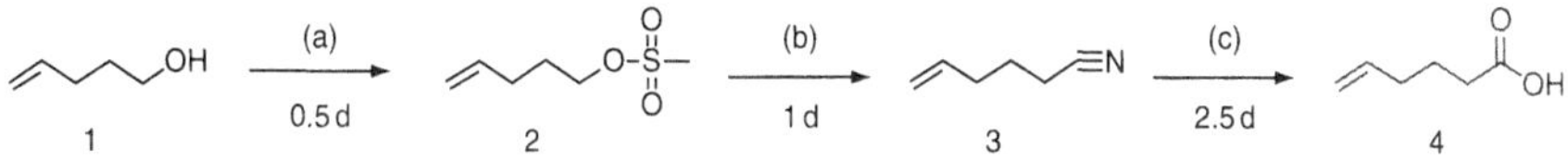

SCHEME 1 Synthesis of the N-1 acid, 4. Reagents and conditions: (a) methanesulfonyl chloride (1.15 equiv.), triethylamine (1.2 equiv.), dichloromethane, 0–20 °C; (b) sodium cyanide (2 equiv.), dimethylsulfoxide, 100 °C reflux; (c) 6 M sodium hydroxide (10 equiv.), methanol, water, reflux 110 °C.

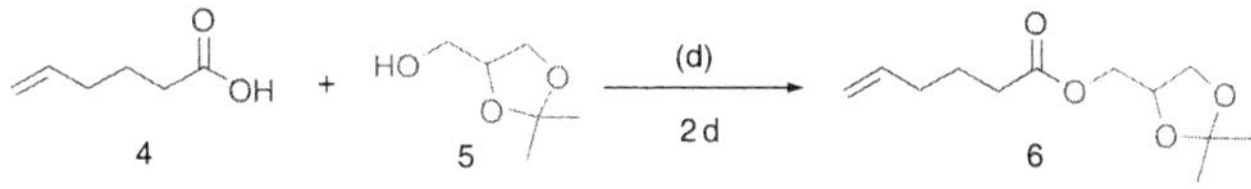

SCHEME 2 Preparation of coupling partner 1, the N-1 acetonide, **6**. Reagents and conditions: (d) *rac*-solketal (1.3 equiv.), *N,N′*-dicyclohexylcarbodiimide (1 equiv.), 4-dimethylamino-pyridine (0.1 equiv.), dichloromethane, 0 °C.

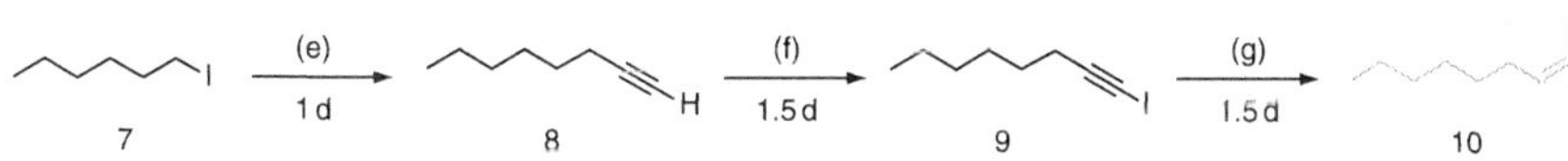

SCHEME 3 Synthesis of coupling partner 2, the *cis*-vinyl iodide, **10**. Reagents and conditions: (e) lithium acetylide, dimethysulfoxide, 10 °C; (f) n-butyllithium (1 equiv.), iodine (1 equiv.), dry tetrahydrofuran, −80 to −50 °C; (g) borane-dimethylsulfide (1.1 equiv.), cyclohexene (2.2 equiv.), glacial acetic acid (16 equiv.), dry diethyl ether, 0 °C.

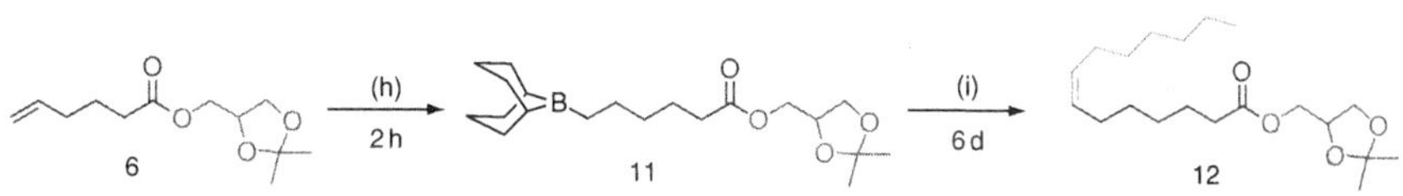

SCHEME 4 The Suzuki-Miyaura coupling reaction—preparation of the N.T acetonide, **12**. Reagents and conditions: (h) 9-borobicyclononane (1.5 equiv.), dry tetrahydrofuran, 5 °C; (i) 10, cesium carbonate (100 mol%), palladium (diphenylphosphinoferrocene) dichloride (Pd(dppf) Cl_2, cat.) (5 mol%), triphenylarsine (10 mol%), dimethylformamide, 22 °C.[a]

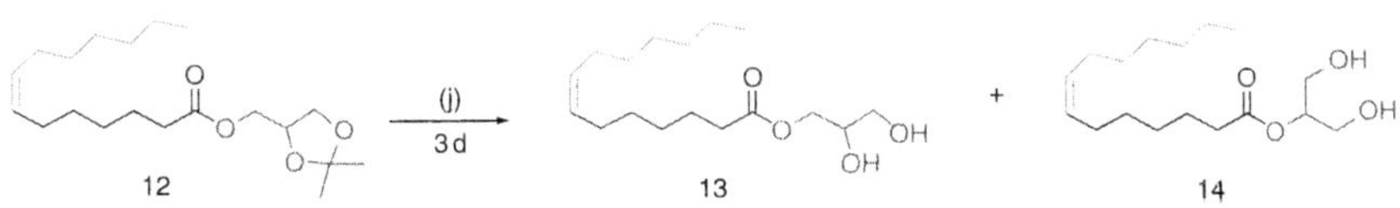

SCHEME 5 Acid hydrolysis of the blocked acetonide. This reaction affords the 1-MAG **13**/2-MAG **14** equilibrium mixture. Reagents and conditions: (j) methanol (96 equiv.), 2 M HCl (4 equiv.), 22 °C. Pure 1-MAG, **13**, is obtained by subsequent recrystallization from diethyl ether/petroleum ether (9:1 v/v) at −22 °C.

[a] A common contaminant in the Suzuki-Miyaura coupling is cyclooctanone. This is one of the materials that must be removed through careful column chromatography. The cyclooctanone has characteristic ^{1}H and ^{13}C NMR peaks that are easy to find if it is present in the 1-MAG acetonide. The cyclooctanone is readily removed after the hydrolysis however, as it has an Rf value much like the 1-MAG acetonide but much larger than a typical 1-MAG.

FIGURE 5 N.T MAG synthesis. The route to 7.7 MAG synthesis is illustrated. Steps involved in the synthesis of coupling partner 1 are described in Schemes 1 and 2. Scheme 3 outlines the synthesis of coupling partner 2. The Suzuki-Miyaura reaction, where the two

is of the highest grade commercially available. Most of the reactions preceding the coupling reaction are straightforward with products easily identified and purified. For a typical synthesis, resulting in ~1 g of N.T MAG, 4.0 cm diameter columns with bed heights of 20 cm are used in the purification of coupling partners 1 and 2 (4-5 g each). In contrast, the Suzuki-Miyaura coupling reaction uses Pd(dppf)Cl_2 as a catalyst and this gives rise to a black sludge containing the product. Separating the N.T acetonide **12** from this suspension requires extensive and repeated column chromatography. For a standard 1 g N.T MAG preparation, the acetonide must be passed up to four times through 7.5 cm diameter columns having a bed height of 35 cm with the product collected in 125 ml fractions. While the column is running, elution is monitored by thin layer chromatography (TLC). Fractions containing the product are combined and concentrated using rotoevaporation. Wherever possible the product is protected from exposure to light and is stored in an atmosphere of nitrogen or argon at −20 °C. Proof that the correct product has been obtained in reasonable purity involves ^{1}H and ^{13}C nuclear magnetic resonance spectroscopy (NMR).

For a typical MAG synthesis, as described above, the final yield of recrystallized 1-MAG is of order 1 g. Reduced yields arise due to incomplete coupling of the two partners, deprotection of the acetonide, **12**, and to losses at the chromatographic and recrystallization steps. The largest of these occur during recrystallization of **13**. Percentage yields for many of the steps in Fig. 5 have been reported (Coleman et al., 2004).

The synthesis, as presented, is limited to the generation of MAGs where $N \geq 5$ and $T < 4$. In the case of N.T MAGs where $N = 4$ for example, the N-1 acid (acrylic acid, $CH_2{=}CH{-}COOH$) is required. Since the latter is conjugated, the preparation of the corresponding trialkylborane required in step (h) (Fig. 5) is not possible. For $T < 4$, the starting alkynes are either gases (propyne) or low boiling point liquids (butyne, b.p. = 8.1 °C) making them difficult to work with. The upper limits for N and T are dictated by the commercial availability of the corresponding starting materials or precursors. Ultimately, of course, utility is limited by the physicochemical properties of the final MAGs. To date, MAGs in the range $5 \leq N, T \leq 13$ that include 5.13 MAG and 13.5 MAG have been synthesized using this methodology (Coleman et al., 2004; Misquitta, 2006; Muthusubramaniam, 2004).

---

coupling partners are combined, is described in Scheme 4. Hydrolysis of the acetonide-protected MAG producing the N.T MAG is shown in Scheme 5. Reagents used and conditions prevailing at each step are summarized. Steps are indicated by arrows and by bold lower case letters while reactants and products are identified by bold numbers. The approximate time in days required to complete each step and to purify intermediates and products is given below the arrows.

B. Lipid Purity and Quality Control

1. Removing the 2-MAG

The final step in the synthesis is the acid hydrolysis of the blocked N.T MAG. The immediate product is the required 1-MAG where the acyl chain is at the *sn*-1 and *sn*-3 positions of glycerol. However, transesterification gives rise to the 2-isomer, among other products, as discussed below. For monoolein, the equilibrium molar ratio of 1-/2-isomer is 88/12 (Coleman et al., 2004; Murgia et al., 2002). Currently, our standard procedure is to produce N.T MAG of highest purity for use in subsequent *in meso* trials and this requires a final recrystallization step that can typically reduce contaminating 2-MAG to <2.5 mol%. Recrystallization is carried out in diethyl ether/petroleum ether (9:1 v/v.) under N_2 at −20 °C which is achieved using a liquid N_2/acetonitrile slush bath.[3] The 1-MAG crystallizes and the solvent containing the 2-MAG is removed under N_2. Recrystallization can be repeated to improve the purity of the final product but this is done at considerable loss of the 1-MAG. For the bulk of the N.T MAGs produced in the laboratory to date a single careful crystallization has sufficed.

It is important to stress that we have not established that this purification step is required in order for the *in meso* method to work. Indeed, partial or complete return to the original equilibrium mix of 1- and 2-MAG will occur during the time in which the lipid is left fully hydrated in the crystallization bolus awaiting crystal growth. It is entirely possible, but not tested, that the compositional change with time facilitates crystallogenesis (Caffrey, 2008b).

2. *cis/trans* Olefinic Purity

The Suzuki-Miyaura coupling reaction is designed to preserve the *cis*-double bond geometry of the T + 1 vinyl iodide, **10**, which in turn is rigorously controlled by the dicyclohexylborane-mediated reduction (step (g), Fig. 5), in the final N.T MAG. In a separate study by Coleman et al. (2004) the *cis* and *trans* makeup of the product of the coupling reaction was established by ^{13}C NMR measurements on the corresponding acetonide **12**. In this work, both *cis* and *trans* isomers were prepared and used to verify ^{13}C NMR signatures. These served to validate the *cis* and *trans* selectivity of the synthesis. Typically, the *trans* contamination level is below 1 mol% (Coleman et al., 2004). We have not ascertained the degree to which *trans* contamination affects *in meso* crystallogenesis. It is expected to have a

[3]The temperature at which recrystallization occurs is dependent on the identity of the N.T MAG. The shorter the chain and the closer to the middle of the chain is the olefinic bond, the lower is that temperature. Thus, the recrystallization temperatures must be established experimentally for each lipid.

profound effect when levels are high given that the *trans* form of the MAGs behave more like their saturated homologs (Chung & Caffrey, 1995; Lutton, 1965).

3. Thin Layer Chromatography

Reaction progress during synthesis is conveniently and sensitively followed by TLC performed with silica coated glass plates with visualization using phosphomolybdic acid and subsequent heating for color development. This gives rise to dark spots on a yellow background and product is typically allowed to form until staining intensity for a given plate loading remains constant with time and/or reactant has disappeared. In some cases, the reaction is stopped as soon as product yield reaches a maximum so as to prevent product degradation. This is relevant especially during the deprotection of **12** (step (j), Fig. 5) where the corresponding fatty acid methyl ester will become the major product if the reaction is not stopped in time.

The purity of the final MAG is evaluated by TLC on binder-free Adsorbosil plates followed by charring on a hotplate with 4.2 M sulphuric acid for visualization. This reveals a black spot on a white background and can be used to quantify purity. Typically, this involves running the final product at increasing loadings from 1 to $\geq$100 μg per spot in three solvent systems chosen to give the MAG an R_f value ideally in the vicinity of 0, 0.5, and 1, respectively (Fig. 6). By comparing spot intensities it is possible to provide estimates of purity to 99% and above.

4. Nuclear Magnetic Resonance

^{1}H and ^{13}C NMR is used throughout the synthesis, and in parallel with TLC, to monitor progress and to verify that the correct intermediates and product have been synthesised. Spectra of starting materials and of products at each step are compared and the expected resonances are monitored for their orderly appearance and disappearance.

Final purity is assessed based on ^{1}H and ^{13}C NMR spectra of the type shown in Fig. 7. All forms of contaminants are looked for at this stage with particular attention paid to the *trans* and 2-MAG isomers. A possible by-product that arises from incomplete activation of the N-1 acetonide **6** to the trialkylborane **11** is the so-called N-1 MAG. This molecule, which is the deprotected form of **6** is visible in the corresponding ^{1}H NMR spectra as three peaks centred at ~5.0 ppm arising from the two hydrogens on the terminal olefinic carbon.[4] Based on the spectra in Fig. 7, the final 7.7 MAG was reported to have a *trans*

[4]The chemical shifts of the protons, especially in the headgroup region of the lipid, are concentration dependant. Self-assembly/aggregation in the apolar solvent used for the NMR experiment may be responsible for such behavior.

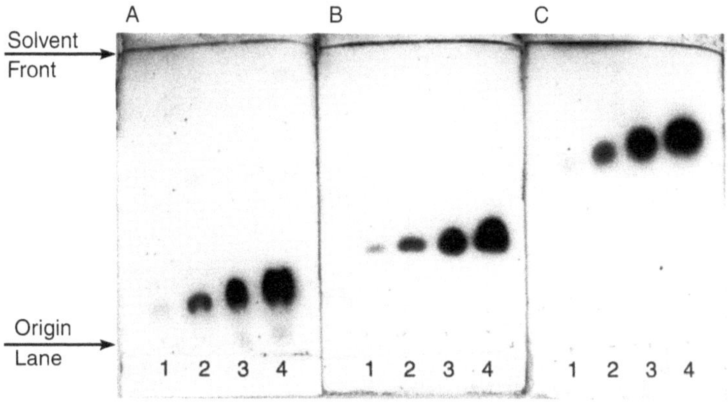

FIGURE 6 Thin layer chromatography of 7.7 MAG. Lipid loadings are 1, 10, 50, and 100 μg in lanes 1–4, respectively. Development solvent systems used are as follows: (A) chloroform: acetone, 96:4 (%, v/v). (B) hexanes:acetone:ethyl acetate, 74:25:1 (%, v/v). (C) hexanes:acetone: ethyl acetate, 50:25:25 (%, v/v). Spots were visualized by spraying with 4.2 M sulfuric acid followed by charring on a hotplate at 260 °C.

content of <1 mol% and a 2-MAG content below 2.5 mol%. The former was verified by the lack of a signal from the *trans*-olefinic carbons at ~130.5–131.0 ppm in the ^{13}C spectra, while the latter was calculated from the integration of ^{1}H NMR (C_6D_6) signals from the combined 1- and 2-MAGs (δ 5.5–5.6 ppm, CH=CH) and unique signals from the 2-MAG (quintet at δ 5.1 ppm, secondary CHOCOR). This is considered to be a high quality lipid.

5. Electronic Absorption and Fluorescence and a Trace Contaminant

The ultimate goal of *in meso* crystallogenesis is to crystallize membrane proteins for the purpose of solving structures at high resolution. However, it is also important to monitor the stability and functional activity of such target proteins reconstituted into the lipidic mesophase as a prelude to crystallization and for related applications. Electronic absorption and fluorescence have been used successfully in such situations (Cherezov et al., 2006; Liu & Caffrey, 2005). The fact that the lipidic cubic phase can be prepared in a form that is optically transparent means that it is nicely suited to spectroscopic measurements. It is unfortunate, however, that the spectroscopic signatures of interest are often in the UV region of the spectrum where small amounts of contaminants in the lipid contribute to background. The problem is exacerbated by the fact that the cubic phase itself represents 1–2 M lipid. Thus, minor contaminants in the lipid that are not obvious by NMR or TLC are clearly visible in the UV spectrum. We have found that both commercially available and synthetic N.T MAGs show strong absorption in this region (Fig. 8). This, in turn,

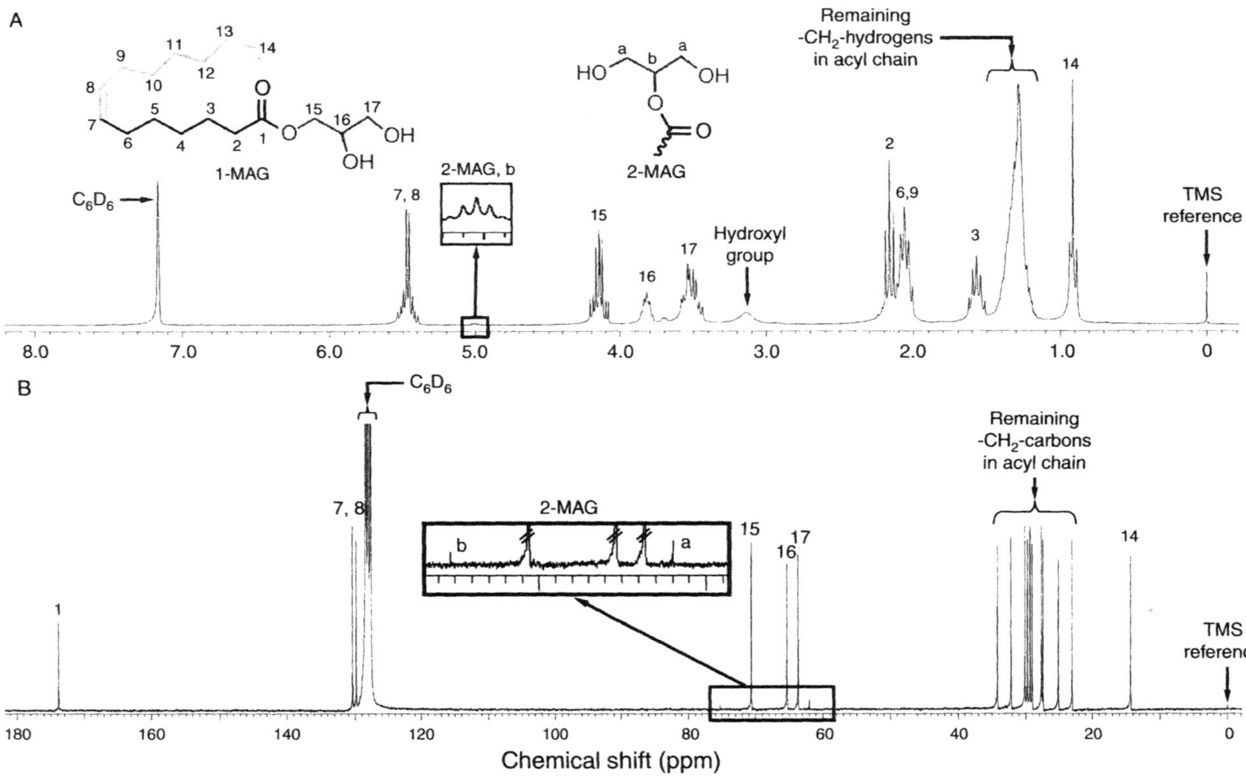

FIGURE 7 NMR spectra of 7.7 MAG. (A) ^{1}H and (B) ^{13}C NMR spectra of 7.7 MAG. Data were collected on a 270 MHz Jeol NMR spectrometer using a sample containing 15 mg MAG in 0.6 ml deuterated benzene (C_6D_6). The sample includes both 1- and 2-MAGs and the corresponding structures or partial structures are shown in (A). Expanded views of tell-tale resonances from 2-MAG are shown in insets. Protons in 1-MAG and relevant carbons in 2-MAG are identified by numerals and letters in the respective structures and corresponding spectra. Trimethylsilane (TMS) was used as the internal standard.

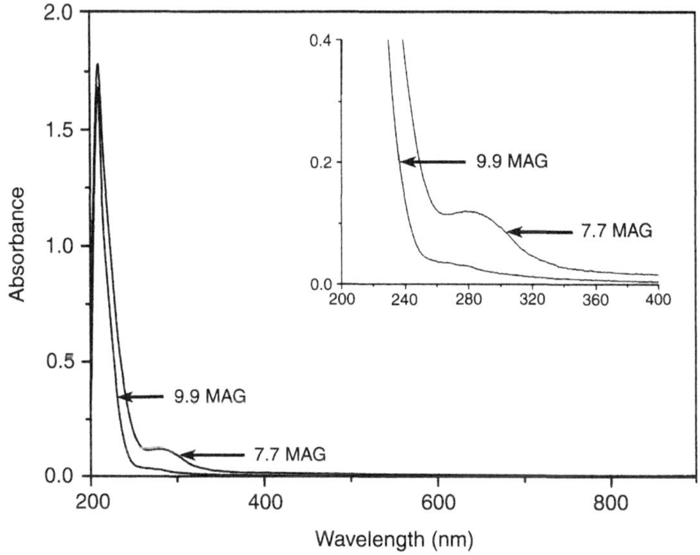

FIGURE 8 UV-visible absorption of 7.7 and 9.9 MAGs. Absorbance in the UV region by the synthetic lipids is a problem that impacts on its use in spectroscopic measurements of reconstituted membrane proteins in the cubic phase, where the lipid is present at a concentration of ~2 M. In this figure, 7.7 MAG, a product of the aforementioned synthesis, is compared against the commercially available 9.9 MAG (monoolein, NuChek Prep). An expanded view of the data in the vicinity of 280 nm is shown in the inset. Each sample is composed of 14 mM lipid in methanol with data collected in a 1 cm path length quartz cuvette. Spectra have been corrected for background absorbance of the solvent.

creates problems for related spectroscopic measurements such as fluorescence and circular dichroism. In the former case, the inner filter effect must be corrected for when absorbance exceeds 0.1. With CD, a failure to record a reasonable signal because of the high background obtained at the lower wavelengths (Liu & Caffrey, 2005). In our experience, the problem of background absorbance and fluorescence is much greater with synthetic compared to commercial MAGs (Fig. 8). Efforts are underway to identify the source of the background with a view to devising synthesis and/or purification strategies that reduce the burden. This will be essential for proposed functional assays with reconstituted membrane proteins where the effects of acyl chain length and olefin position along the chain are to be evaluated.

6. Mass Spectrometry (MS)

MS allows for the rapid, sensitive, and facile determination of the exact molecular weight. The technique is amenable to a wide variety of molecular types, including lipids. In the current application, electrospray ionization mass spectrometry (ESI-MS) is used to confirm the synthesis of new intermediates and MAGs based on the correct and complete assignment of its mass-to-charge ratio (*m/z*). Electrospray ionization is known as a "soft" ionization method as the sample is rendered charged by the relatively gentle addition or removal of a proton with little extra energy remaining to result in fragmentation of the sample ions. In ESI-MS, MAGs are detected as singly charged molecular-related ions, usually protonated (M + H) or deprotonated molecular $(M - H)^-$ species in positive and negative ionization modes, respectively. Analytes may also be ionized by the addition of sodium, potassium, and ammonium ions (Fig. 9). Other adduct ions encountered in negative ion mode include acetate, formate, and chloride. Furthermore, molecular ions may also appear with a combined mass of the parent molecular ion and solvent, as illustrated in Fig. 9. ESI-MS is not capable of quantifying 1-/2- or *cis/trans* isomers present after the synthesis. As such lipid chemical purity is generally assigned from inspection of the corresponding NMR spectra in conjunction with TLC analyses.

C. Time Commitment

By the methods outlined in Section VI.A, synthesis and purification of 1 g N.T 1-MAG, beginning with the N-2 alcohol and the T-1 iodoalkane, takes in total 3–4 weeks. The time required per step is detailed in Fig. 5. At least half of the time is devoted to the purification of intermediate and final products. Purification is carried out by the standard synthetic organic chemistry methodologies of gravity and flash chromatographies, distillation, and crystallization. The synthetic reactions employed in the synthesis are relatively mild and, with care, can be successfully run overnight without difficulty or danger. Some reactions (Scheme 5, Fig. 5), however, require regular monitoring by TLC so as to anticipate product degradation.

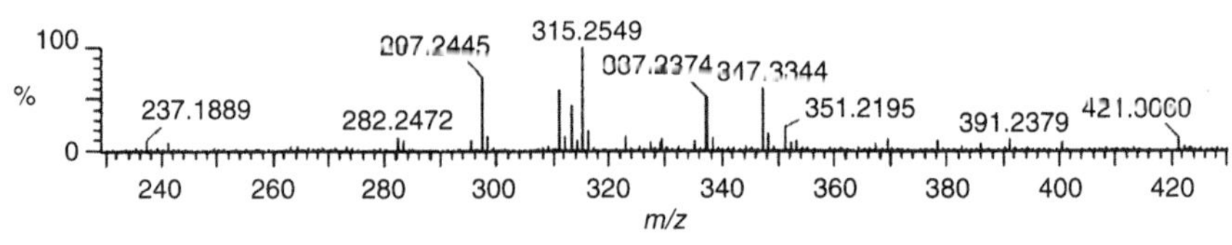

FIGURE 9 Mass spectrum of 7.8 MAG. The spectrum was recorded in the electrospray positive mode on a sample composed of 7.8 MAG in methanol. Species detected with (*m/z*) values of 315.2549, 337.2374, and 347.3344 correspond respectively, to $(M + H^+)$, $(M + Na^+)$, and $(M + CH_3OH + H^+)$. The peak at $(m/z) = 297.2445$ has tentatively been attributed to a dehydrated species that may form through fragmentation.

VII. CONCLUSIONS

In meso crystallogenesis of membrane proteins takes place in a mesophase that is approximately 50% lipid by weight. The lipid forms the curved bilayer into which the membrane proteins reconstitute initially and from which they phase separate in a process that leads to nucleation and crystal growth. The tendency of a protein to partition between the mesophase and the ordered crystal is profoundly sensitive to the chemical and physical identity of the lipid forming the mesophase membrane. It is not unreasonable therefore that the lipid should be considered and exploited as a parameter that can be adjusted to facilitate crystallization.

To be an adjustable parameter however requires that a range of different lipid types is available for use in such applications. In this chapter, we have demonstrated that lipids can be used to form the hosting mesophase and as additives, in both cases acting to facilitate crystal growth. The major focus here however has been on MAGs which neatly serve the role of creating the hosting mesophase. By establishing the temperature-composition phase behavior of these assorted hydrated MAGs, we learn how they function at a molecular level in support of *in meso* crystallogenesis. This information is used subsequently to rationally design MAGs that are capable of growing crystals of diverse membrane proteins under a range of experimental conditions.

The crystallogenesis program described in this chapter and currently underway in the author's laboratory very much relies on the availability of MAGs with different physical and chemical characteristics. This need is met by a modular approach to the synthesis of *cis*-mono-olefinic 1-MAGs in gram quantities and of a quality suitable for *in meso* crystallogenesis. In parallel with the crystal growth studies, as new MAGs are produced, they are subjected to mesophase characterization primarily by SAXS. The phase microstructure and propensity information so derived is, in turn, used to more rationally approach membrane protein crystallization.

While the lipid synthesis program outlined in this chapter is integral to and supports the membrane structural and functional biology work underway in the author's laboratory, challenges remain. It has been noted that lipids produced are of a quality that enable membrane protein crystallogenesis. However, these same lipids fall short when it comes to spectroscopic purity which is key to the functional characterization of membrane proteins reconstituted *in meso*. Accordingly, effort must be devoted to implementing a synthesis and/or a purification route that yields MAGs with significantly lower absorbtivities in the UV region. As esters, MAGs too are intrinsically unstable in an aqueous environment; they succumb to hydrolysis and transesterification reactions that occur during what can be a protracted period of crystallization. Assuming that such chemical changes in the hosting lipid do not impact on nucleation or crystal growth then, in the interest of simplifying the crystallogenesis environment, it is preferred to work with a chemically stable lipid. Such lipids are in development in the author's laboratory. Finally, there is the need to provide these novel lipids, that are not available commercially, to the *in meso* crystallogenesis community. As word spreads, it is expected that usage of these specialized lipids will grow. It will be important that a supplier is identified who is willing and capable of producing and distributing quality lipid at a reasonable price and in a timely manner.

Acknowledgments

There are many who contributed to this work and most are from the Caffrey and Hart groups, both past and present members. In particular, we wish to acknowledge the contributions of V. Cwynar, J. Lee, A. Maher, J. Mohan, and R. Tiedt. To all we extend our warmest thanks and appreciation. This work was supported in part by grants from Science Foundation Ireland (07/IN.1/B1836) and the National Institutes of Health (GM75915).

References

Ai, X., & Caffrey, M. (2000). Membrane protein crystallization in lipidic mesophases. Detergent effects. *Biophysical Journal, 79*, 394–405.

Briggs, J. (1994). *The phase behavior of hydrated monoacylglycerols and the design of an X-ray compatible scanning calorimeter.* The Ohio State University, Columbus, OH.

Briggs, J., & Caffrey, M. (1994a). The temperature-composition phase diagram and mesophase structure characterization of monopentadecenoin in water. *Biophysical Journal, 67*, 1594–1602.

Briggs, J., & Caffrey, M. (1994b). The temperature-composition phase diagram of monomyristolein in water: Equilibrium and metastability aspects. *Biophysical Journal, 66*, 573–587.

Briggs, J., Chung, H., & Caffrey, M. (1996). The temperature-composition phase diagram and mesophase structure characterization of the monoolein/water system. *Journal of Physics II France, 6*, 723–751.

Caffrey, M. (2000). A lipids eye view of membrane protein crystallization in mesophases. *Current Opinion in Structural Biology, 10*, 486–497.

Caffrey, M. (2003). Membrane protein crystallization. *Journal of Structural Biology, 142*, 108–132.

Caffrey, M. (2008a). Membrane protein crystallisation using lipidic mesophases. A practical demonstration from the Caffrey laboratory. http://www.caffreylabs.ul.ie/in%20meso%20demo%2031 1008.pdf.

Caffrey, M. (2008b). On the mechanism of membrane protein crystallization in lipidic mesophases. *Crystal Growth and Design, 8*, 4244–4254.

Caffrey, M. (2009). Crystallizing membrane proteins for structure determination: Use of lipidic mesophases. *Annual Review of Biophysics, 38*, 29–51.

Cherezov, V., & Caffrey, M. (2006). Picolitre-scale crystallization of membrane proteins. *Journal of Applied Crystallography, 39*, 604–609.

Cherezov, V., & Caffrey, M. (2007). Membrane protein crystallization in lipidic mesophases. A mechanism study using X-ray microdiffraction. *Faraday Disc, 136*, 188–205.

Cherezov, V., Clogston, J., Misquitta, Y., Abdel Gawad, W., & Caffrey, M. (2002). Membrane protein crystallization in meso. Lipid type-tailoring of the cubic phase. *Biophysical Journal, 83*, 3393–3407.

Cherezov, V., Clogston, J., Papiz, M., & Caffrey, M. (2006). Room to move. Crystallizing membrane proteins in swollen lipidic mesophases. *Journal of Molecular Biology, 357*, 1605–1618.

Cherezov, V., Liu, J. C., Griffith, M., Hanson, M. A., & Stevens, R. C. (2008). LCP-FRAP assay for prescreening membrane proteins for *in meso* crystallization. *Crystal Growth and Design, 8*, 4307–4315.

Cherezov, V., Liu, W., Derrick, J., Luan, B., Aksimentiev, A., Katrich, V., et al. (2008). *In meso* crystal structure and docking simulations suggest an alternative proteoglycan binding site in the OpcA outer membrane adhesin. *Proteins, 71*, 24–34.

Cherezov, V., Peddi, A., Muthusubramaniam, L., Zheng, Y. F., & Caffrey, M. (2004). A robotic system for crystallizing membrane and soluble proteins in lipidic mesophases. *Acta Crystallographica. Section D, 60*, 1795–1807.

Cherezov, V., Rosenbaum, D. M., Hanson, M. A., Rasmussen, S. G. F., Thian, F. S., Kobilka, T. S., et al. (2007). High-resolution crystal structure of an engineered human β2-adrenergic G protein coupled receptor. *Science, 318*, 1258–1265.

Cherezov, V., Yamashita, E., Liu, W., Zhalnina, M., Cramer, W. A., & Caffrey, M. (2006). *In meso* structure of the cobalamin transporter, BtuB, at 1.95 Å resolution. *Journal of Molecular Biology, 364*, 716–734.

Chung, H., & Caffrey, M. (1995). Polymorphism, mesomorphism, and metastability of monoelaidin in excess water. *Biophysical Journal, 69*, 1951–1963.

Clogston, J. (2005). *Applications of the lipidic cubic phase: From controlled release and uptake to in meso crystallization of membrane proteins*. The Ohio State University, Columbus, OH.

Clogston, J., Graciun, G., Hart, D. J., & Caffrey, M. (2005). Controlling release from the lipidic cubic phase by selective alkylation. *Journal of Controlled Release, 107*, 97–111.

Clogston, J., Rathman, J., Tomasko, D., Walker, M., & Caffrey, M. (2000). Phase behavior of a monoacylglycerol: (Myverol 18-99K)/water system. *Chemistry and Physics of Lipids, 107*, 191–220.

Coleman, B. E., Cwynar, V., Hart, D. J., Havas, F., Mohan, J. M., Patterson, S., et al. (2004). Modular approach to the synthesis of unsaturated 1-monoacyl glycerols. *Synlett, 8*, 1339–1342.

Gordeliy, V. I., Labahn, J., Moukhametzianov, R., Efremov, R., Granzin, J., Schlesinger, R., et al. (2002). Molecular basis of transmembrane signalling by sensory rhodopsin II-transducer complex. *Nature, 419*, 484–487.

Hanson, M. A., Cherezov, V., Griffith, M. T., Roth, C. B., Jaakola, V.-P., Chien, E. Y. T., et al. (2008). A specific cholesterol binding site is established by the 2.8 Å structure of the human β2-adrenergic receptor. *Structure (London, England: 1993), 16*, 897–905.

Hyde, S. T., Andersson, S., Blum, Z., Lidin, S., Larsson, K., Landh, T., et al. (1997). *The language of shape*. Elsevier Science B.V., Amsterdam.

Israelachvili, J. N., Mitchell, D. J., & Ninham, B. W. (1977). Theory of self-assembly of lipid bilayers and vesicles. *Biochimica et Biophysica Acta, 470*, 185–201.

Jaakola, V-P., Griffith, M. T., Hanson, M. A., Cherezov, V., Chien, E. Y. T., Lane, J. R., et al. (2008). The 2.6 Angstrom crystal structure of a human A_{2A} adenosine receptor bound to an antagonist. *Science, 322*, 1211–1217

Landau, E., & Rosenbusch, J. P. (1996). Lipidic cubic phases: A novel concept for the crystallization of membrane proteins. *Proceedings of the National Academy of Sciences of the United States of America, 93*, 14532–14535.

Liu, W., & Caffrey, M. (2005). Gramicidin structure and disposition in highly curved membranes. *Journal of Structural Biology, 150*, 23–40.

Lutton, E. S. (1965). Phase behavior of aqueous systems of monoglycerides. *Journal of American Oil Chemists' Society, 42*, 1068–1070.

Lyons, J., Soulimane, T., & Caffrey, M. (2008). Unpublished work.

Michel, H. (1983). Crystallization of membrane proteins. *Trends in Biochemical Sciences, 8*, 56–59.

Misquitta, L. V., Misquitta, Y., Cherezov, V., Slattery, O., Mohan, J. M., Hart, D. J., et al. (2004a). Membrane protein crystallization in lipidic mesophases with tailored bilayers. *Structure (London, England: 1993), 12*, 2113–2124.

Misquitta, Y. (2006). *The rational design of monoacylglycerols for use as matrices for the crystallisation of membrane proteins.* The Ohio State University, Columbus, OH.

Misquitta, Y., & Caffrey, M. (2001). Rational design of lipid molecular structure: A case study involving the C19:1c10 monoacylglycerol. *Biophysical Journal, 81*, 1047–1058.

Misquitta, Y., & Caffrey, M. (2003). Detergents destabilize the cubic phase of monoolein. Implications for membrane protein crystallization. *Biophysical Journal, 85*, 3084–3096.

Misquitta, Y., Cherezov, V., Havas, F., Patterson, S., Mohan, J. M., Wells, A. J., et al. (2004b). Rational design of lipid for membrane protein crystallization. *Journal of Structural Biology, 148*, 169–175.

Munro, S. (1998). Localization of proteins to the Golgi apparatus. *Trends in Cell Biology, 8*, 11–15.

Murgia, S., Monduzzi, M., Ljusberg-Wahren, H., & Nylander, T. (2002). Acyl migration and hydrolysis in monoolein based systems. *Progress in Colloid and Polymer Science, 120*, 41–46.

Muthusubramaniam, L. (2004). *Automation of the in meso method for membrane protein crystallisation and phase behavior and crystallisation studies on the 5.13 and 13.5 monoacylglycerols.* The Ohio State University, Columbus, OH.

Paas, Y., Cartaud, J., Recouvreur, M., Grailhe, R., Dufresne, V., Pebay-Peyroula, E., et al. (2003). Electron microscopic evidence for nucleation and growth of 3D acetylcholine receptor microcrystals in structured lipid-detergent matrices. *Proceedings of the National Academy of Sciences of the United States of America, 100*, 11309–11314.

Qiu, H., & Caffrey, M. (1999). Phase behavior of the monoerucin/water system. *Chemistry and Physics of Lipids, 100*, 55–79.

Qiu, H., & Caffrey, M. (2000). Phase diagram of the monoolein/water system: Metastability and equilibrium aspects. *Biomaterials, 21*, 223–234.

Qutub, Y., Reviakine, I., Maxwell, C., Navarro, J., Landau, E. M., & Vekilov, P. G. (2004). Crystallization of transmembrane proteins in cubo: Mechanisms of crystal growth and defect formation. *Journal of Molecular Biology, 343*, 1243–1254.

Raman, P., Cherezov, V., & Caffrey, M. (2006). The membrane protein data bank. *Cellular and Molecular Life Sciences, 63*, 36–51. http://www.mpdb.ul.ie/

Slattery, O., Caffrey, M., & Soulimane, T. (2008). Crystallisation and preliminary X-ray diffraction analysis of caa_3-cytochrome c oxidase from *Thermus thermophilus*. *Biochimica et Biophysica Acta, S1*, S74.

Wadsen, P., Wöhri, A. B., Snijder, A., Katona, G., Gardiner, A. T., Cogdell, R. J., et al. (2006). Lipidic sponge phase crystallization of membrane proteins. *Journal of Molecular Biology, 364*, 44–53.

CHAPTER 5

Practical Aspects of Membrane Proteins Crystallization in Bicelles

Salem Faham,* Rachna Ujwal,* Jeff Abramson,* and James U. Bowie[†]

*Department of Physiology, Division of Molecular Medicine, David Geffen School of Medicine, University of California, Los Angeles, California 90095

[†]Department of Chemistry and Biochemistry, UCLA-DOE center for Genomics and Proteomics, Molecular Biology Institute, University of California, Los Angeles, California 90095

I. OVERVIEW

Structure determination of membrane proteins remains a significant challenge. One of the important bottlenecks is growing high quality crystals. Crystallization of membrane proteins directly from detergents is by far the most popular approach; although, lipidic cubic phase crystallization is attracting increasing use. We introduced the use of bicelles as an alternative method for the crystallization of membrane proteins. Bicelles are a mixture of a detergent and a lipid, and can be described as a compromise between the two media with beneficial aspects from both. Membrane proteins reconstituted in bicelles are maintained in a native like bilayer environment and can be manipulated with almost the same ease as detergent solublized membrane

Current Topics in Membranes, Volume 63

1063-5823/09 $35.00
DOI: 10.1016/S1063-5823(09)63005-2

proteins, making it compatible with standard high-throughput screening. An increasing number of recent bicelle crystallization successes support the expanded use of this method, including β2-adrenergic receptor, the voltage-dependent anion channel and xanthorhodopsin. Here we describe the bicelle method, its properties, advantages and disadvantages, and consider how to achieve future progress with the bicelle method.

II. INTRODUCTION

Membrane proteins have traditionally been crystallized from detergents. This is a convenient approach since membrane proteins are generally purified in detergent solubilized form and crystallization can be performed using standard techniques. Many detergent solubilized membrane proteins resist crystallization, or produce low quality crystals, however. Detergents are not always an adequate substitute for a bilayer, often resulting in poor stability, protein aggregation or other heterogeneity. Moreover, detergent micelles shield the hydrophobic surface of the membrane protein from the solvent, reducing the free surface available for crystal contacts. In the absence of a well-ordered soluble domain, crystal contacts are often forced to form in the less rigidly structured loop regions, which can lead to crystal disorder (Carpenter, Beis, Cameron, & Iwata, 2008).

To mitigate the problem of limited surface area available for crystal contacts, the antibody method was developed, where a bound Fv or Fab fragment provides a large stable polar domain for crystal formation (Ostermeier, Iwata, Ludwig, & Michel, 1995). The antibody method has been successful in some cases, but it requires the identification and production of an antibody, a laborious and technically challenging process that is hard to pursue on a routine basis. In addition, there may be concerns that the bound antibody distorts the structure (Lee, Lee, Chen, & MacKinnon, 2005). Another approach to extend the polar surface area is the use of soluble fusion partners (Cherezov et al., 2007; Prive et al., 1994) an approach that proved successful for β2 adrenergic receptor. This method requires extensive knowledge about the protein of interest in order to artfully position the polar domain with minimal disorder.

In some cases conformational changes are observed depending on whether a polypeptide is in a micellar or bilayer environment (Andersson & Maler, 2002; Chou, Kaufman, Stahl, Wingfield, & Bax, 2002; Poget, Cahill, & Girvin, 2007). Thus it should be advantageous to crystallize membrane proteins in lipidic media because the bilayer environment better simulates their native environment and is more likely to preserve their native structure.

Membrane proteins can be influenced by the overall properties of a lipid bilayer, or by specific binding to individual lipids. Ordered lipids have been observed in many membrane protein structures such as bacteriorhodopsin (Luecke, Schobert, Richter, Cartailler, & Lanyi, 1999), yeast cytochrome bc_1 complex (Lange, Nett, Trumpower, & Hunte, 2001), and cytochrome *c* oxidase (Shinzawa-Itoh et al., 2007). Lipids can also play an important role in crystallization (Hunte & Richers, 2008) as was observed with yeast cytochrome bc_1, cyanobacterial cytochrome *b6f* complex (Kurisu, Zhang, Smith, & Cramer, 2003), LacY (Guan, Smirnova, Verner, Nagamori, & Kaback, 2006), and the major light-harvesting complex of photosystem II (Liu et al., 2004).

In 1996 the lipid cubic phase method was introduced for the crystallization of membrane proteins (Landau & Rosenbusch, 1996), demonstrating for the first time that crystals could be obtained from a protein ensconced in a bilayer. The cubic phase is an interconnected three-dimensional lipid bilayer formed by monoolein. Membrane proteins can be incorporated into the cubic phase, where they can diffuse in three dimensions and feed crystal nuclei. A great advantage of the lipid cubic phase is that the protein remains in a more natural bilayer environment. The method has been advanced by others, most notably Caffrey and co-workers, and has an increasing number of success stories to recommend it (Cherezov, Clogston, Papiz, & Caffrey, 2006). A disadvantage of the method is the extremely high viscosity of the lipid cubic phase making specialized apparatus necessary (Cheng et al., 1998) and crystal manipulation problematic (Nollert & Landau, 1998).

Because of some of the technical difficulties associated with the cubic phase method, we explored the possibility of finding a more convenient lipidic medium for the crystallization of membrane proteins. We identified bicelles, which are composed of a lipid/amphiphile mixture, as an attractive medium for the crystallization of membrane proteins. The bicelle medium maintains a beneficial bilayer environment, and if maintained below the gel transition temperature (~30 °C varies depending on composition), it can be easily pipetted so that standard crystallization setups can be used (Fig. 1A). We were successful in crystallizing bacteriorhodopsin in two different bicelle formulations (Fig. 1B) (Faham & Bowie, 2002; Faham et al., 2005), producing different crystal forms. Recently, new membrane protein structures have been determined from crystals grown in bicelles. These are the β2-adrenergic receptor (Rasmussen et al., 2007), the voltage dependent anion channel (VDAC) (Ujwal et al., 2008) and xanthorhodopsin (Luecke et al., 2008). These results argue that bicelle crystallization should be a standard weapon in the membrane protein crystallization arsenal. Here we describe practical aspects of performing bicelle crystallization.

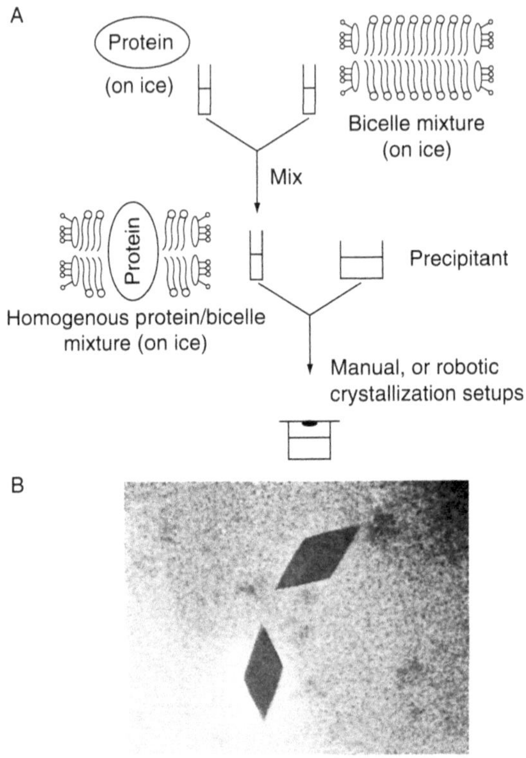

FIGURE 1 *Bicelle crystallization protocol.* (A) The basic steps. The bicelle method can be performed with any standard crystallization setups. (B) Crystals of bacteriorhodopsin grown from bicelles.

III. BICELLES

Bicelles are disc-like micelles that form in particular mixtures of a phosphotidylcholine (PC) lipid including dimyristoyl-phosphatidylcholine (DMPC), ditridecanoyl-phosphatidylcholine (DTPC), dilauroyl-phosphatidylcholine (DLPC), and others along with lipid additives and an amphiphile (i.e., dihexanoyl-phosphatidylcholine, DHPC or 3-[(3-cholamidopropyl) dimethylammonio]-2-hydroxy-1-propanesulfonate, CHAPSO). The bicelle discs are patches of lipid bilayers each with detergent molecules protecting the apolar edges of the bilayer. Bicelles can be described as a compromise between a strictly lipidic medium and a detergent medium offering some of the benefits of both. These benefits are ease of use, like detergents, while maintaining a more native like bilayer environment, like the lipid cubic phase.

Membrane proteins can be incorporated into bicelles easily, and various examples show that membrane proteins preserve their activity in bicelles (Czerski & Sanders, 2000; Sanii, Schill, Moran, & El-Sayed, 2005).

The properties of bicelles vary depending on their formulations. It has been observed that the diameter of the bicelle increases as the ratio of long chain (PC) lipid to detergent increases (Tiburu, Moton, & Lorigan, 2001). The diameter of 2:1 DMPC/DHPC bicelle discs at 20% (w/v) total lipid concentration has been reported in the range of 20–50 nm diameter (Barbosa-Barros, De la Maza, Walther, Estelrich, & Lopez, 2008), while at 1:2 DMPC/DHPC the diameter of bicelle discs is ~8-10 nm (Vold, Prosser, & Deese, 1997). Divalent cations (Ca^{2+} and Mg^{2+}) promote larger diameter size compared to monovalent cations (Na^+ and K^+) (Arnold, Labrot, Oda, & Dufourc, 2002). It was also shown that bicelles prepared with DLPC (12 carbons) lipid have larger diameter than those containing DMPC or dipalmitoyl-phosphatidylcholine (DPPC) lipids (14 or 16 carbons) (Lind, Nordin, & Maler, 2008). Bicelles generally have a thickness of ~40 Å (Luchette et al., 2001).

Bicelles are of importance in the field of NMR due to their predisposition to partially align themselves in a magnetic field (Sanders & Prosser, 1998). Because of this interest, characterization of the properties of bicelles, effects of additives on bicelles, and testing alternative ingredients is growing at a steady pace.

IV. PHASE BEHAVIOR OF BICELLES

Bicelles undergo a transition from a liquid phase to a gel phase as the temperature is raised above a transition temperature (T_m) (Fig. 2). Some of the early characterizations describe the liquid phase as a bicellar phase and the gel phase as a perforated lamellar phase, also described as the Swiss cheese model (Prosser, Hwang, & Vold, 1998). Although the detailed phase behavior of bicelles is still debated, there is an agreement that the phase behavior of bicelles is more complex than some of the initial models. Recent analysis describes four states (Triba, Warschawski, & Devaux, 2005), two bicellar and two lamellar. Worm like topology has also been observed (Nieh et al., 2004), also described as branched flat cylindrical micelles (van Dam, Karlsson, & Edwards, 2004).

The transition temperature, T_m, depends on the ingredients of the bicelle mixture, for example, DMPC/CHAPSO forms a gel above 32 °C, while DTPC/CHAPSO is reported to form a gel above 12 °C (Ottiger & Bax, 1998). As a result the liquid bicelle phase can have a narrow temperature range as in the case for DTPC/CHAPSO, or it can have a wide range.

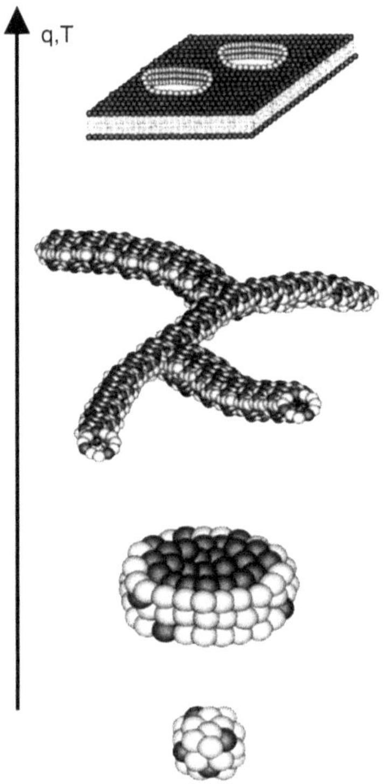

FIGURE 2 *Different phases that can be obtained with bicelle-forming mixtures.* These are from the bottom: small bicelles, or globular micelles; large bicelles; worm like branched flat cylindrical micelles; and perforated lamellar sheet, also described as the swiss cheese model. These transitions occur as the temperature (T) and the ratio DMPC/DHPC (q) are increased. This figure is adapted from van Dam et al. (2004).

For example, when DMPC/DHPC bicelles are doped with unsaturated lipids the range of the liquid phase is expanded from 13 to 55 °C (Triba, Devaux, & Warschawski, 2006). The perforated lamellar phase may have particular advantages for crystal growth as the proteins could be pre-organized in layers.

V. EXAMPLES OF MEMBRANE PROTEINS CRYSTALLIZED IN BICELLES

Bacteriorhodopsin (bR) was the first membrane protein to be crystallized from bicelles. Purple membrane was suspended in water to a bR concentration of ~10 mg/ml and was mixed in a 4:1 ratio with 40% (3:1 DMPC: CHAPSO), giving ~8 mg/ml bR in 8% bicelles. It was crystallized at 37 °C in 2.8 M NaPO4, pH 3.7, and 180 mM hexandiol. The quality of these crystals was sufficient to determine the structure to 1.8 Å resolution. bR was subsequently crystallized again at room temperature in both DMPC/CHAPSO, and in DTPC/CHAPSO. The crystallization condition in DMPC/CHAPSO at room temperature contained 2.45 M NaH_2PO_4 pH 3.7, 180 mM hexanediol, and 3.5% triethyleneglycol in a 8% 2.8:1 (DMPC: CHAPSO) bicelle mixture. These crystals diffracted to 2.2 Å and produced a new crystal form (*C222₁*). The crystallization condition in DTPC/CHAPSO at room temperature is 100 mM sodium formate pH 4.3, 28.5% PEG 2K, 280 mM ammonium sulfate, and 180 mM hexanediol in 8% 3:1 (DTPC: CHAPSO). These crystals diffracted to 1.8 Å and belonged the same *$P2_1$* crystal form obtained at 37 °C (Faham et al., 2005).

A major development was the report of the structure determination of β2-adrenergic receptor (β2AR), a G-protein coupled receptor, from crystals grown in bicelles. β2AR was crystallized in complex with a Fab fragment in bicelles at room temperature.

The β2AR protein was mixed in stoichiometric excess of the Fab fragment. The complex was purified on a gel filtration column in a solution of 10 mM HEPES pH 7.5, 100 mM NaCl, 0.1% dodecylmaltoside, and 10 mM carazolol. The purified β2AR-Fab complex was concentrated to ~60 mg/ml. The complex was mixed with 10% bicelles (3:1 DMPC:CHAPSO in 10 mM HEPES, pH 7.5, 100 mM NaCl) at a 1:5 (protein:bicelle) ratio giving ~10 mg/ml protein in 8.3% bicelles. Initial crystallization leads were identified using multiple 96-well sitting drop screens from Nextal (Qiagen). After optimization, crystals were grown at 22 °C in hanging-drop format over a reservoir solution of 1.85–2.0 M ammonium sulfate, 180 mM sodium acetate, 5 mM EDTA, 100 mM MES or HEPES, pH 6.5–7.5. These crystals diffracted anisotropically to 3.4/3.7 Å resolution (Rasmussen et al., 2007).

Recently, the structure of the VDAC was determined to high resolution from crystals grown from bicelles, the first β-barrel membrane protein to be crystallized from bicelles. Purified murine VDAC1 protein at a concentration of 15 mg/ml was mixed in a 4:1 protein:bicelle ratio with 35% (2.8:1 DMPC: CHAPSO) bicellar solution, giving 12 mg/ml mVDAC1 in 7% bicelles. Initial crystallization trials were performed using the Mosquito nanoliter liquid handling robot (TTP labtech). Almost 700 conditions from commercially available screens were tested. Crystals were observed in three conditions (Fig. 3A,B,C): (1) 0.1 M Tris-HCl (pH 8.5), 15% MPD; (2) 0.1 M Tris-HCl (pH 7.5), 15% Ethanol, 0.3 M NaCl; and (3) 0.1 M Tris-HCl (pH7.5), 9% Isopropanol, 0.3 M NaCl. The proteinacious nature of the crystals was confirmed by UV microscopy (Fig. 3A–C). After optimization by high throughput screening of additives, the best crystals grew in 18–20% MPD, 0.1 M Tris-HCl (pH 8.5) with 10% PEG400 added only to the drop.

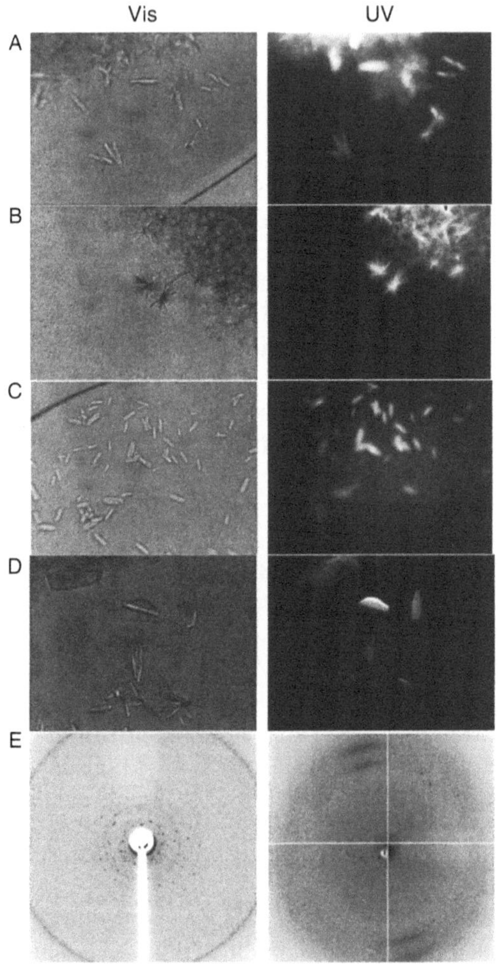

FIGURE 3 *VDAC crystals.* (A)–(D) Pictures of murine VDAC1 crystals taken with visible (left panels) and UV microscopes (right panels). (A) Initial VDAC crystals from an ethanol precipitant condition. (B) Initial VDAC crystals from an isopropanol precipitant condition. (C) Initial VDAC crystals from an MPD precipitant condition. (D) Optimized VDAC crystals from an MPD precipitant condition. (E) Diffraction of VDAC crystals. Left panel shows initial diffraction of crystals obtained from the MPD condition to ~6.0 Å, and right panel shows the diffraction of the optimized VDAC crystals from the MPD condition to ~2.2 Å.

The initial crystals diffracted to ~6.0 Å while the optimized crystals diffracted to 2.2 Å (Fig. 3D,E) (Ujwal et al., 2008). Initial screens as well as optimization additives screens took full advantage of the ability to automate crystallization setups while using bicelles.

Also recently the structure of xanthorhodopsin was determined using crystals grown from bicelles. Xanthorhodopsin is a light activated proton pump with two chromophores: a retinal and the carotenoid salinixanthin. A membrane fraction enriched in xanthorhodopsin was prepared and resuspended in 30 mM phosphate (pH 5.6) and 1 mM sodium azide and concentrated to contain ~5 mg/ml xanthorhodopsin. The protein was mixed with bicelle-type medium made of 16.7% DMPC in 20% Nonylmaltoside (NM) in 1:3 (bicelle:protein) ratio followed by vortexing, and incubating overnight at 4 °C. Crystals were grown at 22 °C over 4–5 months in sitting drops, containing 10 μL of solubilized xanthorhodopsin, 3 μL of 3 M sodium phosphate (pH 5.6), and 2 μL of 2.5 mM sodium azide. The reservoirs contained 2.5–3.0 M sodium phosphate (pH 5.6).

A summary of successful bicelle crystallization conditions are in Table I. This small collection of proteins shows that colored, colorless, alpha helical, and beta sheeted membrane proteins can be crystallized in bicelles. Bicelles are also compatible with soluble domains and protein complexes as demonstrated by the β2AR/Fab crystals. Precipitant conditions include salts, PEGs, and organics. The diversity of the crystallization conditions shows that many precipitants are suitable for screening with the bicelle method. Moreover,

TABLE I

Summary of Crystallization Conditions for Membrane Proteins Crystallized in Bicelles

Protein	Bicelle formulation	Crystallization conditions
Bacteriorhodopsin (~8 mg/ml) *P2₁* crystal form	8% (2.8:1) DMPC: CHAPSO at 37°C	2.8 M NaH_2PO_4 pH 3.7, 180 mM hexanediol
Bacteriorhodopsin (~8 mg/ml) *P2₁* crystal form	8% (3:1) DTPC: CHAPSO	100 mM sodium format pH 4.3, 28.5% PEG2K, 280 mM ammonium sulfate, 180 mM Hexanediol
Bacteriorhodopsin (~8 mg/ml) *C222₁* crystal form	8% (2.8:1) DMPC: CHAPSO	2.45 MNaH2PO4 pH 3.7, 180 mM hexanediol, 3.5% triethyleneglycol
β2 Adrenergic receptor/Fab complex (~10 mg/ml)	8.3% (3:1) DMPC: CHAPSO	1.85–2.0 M ammonium sulfate, 180 mM sodium acetate, 5 mM EDTA, 100 mM MES or HEPES pH 6.5–7.5
Voltage dependent anion channel (12 mg/ml)	7% (2.8:1) DMPC: CHAPSO	18–20% MPD, 0.1 M Tris-HCl, pH 8.5, with 10% PEG400 PEG added only to the drop
Xanthorhodopsin (3.75 mg/ml)	4.2% DMPC, 5% NM	2.5–3.0 M sodium phosphate pH 5.6, 2.5 mM sodium azide

given the interesting phase behavior of bicelle formulations, the same screens can be performed in different phases simply by altering the temperature, providing an increased diversity of crystallization conditions.

VI. CRYSTAL PACKING

It was recognized early on that membrane proteins can form two types of crystals categorized based on the type of crystal packing observed, referred to as type I and type II (Ostermeier & Michel, 1997). Type I crystals result from the stacking of 2D crystals or 2D layers. In each layer the membrane protein molecules are oriented as if in a bilayer stacking side by side with the hydrophobic surface providing substantial crystal contacts. Crystals grown from lipidic media tend to produce type I crystals, because bilayers support the formation of 2D layers. These crystals usually grow well in two dimensions with possibly slower growth in the third dimension producing thin plates. Developments of microfocused beamlines at synchrotrons can be critical for structure determination from some of these very thin crystals. Indeed, crystals grown from bicelles tend to belong to type I crystals, including VDAC, xanthorhodopsin, and both crystal forms of bR (Fig. 4). The β2AR crystal packing is more difficult to classify due to lack of clear crystal contacts in one direction, and the presence of the Fab fragment which forms the majority of the crystal contacts. But even in this case the packing follows the stacking of layers pattern. Type II crystals are more commonly observed with delipidated, detergent solublized membrane proteins. In this case the hydrophobic surface is shielded and most of the crystal contacts are formed by the hydrophilic surface of the membrane protein. Strategies that involve screening many detergents and the selection of the appropriate additives to modulate the micelle size can be crucial for type II crystals. Strategies that include the screening of lipid media can offer a different surface for crystal contacts and thus can lead to completely different crystal forms.

VII. PRACTICAL INFORMATION

Crystallization from bicelles, as with other lipidic media is performed in three steps. The first step is the preparation of the lipidic medium. The second step is the incorporation of the protein into the lipidic medium. And the third step is the setup of crystallization trials. A quick reference guide for Steps 1 and 2 is provided in Table II.

Step 1: Making the bicelle mixture. Since bicelles can be formed by diverse ingredients and in a wide range of concentrations it is desirable to have a good starting point for initial screens. Here we provide an example of a good

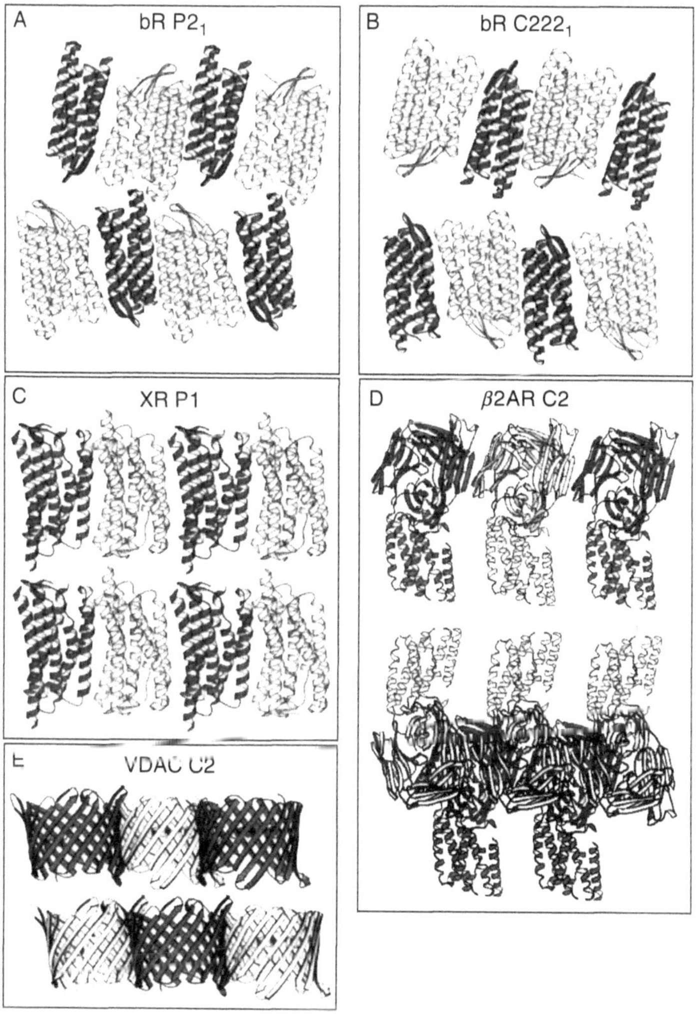

FIGURE 4 *Crystal packing using bicelle method.* Crystals obtained using the bicelle method generally belong to type I crystals. (A) Packing of bacteriorhodopsin in $P2_1$ crystals. (B) Packing of bacteriorhodopsin in $C222_1$ crystals. (C) Packing in xanthorhodopsin crystals. (D) Packing in β2AR-FAB crystals. The majority of crystal contacts are formed by the Fab fragments. (E) Packing of VDAC.

starting point for initial crystal screens in bicelles. To prepare 1.0 ml of a 35% bicelle solution at 2.8:1 (DMPC:CHAPSO), one will need to mix 0.263 g DMPC, 0.087 g CHAPSO, and add water to a final volume of 1.0 ml. The total weight of bicelle components adds up to 0.35 g (corresponding to 35%),

TABLE II

Quick Guide for Protein/Bicelle Mixture Preparation for Crystal Trials

Step 1: Make the bicelle mixture. The following is a recipe for making 1 ml of a 35% bicelle solution at 2.8:1 (DMPC:CHAPSO):
1. Weigh out 87.35 mg of CHAPSO in a 1.5 ml eppendorf tube
2. Add 650 μl water to the eppendorf tube and vortex until the CHAPSO dissolves
3. Weigh out 262.6 mg of DMPC in a 15 ml falcon tube
4. Add the dissolved CHAPSO to the falcon tube. Vortex
5. Perform cycles of flash freezing using liquid nitrogen, heating to 55 °C, cooling on ice and vortexing to dissolve the DMPC
6. Employ a few short bursts of sonication (1 s) using a microtip sonicator help to get a homogenous mixture. Sonication can lead to froth and should be used only briefly
7. Repeat step 5 until the bicelle mixture forms a clear liquid on ice and a clear gel upon warming to 37 °C
8. The bicelle mixture can be stored for long periods at −20 °C
Step 2: Make the protein/bicelle mixture:
1. When ready to use, thaw the bicelle mixture and place it on ice (to keep it liquid). The mixture should be vortexed after the freeze-thaw to re-establish a homogenous bicelle phase
2. Add the bicelle mixture to the protein (>10 mg/ml) in a 1:4 (bicelle:protein) ratio while keeping everything on ice
3. Mixing can be achieved by simply pipetting the contents up and down until the solution appears homogenous
4. Incubate the protein/bicelle mixture on ice for ~30 min before setting up crystallization trials in any standard format

and the molar ratio of DMPC to CHAPSO is 2.8:1. Mixing to obtain a homogenous solution can take some work. Short bursts of sonication, cycles of heating to ~50 °C and cooling, including flash cooling with liquid nitrogen, and vortexing the cooled liquid phase will produce a clear solution. An indication that the bicelle phase has properly formed occurs when the mixture is clear and a liquid when cooled on ice, but forms a clear gel when warmed to 37 °C. Prepared bicelles can be stored frozen for future use. Bicelles are not stable enough for long term storage at room temperature due to the potential hydrolysis of the phospholipid head group. When retrieving the frozen bicelle mixture, it might be necessary to re-establish the bicelle phase by vortexing, heating, and cooling.

Step 2: Making the protein/bicelle mixture. Once the bicelle solution is prepared, protein purified in detergent can be added to the bicelles by simply pipetting up and down to homogeneity. Further incubation on ice for 30 min is a good general practice to allow for complete incorporation of the protein into the bicelles. For example, if the protein concentration is 15 mg/ml, one can mix it in a 4:1 protein:bicelle ratio, and obtain 12 mg/ml protein concentration in 7% bicelles. Detergent carried along can influence the behavior of bicelles, but the concentration of added detergents are generally much lower than the 7% bicelle concentration, thus the behavior of the bicelle medium generally prevails. For example before addition to bicelles, VDAC protein was present in 0.1% LDAO, β2AR was in 0.1% dodecylmaltoside. Also, in the case of bR, β-Octyl Glucoside was initially used as an additive, further demonstrating that additional detergents can be compatible with the bicelle system. If the bicelle solution is kept on ice it can be pipetted and mixed easily with the protein solution. This is particularly important when working with DTPC/CHAPSO since this mixture forms a gel at room temperature. Preparing a high concentration of bicelle mix (35%) that can be diluted fivefold to a 7% bicelle concentration is useful because it is not necessary to dilute the protein much. However if one is able to obtain a high protein concentration this consideration is not critical and one may benefit from adding less detergent.

Hydrolysis of the ester linkage of the PC lipids can occur over time, particularly at extreme pH values. For example, it was shown that DMPC/DHPC bicelle character is lost after 2 weeks at pH 4. Ether-linked lipids are found in the cell membranes of archeal microorganisms and are able to withstand extreme conditions (Koga & Morii, 2005). Similarly, PC lipids that replace the ester linkage with an ether linkage can form bicelles and are resistant to hydrolysis (Aussenac, Lavigne, & Dufourc, 2005). These lipids are much more expensive, however. In the case of bR, the crystallization condition is at pH of ~3.7, but lipid hydrolysis does not appear to inhibit the production of high quality bR crystals (Faham & Bowie, 2002).

Step 3: Crystal screening. Essentially any standard crystallization setups can be used with bicelles including robotics. We have tested both hanging and sitting drop formats from a variety of suppliers. It is possible that bicelle-specific crystallization screens can be developed, but for now, standard commercially available crystallization screens can be used.

Visualization can be more challenging when using lipid media. The lipidic media can be slightly opaque, or certain lipidic phases can form that appear crystalline. Indeed, the majority of early successful crystallizations in lipidic media were of colored membrane proteins (Nollert, 2004). To overcome this challenge we utilize a UV microscope that detects tryptophan fluorescence. Thus, proteinaceous features can be distinguished from other spurious phenomena that occur in crystallization trials. Without the benefit of a positive identification by a UV microscope, false leads may distract efforts away from pursuing a real lead. Alternatively, a positive identification can be achieved by preparing each tray for both the protein/bicelle mixture and bicelle mixture alone.

Crystal extraction and freezing is relatively straightforward when working with bicelles. In contrast to lipid cubic phase media, even the gel phase of bicelles does not prohibit crystal extraction with standard procedures. The gel phase is more like soft glue so it can be manipulated. It also appears that the bicelle mixture itself provides some cryo protection.

Although the number of crystals remains low, we can attempt to draw some conclusions. The range of bicelle ingredients successful for producing membrane protein crystals, although wide and diverse, does seem to show a consensus in the range of ~8% bicelles 2.9:1 DMPC:CHAPSO. The GPCR Opsin has been reported to be more stable in DMPC/CHAPSO than in DMPC/DHPC (McKibbin et al., 2007). If this is a general trend for the stability of other membrane proteins, it could account for the observation of higher success rates in bicelles made with CHAPSO instead of DHPC.

The unique advantage of the bicelle method is that steps 2 and 3 are straightforward and are no different from standard protocols as long as the bicelle mixture is kept liquid at lower temperatures. It is generally accepted that for the crystallization of membrane proteins many detergents should be screened to determine which provides the best medium for successful crystallization. Bicelles can easily be added to such screens. Indeed, bicelles are easier to screen than alternate detergents because it does not require an additional purification or a detergent exchange step as the detergent solubilized protein can simply be added to a bicelle mixture directly. In addition, bicelles can be easily doped with other lipids, which could be particularly advantageous for those proteins that bind specific lipids.

VIII. CONCLUSION

The use of lipidic media has gained acceptance and now includes the cubic and sponge phase methods (see Chapter 4 for detailed explanation of these methods) and the bicelle method. But crystallizations from lipidic media present challenges. These challenges include practical aspects such as the ease of use, incorporation of the membrane protein into the lipidic media, high throughput crystallization trial preparation, crystal observation, and crystal extraction. The bicelle method is well suited to deal with all these challenges and it offers considerable versatility. More work is required to better understand how crystals form in bicelles so the conditions that favor crystallization can be more rationally approached and smaller more focused screens employed. If protein is not limiting, however, it is straightforward to employ large-scale screening using existing formulations. There is little reason for not employing bicelles as part of the screening process.

References

Andersson, A., & Maler, L. (2002). NMR solution structure and dynamics of motilin in isotropic phospholipid bicellar solution. *Journal of Biomolecular NMR, 24*, 103–112.

Arnold, A., Labrot, T., Oda, R., & Dufourc, E. J. (2002). Cation modulation of bicelle size and magnetic alignment as revealed by solid-state NMR and electron microscopy. *Biophysical Journal, 83*, 2667–2680.

Aussenac, F., Lavigne, B., & Dufourc, E. J. (2005). Toward bicelle stability with ether-linked phospholipids: Temperature, composition, and hydration diagrams by 2H and 31P solid-state NMR. *Langmuir, 21*, 7129–7135.

Barbosa-Barros, L., De la Maza, A., Walther, P., Estelrich, J., & Lopez, O. (2008). Morphological effects of ceramide on DMPC/DHPC bicelles. *Journal de Microscopie, 230*, 16–26.

Carpenter, E. P., Beis, K., Cameron, A. D., & Iwata, S. (2008). Overcoming the challenges of membrane protein crystallography. *Current Opinion in Structural Biolology, 18*, 581–586.

Cheng, A., Hummel, B., Qiu, H., & Caffrey, M. (1998). A simple mechanical mixer for small viscous lipid-containing samples. *Chemistry and Physics of Lipids, 95*, 11–21.

Cherezov, V., Clogston, J., Papiz, M. Z., & Caffrey, M. (2006). Room to move: Crystallizing membrane proteins in swollen lipidic mesophases. *Journal of Molecular Biology, 357*, 1605–1618.

Cherezov, V., Rosenbaum, D. M., Hanson, M. A., Rasmussen, S. G., Thian, F. S., Kobilka, T. S., et al. (2007). High-resolution crystal structure of an engineered human beta2-adrenergic G protein-coupled receptor. *Science, 318*, 1258–1265.

Chou, J. J., Kaufman, J. D., Stahl, S. J., Wingfield, P. T., & Bax, A. (2002). Micelle-induced curvature in a water-insoluble HIV-1 Env peptide revealed by NMR dipolar coupling measurement in stretched polyacrylamide gel. *Journal of the American Chemical Society, 124*, 2450–2451.

Czerski, L., & Sanders, C. R. (2000). Functionality of a membrane protein in bicelles. *Analytical Biochemistry, 284*, 327–333.

Faham, S., Boulting, G. L., Massey, F. A., Yohannan, S., Yang, D., & Bowie, J. U. (2005). Crystallization of bacteriorhodopsin from bicelle formulations at room temperature. *Protein Science, 14*, 836–840.

Faham, S., & Bowie, J. U. (2002). Bicelle crystallization: A new method for crystallizing membrane proteins yields a monomeric bacteriorhodopsin structure. *Journal of Molecular Biology, 316*, 1–6.

Guan, L., Smirnova, I. N., Verner, G., Nagamori, S., & Kaback, H. R. (2006). Manipulating phospholipids for crystallization of a membrane transport protein. *Proceedings of the National Academy of Sciences of the United States of America, 103*, 1723–1726.

Hunte, C., & Richers, S. (2008). Lipids and membrane protein structures. *Current Opinion in Structural Biolology, 18*, 406–411.

Koga, Y., & Morii, H. (2005). Recent advances in structural research on ether lipids from archaea including comparative and physiological aspects. *Bioscience, Biotechnology, and Biochemistry, 69*, 2019–2034.

Kurisu, G., Zhang, H., Smith, J. L., & Cramer, W. A. (2003). Structure of the cytochrome *b6f* complex of oxygenic photosynthesis: Tuning the cavity. *Science, 302*, 1009–1014.

Landau, E. M., & Rosenbusch, J. P. (1996). Lipidic cubic phases: A novel concept for the crystallization of membrane proteins. *Proceedings of the National Academy of Sciences of the United States of America, 93*, 14532–14535.

Lange, C., Nett, J. H., Trumpower, B. L., & Hunte, C. (2001). Specific roles of protein-phospholipid interactions in the yeast cytochrome *bc1* complex structure. *The EMBO Journal, 20*, 6591–6600.

Lee, S. Y., Lee, A., Chen, J., & MacKinnon, R. (2005). Structure of the KvAP voltage-dependent K+ channel and its dependence on the lipid membrane. *Proceedings of the National Academy of Sciences of the United States of America, 102*, 15441–15446.

Lind, J., Nordin, J., & Maler, L. (2008). Lipid dynamics in fast-tumbling bicelles with varying bilayer thickness: Effect of model transmembrane peptides. *Biochimica et Biophysica Acta,, 1778*, 2526–2534.

Liu, Z., Yan, H., Wang, K., Kuang, T., Zhang, J., Gui, L., et al. (2004). Crystal structure of spinach major light-harvesting complex at 2.72 A resolution. *Nature, 428*, 287–292.

Luchette, P. A., Vetman, T. N., Prosser, R. S., Hancock, R. E., Nieh, M. P., Glinka, C. J., et al. (2001). Morphology of fast-tumbling bicelles: A small angle neutron scattering and NMR study. *Biochimica et Biophysica Acta, 1513*, 83–94.

Luecke, H., Schobert, B., Richter, H. T., Cartailler, J. P., & Lanyi, J. K. (1999). Structure of bacteriorhodopsin at 1.55 Å resolution. *Journal of Molecular Biology, 291*, 899–911.

Luecke, H., Schobert, B., Stagno, J., Imasheva, E. S., Wang, J. M., Balashov, S. P., et al. (2008). Crystallographic structure of xanthorhodopsin, the light-driven proton pump with a dual chromophore. *Proceedings of the National Academy of Sciences of the United States of America, 105*, 16561–16565.

McKibbin, C., Farmer, N. A., Jeans, C., Reeves, P. J., Khorana, H. G., Wallace, B. A., et al. (2007). Opsin stability and folding: Modulation by phospholipid bicelles. *Journal of Molecular Biology, 374*, 1319–1332.

Nieh, M. P., Raghunathan, V. A., Glinka, C. J., Harroun, T. A., Pabst, G., & Katsaras, J. (2004). Magnetically alignable phase of phospholipid "bicelle" mixtures is a chiral nematic made up of wormlike micelles. *Langmuir, 20*, 7893–7897.

Nollert, P. (2004). Lipidic cubic phases as matrices for membrane protein crystallization. *Methods, 34*, 348–353.

Nollert, P., & Landau, E. M. (1998). Enzymic release of crystals from lipidic cubic phases. *Biochemical Society Transactions, 26*, 709–713.

Ostermeier, C., Iwata, S., Ludwig, B., & Michel, H. (1995). Fv fragment-mediated crystallization of the membrane protein bacterial cytochrome *c* oxidase. *Nature Structural Biology, 2*, 842–846.

Ostermeier, C., & Michel, H. (1997). Crystallization of membrane proteins. *Current Opinion in Structural Biolology, 7*, 697–701.

Ottiger, M., & Bax, A. (1998). Characterization of magnetically oriented phospholipid micelles for measurement of dipolar couplings in macromolecules. *Journal of Biomolecular NMR, 12*, 361–372.

Poget, S. F., Cahill, S. M., & Girvin, M. E. (2007). Isotropic bicelles stabilize the functional form of a small multidrug-resistance pump for NMR structural studies. *Journal of the American Chemical Society, 129*, 2432–2433.

Prive, G. G., Verner, G. E., Weitzman, C., Zen, K. H., Eisenberg, D., & Kaback, H. R. (1994). Fusion proteins as tools for crystallization: The lactose permease from *Escherichia coli*. *Acta Crystallogrphica Section D, Biological Crystallogrphy, 50*, 375–379.

Prosser, R. S., Hwang, J. S., & Vold, R. R. (1998). Magnetically aligned phospholipid bilayers with positive ordering: A new model membrane system. *Biophysical Journal, 74*, 2405–2418.

Rasmussen, S. G., Choi, H. J., Rosenbaum, D. M., Kobilka, T. S., Thian, F. S., Edwards, P. C., et al. (2007). Crystal structure of the human beta2 adrenergic G-protein-coupled receptor. *Nature, 450*, 383–387.

Sanders, C. R., & Prosser, R. S. (1998). Bicelles: A model membrane system for all seasons? *Structure, 6*, 1227–1234.

Sanii, L. S., Schill, A. W., Moran, C. E., & El-Sayed, M. A. (2005). The protonation-deprotonation kinetics of the protonated Schiff base in bicelle bacteriorhodopsin crystals. *Biophysical Journal, 89*, 444–451.

Shinzawa-Itoh, K., Aoyama, H., Muramoto, K., Terada, H., Kurauchi, T., Tadehara, Y., et al. (2007). Structures and physiological roles of 13 integral lipids of bovine heart cytochrome *c* oxidase. *The EMBO Journal, 26*, 1713–1725.

Tiburu, E. K., Moton, D. M., & Lorigan, G. A. (2001). Development of magnetically aligned phospholipid bilayers in mixtures of palmitoylstearoylphosphatidylcholine and dihexanoylphosphatidylcholine by solid-state NMR spectroscopy. *Biochimica et Biophysica Acta, 1512*, 206–214.

Triba, M. N., Devaux, P. F., & Warschawski, D. E. (2006). Effects of lipid chain length and unsaturation on bicelles stability. A phosphorus NMR study. *Biophysical Journal, 91*, 1357–1367.

Triba, M. N., Warschawski, D. E., & Devaux, P. F. (2005). Reinvestigation by phosphorus NMR of lipid distribution in bicelles. *Biophysical Journal, 88*, 1887–1901.

Ujwal, R., Cascio, D., Colletier, J. P., Faham, S., Zhang, J., Toro, L., et al. (2008). The crystal structure of mouse VDAC1 at 2.3 A resolution reveals mechanistic insights into metabolite gating. *Proceedings of the National Academy of Sciences of the United States of America, 105*, 17742–17747.

van Dam, L., Karlsson, G., & Edwards, K. (2004). Direct observation and characterization of DMPC/DHPC aggregates under conditions relevant for biological solution NMR. *Biochimica et Biophysica Acta, 1664*, 241–256.

Vold, R. R., Prosser, R. S., & Deese, A. J. (1997). Isotropic solutions of phospholipid bicelles: A new membrane mimetic for high-resolution NMR studies of polypeptides. *Journal of Biomolecular NMR, 9*, 329–335.

CHAPTER 6

Membrane Protein Crystallization: Approaching the Problem and Understanding the Solutions

Mads Gabrielsen,* Alastair T. Gardiner,* Petra Fromme,† and Richard J. Cogdell*

*Division of Molecular and Cellular Biology, Faculty of Biological Life Sciences, Glasgow Biomedical Research Centre, University of Glasgow, 120 University Place, Glasgow G12 8TA, UK
†Department of Chemistry and Biochemistry, Arizona State University, Tempe, Arizona 85287-1604

Current Topics in Membranes, Volume 63

1063-5823/09 $35.00
DOI: 10.1016/S1063-5823(09)63006-4

I. OVERVIEW

To the uninitiated, the process of crystallizing a membrane protein may appear rather daunting, due partly to the startling discrepancy between the number of soluble protein structures deposited in the Protein data base (PDB) compared with that for membrane proteins. As expression systems, vectors, purification, and refolding strategies for membrane proteins become more sophisticated for, and targeted towards, membrane proteins, ever increasing numbers are becoming available that are of sufficient purity/stability and are ready to undergo screening in crystallization trials. This chapter sets out to demystify membrane protein crystallization and present the process in a logical, accessible, and systematic manner for researchers unfamiliar to the field. By detailing the methods and techniques available, hopefully, this will provide a general understanding of membrane protein crystallization and allow rational and informed decisions to be made, either when crystals have been obtained or how to proceed if they have not.

II. INTRODUCTION

This chapter provides a brief overview of the main problems associated with crystallizing membrane proteins followed by an annotated guide to approach the problem of crystallizing a membrane protein of choice. Although the major bottleneck when determining the 3-D structure of a membrane protein is to obtain a sufficient quantity of pure, stable protein, it can still be quite challenging to grow sufficiently large, well ordered crystals suitable for X-ray crystallography. There has been, however, consistent progress in this endeavor over the past few years and the task of using X-ray crystallography to determine the structure of membrane proteins is now considerably less daunting than it was a few years ago. A great deal of useful information is now obtainable from the internet, for example, at the time of writing the web sites of Stephen White at UC Irvine and the membrane protein data bank (Raman, Cherezov, & Caffrey, 2006) contain a regularly updated list of all membrane proteins structures determined to high resolution.

This chapter focuses on 3-D crystallization. For 2-D crystallization, readers are referred to the following excellent reviews (Kuhlbrandt, 2003; Miller, 1991; Ringler, Heymann, & Engel, 2000).

III. THE BASIC CRYSTALLIZATION PROCESS

The process of crystallizing a membrane protein is, in principle, very similar to that involved in crystallizing a water soluble protein except that the presence of solubilizing detergent molecules introduces an added layer of complication. Proteins form crystals when they are induced to precipitate slowly and in an ordered manner. The crystallization process is usually initiated by the addition of various precipitating agents. These can be salts, polymers like polyethylene glycols, or various other organic molecules, which are often alcohols such as 2-methyl-2,4-pentanediol (MPD). Precipitants reduce the solubility of the protein solution to the point that the solubility borderline is reached, leading to a supersaturated solution. When the supersaturation reaches a certain threshold, the protein-protein aggregates (or microcrystallites) are at the critical radius and the protein comes out of solution. Vapor diffusion is the most commonly used method to reach the supersaturated phase. In this way the concentration of protein is increased through the gradual abstraction of water using a hydroscopic reservoir solution. Crystallization can also be achieved by reducing the ionic strength by dialysis (Fromme, 2003; Witt et al., 1992). At very low ionic strength, the surface of the protein is depleted of counterions, which then reduces the protein's solubility and increases the electrostatic interactions between charged residues of the proteins (see Chapter 7 for more details about this crystallization method).

When membrane proteins were first used in crystallization trials a major problem was encountered. Addition of the precipitants often caused a reaction with the detergents used that resulted in a phase separation, illustrated in Fig. 1A. In this case the detergent molecules have been excluded from the aqueous phase and a two phase system is formed that contains two aqueous nonmiscible phases; the form that is detergent rich forms an "oily layer" or "oily drops" on the surface of the crystallization trial. This often occurs at concentrations of the precipitant below those required to precipitate the protein thereby preventing crystal formation. Phase separation results from early trials led to the belief that it was impossible to crystallize membrane proteins. However, two people (Drs. Hartmut Michel and Michael Garavito) independently discovered that if certain amphiphilic, small molecules were added to the crystallization mixture, the point at which phase separation occurs moved above the critical precipitation point allowing crystallization to take place (Fig. 1B). Using this approach, Michel was able to crystallize reaction centers from a purple bacterium (Michel, 1982) and Garavito was able to crystallize porin (Garavito & Rosenbusch, 1980). The effect of the different amphiphiles on identical crystallization conditions is illustrated in Fig. 1C and D. It should be noted that the "reputation" of phase separation in membrane protein crystallization has changed. Structured precipitation along a phase separation borderline can often be regarded as an indication that the protein may crystallize under slightly modified conditions. It has also been observed that some membrane proteins concentrate and crystals grow in the detergent rich phase (Yurkova, Demin, & Abdulaev, 1990;

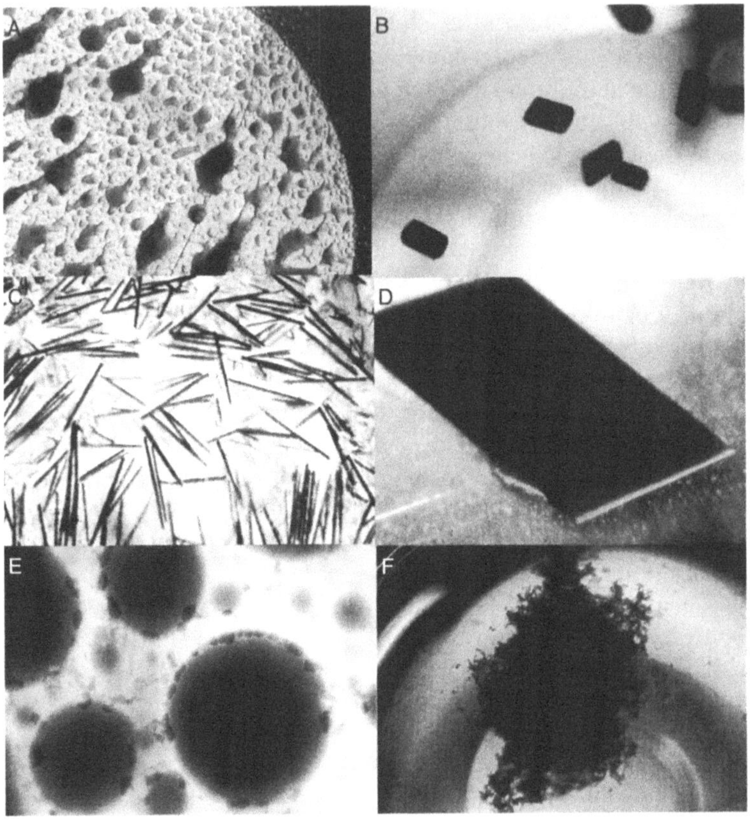

FIGURE 1 Crystallization trial outcomes. (A) Drop showing phase separation between the hydrophobic, detergent rich areas, and the aqueous phase, (B) membrane protein crystals, in this case a purple bacterial photosynthetic LH2 antenna complex from *Rhodopseudomonas palustris*, (C) LH2 antenna complex crystals from *Rhodopseudomonas acidophila* strain 10050 grown with the amphiphile heptanetriol, (D) LH2 antenna complex crystals from *R. acidophila* strain 10050 grown with the amphiphile benzamidine, (E) LH2 antenna complex crystals from *Rhodopseudomonas cryptolactis* growing in the phase separated, detergent-rich phase, (F) amorphous precipitate.

Fig. 1E). In the case of Photosystem I, the detergent rich phase is formed by protein-detergent micelles and the "oily droplets" convert into crystals themselves (see Chapter 7).

The most common techniques, with a few exceptions, used to crystallize membrane proteins are the same as those used to crystallize water soluble proteins, namely, vapor diffusion, microdialysis, and batch techniques (Ducruix & Giege, 1992). Vapor diffusion is the most common approach since it is technically the easiest, but all three methods have been successfully used to crystallize membrane proteins (Cherezov & Caffrey, 2003; Garavito & Rosenbusch, 1980; Lunin et al., 2006).

IV. VAPOR DIFFUSION

Vapor diffusion experiments can be conducted using sitting or hanging droplets (Ducruix & Giege, 1992). Typical hanging and sitting drop experimental setups are shown in Fig. 2. A small drop of the crystallization mixture, containing the membrane protein and precipitant (at a concentration slightly below that required to induce precipitation of the protein) is placed in the central well (or on a cover slip) and the reservoir filled with a larger volume of the precipitant solution. The reservoir wells have a precipitant concentration above that required for precipitation and usually, but not always, contain all the components (detergent, amphiphile, buffer, etc.) present in the drop so that the only parameter varied is the concentration of the precipitant. The vessel is sealed, incubated at constant temperature, and the closed system over time allowed to reach a state of equilibrium with respect to the water concentration in the drop versus reservoir. Diffusion of water vapor from the drop into the reservoir will slowly concentrate the protein-detergent micelle, the empty micelles and the precipitant in the central well (or the hanging drop). If the designated conditions provide a suitable environment, then crystals of the protein will slowly form, or, in other words, the protein will come out of solution as an ordered aggregate (Figs. 1B-D) rather than nonordered, amorphous precipitate (Fig. 1F). The rate of vapor diffusion can be controlled by preparing trays at different temperatures, varying the concentration gradient of the precipitant in the two aqueous compartments or by changing their volumes. Application of silicon oil on the surface of the reservoir can also be used to reduce the speed of equilibration. Hanging drops have an advantage in that the effect of gravity means that crystals are formed preferentially at the bottom of the drop and are not in direct contact with the glass or plastic used. This limits any direct influence of the cover slide or foil material on nucleation and crystal growth and facilitates harvesting of the crystals. However, the drops are obviously size-limited and occasionally prone, due to the presence of detergent, to spreading. Sitting drops have become increasingly popular because they can be conveniently combined with the use of crystallization robots (Hui & Edwards, 2003). If one considers the possible number of variables involved in preparing crystallization trials; protein concentration, pH, type and concentration of precipitant, type and concentration of detergent, type and concentration of additives, etc., then the number of conditions

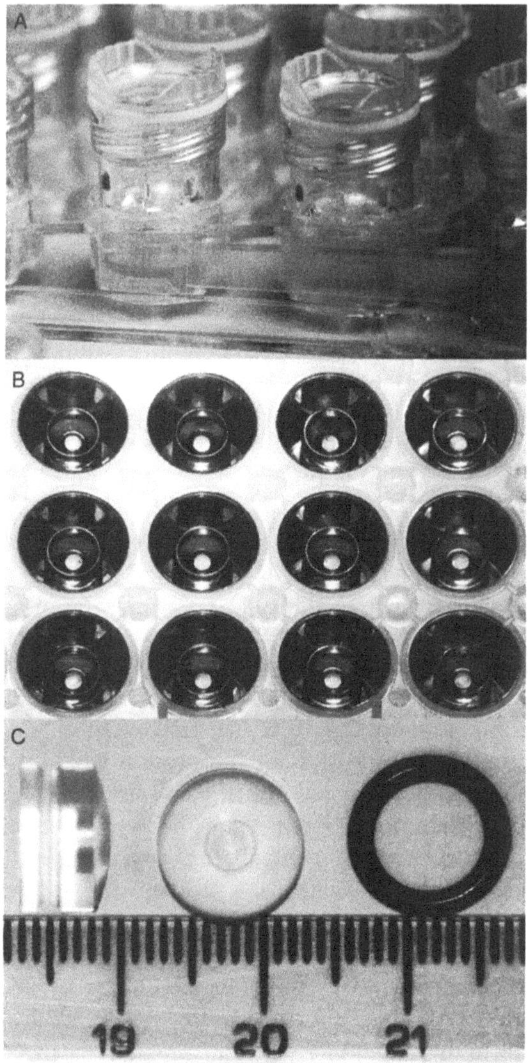

FIGURE 2 Typical crystallization experiments. In order to help those unfamiliar with vapor diffusion and microdialysis visualize the dynamics of the experiments this figure has been provided. (A) Nextal™ (Qiagen) crystallization plates for hanging drop experiments, the protein and precipitant solutions are pipetted onto the lid that can be inverted and screwed onto the base to form a sealed chamber. These plates have the advantage over previous hanging drop preparations in that the lid can be easily unscrewed and opened to allow addition or removal of precipitant. (B) CrysChem™ [NBS Biologicals] sitting drop plates, the protein and precipitant solutions are pipetted onto the central bridge and the chamber sealed using tape. Preparation of high-throughput (HTP)

to investigate quickly becomes overwhelming. The use of crystallization robots (instead of manually preparing experiments) increases one's ability to screen many more possible crystallization conditions using micro- and even nanoliter drop volumes (i.e., requiring much less protein). The vapor diffusion vessels shown in Figs. 2A and B are also convenient for monitoring the crystallization process as they can be easily placed in a microscope without opening the system. A further, yet somewhat understated advantage is the use of plate hotel and imaging systems (Hui & Edwards, 2003). These provide a systematic record of the crystallization trial timeline and facilitate exact comparisons, between, that is, different salt concentrations or amphiphiles, leading in time to a wealth of easily accessible information that can be mined to inform future screening decisions.

V. MICRODIALYSIS

Microdialysis is fundamentally different from vapor diffusion as it allows the free exchange of water and all solutes (buffer, salts, precipitants) that are smaller than the pore size of the membrane. It allows changes for only one parameter at the time (e.g., the ionic strength) with the ability to increase or decrease the parameter without changing the detergent or protein concentration, resulting in much more defined conditions for crystallization. It has been used for the crystallization of Photosystem I from cyanobacteria (Fromme & Witt, 1998; Witt et al., 1992). Despite these advantages microdialysis is not very commonly used as it requires more protein per experiment than vapor diffusion and experiments must be manually prepared. Another important point is that one of the most common precipitants, polyethylene glycol 1000 (PEG-1000), does not easily cross the dialysis membrane of 16 kDa due to the flexible polymer nature of the molecule. Dialysis experiments prepared in the presence of higher molecular weight PEGs essentially mimic vapor diffusion experiments as water moves across a semipermeable membrane rather than through the vapor phase, thereby concentrating the protein-detergent micelles. However, in dialysis experiments buffers and salts are not co-concentrated, allowing the user to discriminate between the precipitating effect of PEGs and salt. An important factor in the rare use of microdialysis is the difficult handling of the typical microdialysis cells shown

experiments is essentially identical except the size of the plate is much smaller. (C) A dialysis button. From left; side view, top view, and the O-ring used to hold on the dialysis membrane. The protein solution to be crystallized is placed in the well and sealed with the dialysis membrane. The precipitant diffuses across the membrane and induces crystallization.

in Fig. 2C. The crystallization mixture is placed in the button, which is then sealed with a small piece of dialysis membrane. The molecular weight cutoff of this membrane must be such that the protein is nonpermeable to the membrane, but the precipitant can freely penetrate (the exception occurs when PEG is used as explained above). The microdialysis cell is floated in a solution where the concentration of the precipitant is higher outside than that inside the cell. Since it is possible to change the outside solution if necessary, the ability exists to vary conditions if crystals do not appear. Microdialysis is, however, less convenient for monitoring the possible crystallization process as the dialysis membrane usually requires removal before crystals can be microscopically detected. Formation of bubbles and membrane leakage are also common problems. An alternate setup for microdialysis that is much easier to handle, requires less protein, allows easy *in situ* visualization of the crystals and can accommodate seeding is described in Chapter 7.

VI. MICROFLUIDICS

Recently, as more groups are targeting "high-hanging fruit" where only small amounts of protein are available for structural trials, the use of microfluidics has been introduced. This technique allows the use of 10 μl of protein solution to prepare 1300 crystallization conditions (Li et al., 2006). The protein and precipitant solutions are introduced at opposite ends of a tube and allowed to contact each other at an interface. Over time, diffusion occurs across this interface and induces the formation of crystals. Microfluidics can be performed under both batch and vapor diffusion conditions. There are now specific robots that are able to prepare trays for microfluidic trials and subsequent visualization of any crystals obtained. This method is excellent for the initial systematic screening of a large matrix of crystallization conditions, However, recovery of crystals is difficult. A detailed description of this method can be found in Hansen, Skordalakes, Berger, and Quake (2002), Lounaci, Rigolet, Casquillas, Huang, and Chen (2006) and in Chapter 10.

VII. COUNTER-DIFFUSION

The method of crystallization by counter-diffusion differs from microfluidics in that there is a liquid-gel interface rather than a liquid-liquid interface. The protein solution is kept in a capillary and inserted into an agarose gel layer then overlaid with the precipitant. The crystals form in the capillary as the precipitant solution slowly diffuses through the gel and into the capillary. Since during the diffusion a wide range of precipitant concentrations occur with time, this naturally allows a single trial to screen for optimal precipitant concentrations in a single run. This method can be automated, but there are also less expensive, "low-tech" solutions that may be used on the lab bench (i.e., "Granada-crystallization-box"; Garcia-Ruiz, Gonzalez-Ramirez, Gavira, & Otalora, 2002). This method was used to optimize crystals of the PSII core complex from *Pisum sativum* (Kuta Smatanova, Gavira, Rezacova, Vacha, & Garcia-Ruiz, 2006).

VIII. NOVEL METHODS USING LIPIDS IN CRYSTALLIZATION

The crystallization methods described previously work best for membrane proteins that are sufficiently robust to survive solubilization and handling in detergent solutions. A membrane protein will generally be most stable when solubilized with the detergent that best mimics the "average" properties of the lipids that comprise the original membrane from which it was obtained. However, some proteins have very specific, defined interactions with particular lipids that detergent molecules cannot replace. Lipids, either are required, to confer enzymatic activity, for example, cytochrome oxidase requires cardiolipin (Awasthi, Chuang, Keenan, & Crane, 1971), or have a stabilizing effect on the protein that facilitates good crystal growth. In this regard, the case of cytochrome b_6f is interesting because crystals suitable for high resolution structure determination were grown using standard vapor diffusion techniques where the protein had been supplemented with the required lipid (Zhang, Kurisu, Smith, & Cramer, 2003).

However, it is also possible to use the properties of phospholipids themselves, in novel ways, to produce crystals of membrane proteins. Examples of this include crystallization in the lipidic cubic phase, a novel method of membrane protein crystallization invented and developed by Rosenbusch and Landau (Chiu et al., 2000), spongy phase crystallization (Wohri et al., 2008) and the unusual liposome reconstitution method used to crystallize the major light-harvesting protein from spinach (Liu et al., 2004). A detailed description of these methods is beyond the scope of this chapter, therefore, interested readers are referred to the following excellent reviews/papers (Caffrey, 2003; Cherezov, Clogston, Papiz, & Caffrey, 2006; Nollert, 2004; Wadsten et al., 2006; Wohri et al., 2008) as well as Chapter 6. For crystallization in lipid phases the membrane proteins are reconstituted in monoacylglycerols (MAG), of which monoolein (MAG 9.9) is the most commonly used. These lipids are either manipulated into the lipidic cubic phase (Pn3m), which is a rather stiff, viscous phase and is represented schematically in Fig. 3A, or by adding an additional solvent to the MAG:H_2O mixture a fluid, spongy phase (L_3) can be formed. The cubic and spongy phases form 3-D structures that contain interconnecting water channels. In these phases

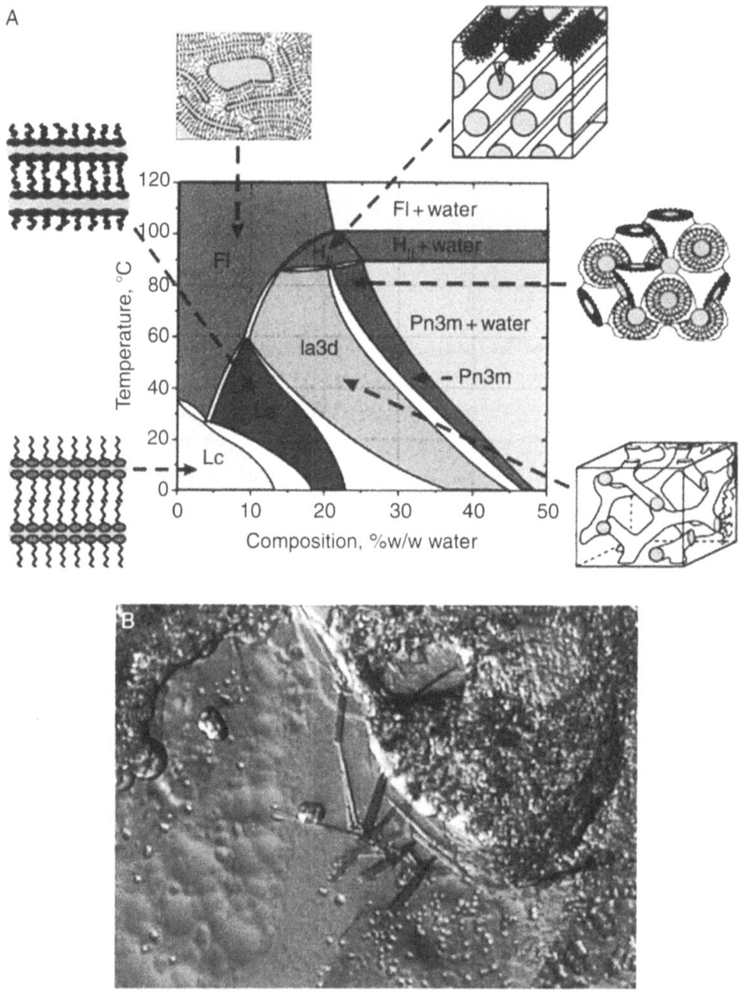

FIGURE 3 Membrane protein crystallization in mesophases. (A) Cartoon representation of the mesomorphism of monoolein as a function of temperature and hydration. Lipidic-lamellar, -hexagonal and -cubic phases can be formed with the cubic Pn3m phase, used for crystallization, only being formed within a relatively narrow window. (B) The lipidic sponge phase (L_3) can be obtained by adding an additional solvent to the monoolein:water system. This panel shows the RC-LH1 "core" antenna complex from *Rhodospeudomonas viridis* crystallized in the lipidic sponge phase.

the membrane proteins are located in the lipid matrices and can diffuse laterally in them. This movement enables the proteins to be delivered along the lipid scaffold and assemble at the nucleation point to form relatively dense, small crystals (Fig. 3B). It has been suggested that a lamellar phase may be involved in the nucleation. The lipidic cubic phases have been successfully used to crystallize bacteriorhodopsin (Chiu et al., 2000) and bacterial photosynthetic antenna complexes (Cherezov et al., 2006; Wadsten et al., 2006). Lipidic cubic phase robots have been developed but these are rather specialized and not yet commercially available. Crystallization using spongy phase has some advantages over that with the lipidic cubic phase. The sponge phase exists as a fluid, making it easier to handle and greater aqueous pore size facilitate incorporation of proteins with larger hydrophilic domains. The ability of the spongy phase to be pipetted also makes it possible to combine spongy phase trials with standard crystallization robots (Wohri et al., 2008), which will undoubtedly increase its usefulness. Thus far the liposome method has only been used successfully once. Plant light-harvesting complexes were reconstituted into microliposomes to form a regular 2-D curved array. These liposomes were then crystallized to form the macrolattice (Liu et al., 2004). It remains to be seen whether this remarkable method will be of general use.

IX. HOW TO OBTAIN CRYSTALS FROM A MEMBRANE PROTEIN OF CHOICE

It is assumed here that a sufficient quantity of pure, native, monodisperse membrane protein is available. It should be emphasized that this assumption is not trivial and for many important membrane proteins it cannot yet be fulfilled. How can crystals be obtained that are suitable for X-ray crystallography and structure determination? As there is now a small but increasing database of membrane proteins that have been crystallized, it has been possible to mine this information to produce sparse matrix screens that incorporate the conditions used thus far (Newstead, Ferrandon, & Iwata, 2008). There are several commercially available screens that provide excellent starting points. Even when using a sparse matrix screen there are many possibilities of how to begin the crystallization process. What concentration of protein should be tried? What detergent should be used? At what temperature should the crystallization trial be conducted? The UK structural genomics effort (MPSi) has developed an easy set of starting conditions to initiate crystallization screening on a novel membrane protein (McLuskey et al., 2008). It is recommended to begin with the protein at a concentration of about 10 mg/ml, solubilized in at least four different detergents (e.g., DDM, β-OG, $C_{12}E_8$, and LDAO) and perform the trials at two different temperatures (4 and 18 °C). The detergents are usually used at concentrations of one to four times their CMC. Beginning trials using different sparse matrix screens will cover a broad range of precipitants, pH, and salts, providing a good indication as to whether the protein is likely to crystallize. In the first instance any "hits" obtained are unlikely to provide highly-diffracting crystals but the conditions can be optimized as described below. If, however, no crystal "hits" are evident, what is the next step? This is the most common problem as membrane proteins rarely lead to crystals in the first screening attempt. The quality of the protein preparation is in most cases the critical factor. It is very important to check for the homogeneity of the protein preparation in respect to the oligomeric state, monodispersity, and proteolytic degradation. Especially for eukaryotic membrane proteins, post-translational modifications such as glycosylation and phosphorylation may be a major factor that must be considered. Heterogeneity and purity of the detergent may also be an important factor. It is very often the case that any given protein, even a water soluble one, may not crystallize. If the protein preparation has been carefully optimized yet no microcrystals are observed, changing the protein source might be another promising option. The chances of obtaining crystals are increased if orthologues (or mutants) of the protein are available. This approach resulted in good crystals of the LH2 complex from purple bacteria. Eight different orthologues were tried yet only one worked (Cogdell & Hawthornthwaite, 1993). If, however, the membrane protein to be crystallized cannot be changed, then other approaches, including a wider screening of possible crystallization conditions, should be attempted. These may be particularly relevant for certain key eukaryotic membrane proteins or proteins that naturally have different conformational states, such as transporters or receptors (Abramson et al., 2003).

Hartmut Michel pioneered the cocrystallization of truncated antibodies with membrane proteins to promote the formation of well-ordered crystals. The antibody fragment binds to the polar region of the membrane protein, thereby increasing the size of the portion of the molecule available to form contacts within the crystal lattice. This method was first used successfully with cloned Fv fragments to obtain the structure of cytochrome oxidase (Dohse et al., 1995) and has been spectacularly applied by MacKinnon with purified Fab fragments to solve the structure of a number of ion channels (Jiang et al., 2003; Lee et al., 2005). Very recently this basic idea, to increase the hydrophilic/hydrophobic proportion of the whole molecule, has been further developed by Stevens (Cherezov et al., 2007; Rosenbaum et al., 2007) to produce a chimeric molecule of a G-protein coupled receptor with T4 lysozyme, which could then be crystallized and its structure determined. In principle, these approaches have the added bonus that, as the structures of the antibody fragment or lysozyme are already known, the structure of the new membrane protein can be solved by a more rapid crystallographic technique known as molecular replacement.

If the protein of choice is either unstable or naturally has several conformational states, crystallization of the native protein may be very difficult if not impossible. In such cases site-directed mutagenesis may prove very useful. For many years Ronald Kaback and others tried to crystallize the lactose permease from *Escherichia coli* with success finally achieved using a mutant that locked the protein into a single conformation (Abramson et al., 2003). The Cambridge group had tried for many years to crystallize G-protein coupled receptors and their recent success was accomplished by systematically mutating nonessential amino acids in the primary sequence of the β-androgenic receptor and assaying which mutations resulted in the expressed protein having an increased thermal stability (Magnani, Shibata, Serrano-Vega, & Tate, 2008; Serrano-Vega, Magnani, Shibata, & Tate, 2008). Such an approach is very labor intensive but if the goal is important enough, systematic mutagenesis can prove very worthwhile.

X. OPTIMIZATION OF INITIAL "HITS"

It is often the case that a membrane protein undergoing crystallization trials with a sparse matrix screen will result in "hits" with small, rather disordered crystals. To go from these small crystals to larger well-ordered ones, that are suitable for X-ray analysis, can be a major hurdle and a frustrating bottleneck. How are initial crystal screen "hits" optimized? Unfortunately, there is no panacea for this problem and it often comes down to systematic, painstaking work. It is important to know the protein and its particular properties. One point that cannot be emphasized highly enough, particularly when robots or imaging systems are unavailable, is the importance of keeping accurate and exhaustive records of the progress of the crystallization experiment. The way the membrane protein is solubilized, purified, and prepared for the crystallization (i.e., if/how the detergent is exchanged) as well as the exact composition of the crystallization conditions, all determine the outcome of the trial (the sparse matrix screen conditions are defined but the buffer solution containing the protein may not be).

However, based on our experience it is possible to outline some strategies that can be attempted for optimization of initial crystal "hits." Clearly, the first attempt to try is fine screening of the crystallization parameters; protein concentration, pH, temperature, type and concentration of salt, and altering the type and concentration of the precipitant (i.e., polyethylene glycol monomethyl ether 2000 (MMePEG 2000) instead of PEG 2000). Almost certainly fine screening will improve the crystals but the effects of different additives (such as heptanetriol or benzamidine) should also be tried in tandem. Indeed, Hampton Research offers a dedicated additive screen for exactly this purpose. Trying a family of detergents where the size of the micelle is systematically varied (Prive, 2007) is often very helpful and is a strategy that has been successfully used by So Iwata (personal communication). The purity of the detergent and its stereochemistry is also an important factor in the stability of the protein and the crystallization process. As mentioned briefly above, the details of the solubilization and purification can be very important for the crystallization of membrane proteins. Researchers are often unaware that the lipid composition of the membrane protein sample may influence the outcome of any crystallization trials. The initial solubilization conditions are obviously the most important parameter that determines the lipid composition of the membrane protein. If these conditions are relatively harsh, all or too many of the annular lipids may be removed resulting in an inactive or unstable protein that is difficult to crystallize. The converse may also hold true, namely that gentle solubilization does not remove enough lipids resulting in a sample that is not sufficiently homogeneous for crystallization. It may be worthwhile, therefore, if no initial crystal "hits" are evident to go back and try different solubilization conditions with the same detergent, that is, temperature, length of solubilization, ratio of detergent to protein. In a similar manner, the methods used to purify the membrane protein will also have a bearing on the lipid composition of the final sample. The use of different types of chromatographic matrices can have a strong influence on which lipids remain bound to the protein. If no "hits" are evident it may be worthwhile, where the possibility exists, to slightly alter the purification protocol of the membrane protein.

XI. SEEDING

A seed crystal can provide the template from which further crystal growth can occur. It is energetically more favorable for crystal growth to occur on an already formed crystal surface than to create a new nucleation event. Seeding, therefore, can be particularly useful where knowledge of the phase diagram for the crystallization process is already known. A generalized phase diagram for protein crystallization is given in Fig. 4. For a comprehensive overview on the use of seeding to improve crystal quality see the review by Bergfors (2003).

At low protein and/or precipitant concentration, the protein will be in the undersaturated zone, labeled U, and no crystals will form. At too high protein and/or precipitant concentrations, zone P, the conditions are essentially too harsh and the protein will come out of solution in an unordered manner as illustrated in Fig. 1F, forming amorphous precipitates in the drops. However, if the protein and precipitant concentrations are such that

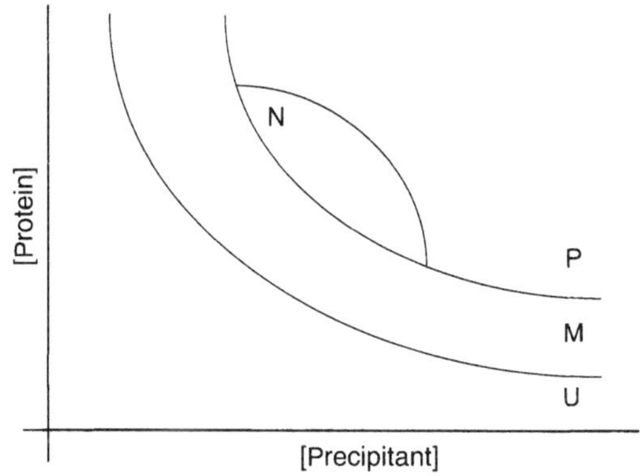

FIGURE 4 Schematic phase diagram for protein crystallization. U, undersaturated zone; M, metastable zone; N, nucleation zone; P, precipitation zone.

they fall into the nucleation zone, N, then spontaneous nucleation, that is, formation of the first ordered aggregates, events will occur and crystal growth can be initiated. There is an additional region in the phase diagram, called the metastable zone, M, where protein and/or precipitant concentrations are not sufficiently saturated to induce nucleation but are high enough to facilitate crystal growth. Therefore, if a seed crystal is added to protein that is in the metastable zone, the seed crystal is then in the correct physical environment to continue growing and, in effect, the nucleation event has been decoupled from subsequent crystal growth. A seed crystal placed in a drop which is just in the undersaturated state rather than the metastable phase will slowly dissolve. Addition of more precipitant into this reservoir solution will bring the conditions into the metastable zone, if this is done at the correct moment the seed crystal will be able to grow. It is apparent, therefore, that a precise knowledge of the boundaries of the phase diagram can provide a useful means to improve and enlarge crystals. In order to obtain this information it is necessary to calculate the molar concentration of each component of the reservoir and in the drop before equilibration occurs. Following equilibration the concentrations in the drop will be the same as those in the reservoir. It is, therefore, possible to set up solutions whose concentrations span the different regions of the phase diagram, that is, undersaturated and metastable. The precipitant range should be set so that the drops remain clear at low concentrations, that is, the equilibrated drop is still in the undersaturated zone, whereas the drops at high concentrations of precipitant contain amorphous precipitate. Obviously drops that contain crystals must have been at some point in the nucleation zone whereas drops that that remain clear, but at concentrations just below those that formed crystals, could possibly be in a metastable zone. From this knowledge and the generalized shape of phase diagram curves given in Fig. 4, the boundaries between each phase can be estimated. The determination of the exact phase boundaries must be determined empirically. Equilibrated drops can be prepared covering a range of protein and precipitant concentrations, into which a singe crystal is then placed. If the crystal immediately begins to dissolve and has disappeared within minutes, then the drop must lie in the undersaturated zone. When the crystal does not dissolve, it can be assumed that the drop is in the metastable zone. If the crystal is placed into an equilibrated drop that is in an undersaturated zone but closer to the boundary with the metastable zone than above, then it will take correspondingly longer to dissolve. In this way, over a range of drops the boundary can be mapped between the undersaturated and metastable zones.

A. Microseeding

Generally, microseeding is the easiest and most reliable form of seeding. A seed crystal is mechanically shattered, using a needle or a specialist tool (such as the Seed Beads™ [Hampton Research]) to produce a fine microcrystalline powder in equilibrated mother liquor. Any large pieces remaining from the original seed crystal are removed and the suspension (as a rule of thumb the seeds should be invisible even when viewed under a microscope) serially diluted to find the optimal concentration. A small quantity ($\sim$0.5 μl) of the seed suspension is then pipetted into a drop preequilibrated in the metastable state. Microseeding has the advantage that the quality of the initial crystal used to produce the seed suspension has no bearing on the outcome of the trial. This method has been used to improve SeMet derivatives of the transporter MgtE for structure determination (Hattori, Tanaka, Fukai, Ishitani, & Nureki, 2007a,b).

The microseeding technique can also be utilized in a broader way, known as microseed matrix seeding. This method uses the seed suspension as an additive when preparing random sparse matrix screens. Sometimes the introduction of the seed crystal allows crystal growth under conditions which would otherwise not produce crystals. This method is very convenient for use with robotic systems, enhancing reproducibility of crystal quality (D'Arcy, Villard, & Marsh, 2007; Walter et al., 2008).

B. Streak Seeding

Streak seeding is similar to microseeding except that a hair is used. A suitable "mother" crystal is gently probed with the hair so that microseeds attach themselves onto it. The hair is then passed through a preequilibrated crystallization drop in the metastable zone. If the seeding has been successful, crystals will subsequently appear along the streak line and possibly some self-nucleated crystals may also appear further away in the drop (Boer, Muller, Fetzner, Lowe, & Romao, 2005; Drory, Mor, Frolow, & Nelson, 2004).

C. Macroseeding

The most obvious difference between macro and microseeding is that with the latter the seed crystal is visible under the microscope and sometimes even visible to the naked eye. Additionally, the quality of the initial seed crystal is of critical importance. If the experiment starts with seed crystals that are not the best quality, then the final crystals will be poor. Conversely, seeding with a good quality seed crystal can lead to the growth of excellent large single crystals. The seed crystal chosen should be single and free from any obvious defects. The important point for macroseeding is that the surface of the seeding crystal has to be partially dissolved in the protein solution. The best way to achieve this goal is to start with a protein/precipitant concentration in the crystallization drop that is close to the phase separation borderline but still unsaturated. In this ideal case the crystal will only partially dissolve. The vapor diffusion chamber is immediately sealed after addition of the seed crystal. The solution must reach the metastable phase before the seed is completely dissolved, which allows the seeding crystal to grow into a large well ordered single crystal. There is one point of caution when PEG is used as precipitant. In this case it is often observed that the crystal and the crystallization drop are covered with a "skin" of PEG and detergent. This skin hinders the dissolution of the crystal, even under conditions that are unsaturated, and makes macroseeding very difficult. If a skin problem is observed the seeding crystals must be harvested in the exponential growth phase before skin formation reoccurs or streak seeding with microcrystals (see above) can be used as an alternate strategy.

If the precise boundary in the phase diagram between the metastable and undersaturated states is known with sufficient precision then macroseeding can be a useful addition to the available techniques. This method has been used successfully on several soluble proteins (Noble et al., 1993; Pylypenko et al., 2008) and the membrane protein Photosystem I (Fromme & Witt, 1998).

D. Heterogeneous Seeding

The examples given above are variations on homogeneous seeding. Heterogeneous seeding uses foreign solid substances such as, but not exclusively, hair or wool to induce nucleation. If these surfaces are regular they may

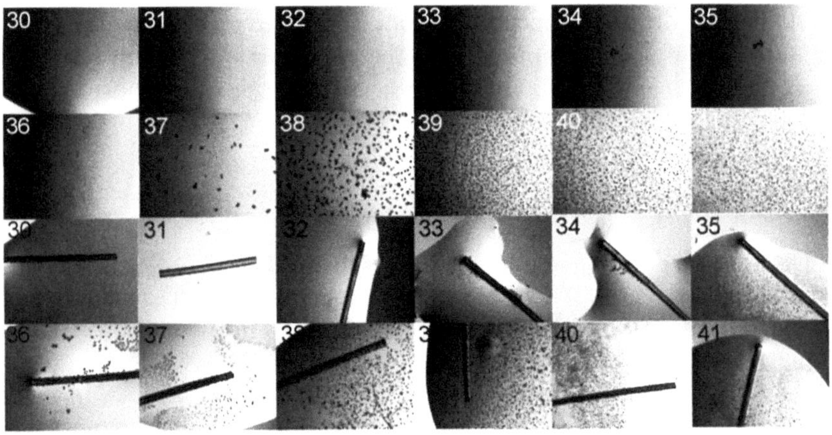

FIGURE 5 Heterologous seeding using a hair. Two identical crystallization trials were set up using increasing concentrations of the precipitant PEG 400 (30-41%). The top experiment shows that crystals start to appear around 36% PEG 400, whereas with the hair present they are visible at 33%. The hair provides a suitable surface that is able to facilitate nucleation and crystal growth.

promote adhesion of the protein molecules and also provide a suitable template to produce an ordered protein layer that then initiates crystal growth. It is energetically more favorable to add to a protein layer than to create a new crystal nucleus. Figure 5 illustrates the effect of a single hair on inducing crystallization of purple bacterial antenna complexes. Crystals appear in drops with lower precipitant concentrations than those with no hair present.

The heterologous seeding effect can sometimes be rather disadvantageous in sitting drop crystallization experiments as crystals sometimes grow only when attached to the plastic well. In such cases, unfortunately, even the minimal amount of force necessary to dislodge the crystal is enough to disrupt the lattice contacts and render the crystal useless for data collection. However, with luck crystals that are formed on the heterogeneous material can be gently harvested and used in a subsequent experiment, with hopefully, the result that a few, well ordered crystals appear that are suitable for diffraction studies (Georgieva, Kuil, Oosterkamp, Zandbergen, & Abrahams, 2007).

XII. IS IT POSSIBLE TO PRESELECT CRYSTALLIZATION CONDITIONS THAT ARE INHERENTLY MORE FAVORABLE FOR CRYSTALLIZATION?

The second virial coefficient, or B_{22} value, is a measurement of how strongly a protein interacts with itself in solution. It has been shown (George & Wilson, 1994) that most soluble proteins that have been crystallized in conditions that fall within a narrow range of B_{22} values (between -1×10^{-4} and -8×10^{-4} mol ml/g^2), which is known as the "crystallization slot." A high B_{22} value indicates good protein solubility and a low tendency to self-aggregate. Therefore, to achieve nucleation, under these conditions, protein concentrations may have to be as high as 100 mg/ml. For lower B_{22} values, the solubility of the protein in the buffer is decreased and it has a greater tendency to precipitate (George & Wilson, 1994; Guo et al., 1999). Determination of the B_{22} value can be achieved through static light scattering (SLS), small angle scattering and sedimentation equilibrium studies. However, the easiest and most cost-effective method, in terms of time and protein is self-interaction chromatography (SIC). SIC involves binding the protein of interest to a resin and creating a column for use with an HPLC instrument (Patro & Przybycien, 1996). The same protein, in solution, is then loaded onto the column, and the retention volume is measured. This retention volume is directly correlated with the B_{22} value. The exact relationship between the retention volume and the B_{22} value is explained in Chapter 9. As the composition of the solution changes (adding precipitants, additives, exchanging detergents, etc.), the protein will interact more or less with itself and the retention volume will change.

Using SIC to calculate the B_{22} value, one can quickly identify conditions that might be more suitable for crystallization, or improve initial crystallization conditions by the addition of small molecules at various concentrations and monitor the changes in the measured value of B_{22} that this causes. Thus far this technique has been predominantly applied to the problem of crystallizing water soluble proteins but recent studies on bacteriorhodopsin and other membrane proteins have shown that the crystallization conditions reside within a similar range of slightly negative B_{22} values, suggesting that weakly attractive interactions are as important for membrane proteins as they are for soluble proteins (Berger, Gendron, Lenhoff, & Kaler, 2006). Preliminary data has been produced indicating that SIC will also be useful with other membrane proteins (M. Gabrielsen, R.J. Cogdell, and L. DeLucas, unpublished data). See Chapter 9 for a comprehensive discussion of the second virial coefficient and self-interaction chromatography.

XIII. CONCLUSIONS

Attempting to crystallize a membrane protein can be a rather daunting prospect. It is hoped that the summary of methods outlined in this chapter will provide a useful guide for tackling this problem. If sufficient purified protein is available then many useful methods are already established for obtaining crystals suitable for subsequent X-ray structure determination.

While preparation of the experiments is no more difficult than that used with soluble proteins, the optimization of crystallization conditions and the growth of sufficiently ordered crystals may well be. Nevertheless, this chapter has outlined new tools and methods developed to aid the optimization process. The presence of detergent adds another variable to the challenge of determining optimal crystallization conditions but this should not be viewed as an insurmountable barrier. Although the number of crystal structures of membrane proteins in the PDB is relatively small, additional research groups are now working on the crystallization of membrane proteins and the number of structures is steadily increasing. The authors hope that this review will encourage more research groups to undertake projects in this very exciting area of structural biology.

Acknowledgments

R.J.C., A.T.G., and M.G. acknowledge the Biotechnology and Biological Sciences Research Council (BBSRC) and the European Membrane Protein Consortium (E-MeP) for funding. Prof. Martin Caffrey is thanked for permission to reproduce Fig. 3A. P.F. is funded by the National Institute of Health (grant nos. GM71619 and GM81490) and the National Science Foundation (grant no. 0417142).

References

Abramson, J., Smirnova, I., Kasho, V., Verner, G., Kaback, H. R., & Iwata, S. (2003). Structure and mechanism of the lactose permease of *Escherichia coli*. *Science, 301*, 610–615.

Awasthi, Y. C., Chuang, T. F., Keenan, T. W., & Crane, F. L. (1971). Tightly bound cardiolipin in cytochrome oxidase. *Biochimica et Biophysica Acta, 226*, 42.

Berger, B. W., Gendron, C. M., Lenhoff, A. M., & Kaler, E. W. (2006). Effects of additives on surfactant phase behavior relevant to bacteriorhodopsin crystallization. *Protein Science, 15*, 2682–2696.

Bergfors, T. (2003). Seeds to crystals. *Journal of Structural Biology, 142*, 66–76.

Boer, D. R., Muller, A., Fetzner, S., Lowe, D. J., & Romao, M. J. (2005). On the purification and preliminary crystallographic analysis of isoquinoline 1-oxidoreductase from Brevundimonas diminuta 7. *Acta Crystallographica. Section F, 61*, 137–140.

Caffrey, M. (2003). Membrane protein crystallization. *Journal of Structural Biology, 142*, 108–132.

Cherezov, V., & Caffrey, M. (2003). Nano-volume plates with excellent optical properties for fast, inexpensive crystallization screening of membrane proteins. *Journal of Applied Crystallography, 36*, 1372–1377.

Cherezov, V., Clogston, J., Papiz, M. Z., & Caffrey, M. (2006). Room to move: Crystallizing membrane proteins in swollen lipidic mesophases. *Journal of Molecular Biology, 357*, 1605–1618.

Cherezov, V., Rosenbaum, D. M., Hanson, M. A., Rasmussen, S. G., Thian, F. S., Kobilka, T. S., et al. (2007). High-resolution crystal structure of an engineered human beta2-adrenergic G protein-coupled receptor. *Science, 318*, 1258–1265.

Chiu, M. L., Nollert, P., Loewen, M. C., Belrhali, H., Pebay-Peyroula, E., Rosenbusch, J. P., et al. (2000). Crystallization in cubo: General applicability to membrane proteins. *Acta Crystallographica. Section D, 56*, 781–784.

Cogdell, R. J., & Hawthornthwaite, A. M. (1993). Preparation, purification and crystallization of purple bacteria antenna complexes. In J. Deisenhofer & J. R. Norris (Eds.), *The photosynthetic reaction centre* (Vol. 1, pp. 23–42). San Diego: Academic Press.

D'Arcy, A., Villard, F., & Marsh, M. (2007). An automated microseed matrix-screening method for protein crystallization. *Acta Crystallographica. Section D, 63*, 550–554.

Dohse, B., Mathis, P., Wachtveitl, J., Laussermair, E., Iwata, S., Michel, H., et al. (1995). Electron transfer from the tetraheme cytochrome to the special pair in the *Rhodopseudomonas viridis* reaction center: Effect of mutations of tyrosine L162. *Biochemistry, 34*, 11335–11343.

Drory, O., Mor, A., Frolow, F., & Nelson, N. (2004). Expression, crystallization and phasing of vacuolar H(+)-ATPase subunit C (Vma5p) of *Saccharomyces cerevisiae. Acta Crystallographica. Section D, 60*, 1906–1909.

Ducruix, A., & Giege, R. (1992). *Crystallization of nucleic acids and proteins: A practical approach.* New York: IRL Press.

Fromme, P. (2003). Crystallisation of photosystem I. In I. Iwata (Ed.), *Methods and results in crystallisation of membrane proteins* (pp. 147–173). La Jolla, California, USA: International University Line.

Fromme, P., & Witt, H. T. (1998). Improved isolation and crystallization of Photosystem I for structural analysis. *Biochimica et Biophysica Acta, 1365*, 175–184.

Garavito, R. M., & Rosenbusch, J. P. (1980). Three-dimensional crystals of an integral membrane protein: An initial X-ray analysis. *Journal of Cell Biology, 86*, 327–329.

Garcia-Ruiz, J. M., Gonzalez-Ramirez, L. A., Gavira, J. A., & Otalora, F. (2002). Granada Crystallisation Box: A new device for protein crystallisation by counter-diffusion techniques. *Acta Crystallographica. Section D, 58*, 1638–1642.

George, A., & Wilson, W. W. (1994). Predicting protein crystallization from a dilute-solution property. *Acta Crystallographica. Section D, 50*, 361–365.

Georgieva, D. G., Kuil, M. E., Oosterkamp, T. H., Zandbergen, H. W., & Abrahams, J. P. (2007). Heterogeneous nucleation of three dimensional protein nanocrystals. *Acta Crystallographica. Section D, 63*, 564–570.

Guo, B., Kao, S., McDonald, H., Asanov, A., Combs, L. L., & Wilson, W. W. (1999). Correlation of second virial coefficients and solubilities useful in protein crystal growth. *Journal of Crystal Growth, 196*, 424–433.

Hansen, C. L., Skordalakes, E., Berger, J. M., & Quake, S. R. (2002). A robust and scalable microfluidic metering method that allows protein crystal growth by free interface diffusion. *PNAS, 99*, 16531–16536.

Hattori, M., Tanaka, Y., Fukai, S., Ishitani, R., & Nureki, O. (2007a). Crystal structure of the MgtE Mg^{2+} transporter. *Nature, 448*, 1072–1075.

Hattori, M., Tanaka, Y., Fukai, S., Ishitani, R., & Nureki, O. (2007b). Crystallization and preliminary X-ray diffraction analysis of the full-length Mg^{2+} transporter MgtE. *Acta Crystallographica. Section F, 63*, 682–684.

Hui, R., & Edwards, A. (2003). High-throughput protein crystallization. *Journal of Structural Biology, 142*, 154–161.

Jiang, Y., Lee, A., Chen, J., Ruta, V., Cadene, M., Chait, B. T., et al. (2003). X-ray structure of a voltage-dependent K^+ channel. *Nature, 423*, 33–41.

Kuhlbrandt, W. (2003). Two-dimensional crystallization of membrane proteins: A practical guide. In C. Hunte, G. V. Jagow, & H. Schägger, (Eds.), *Membrane protein purification and crystallization: A practical guide.* San Diego: Academic Press.

Kuta Smatanova, I., Gavira, J. A., Rezacova, P., Vacha, F., & Garcia-Ruiz, J. M. (2006). New techniques for membrane protein crystallization tested on photosystem II core complex of Pisum sativum. *Photosynthesis Research, 90*, 255–259.

Lee, S. Y., Lee, A., Chen, J., & MacKinnon, R. (2005). Structure of the KvAP voltage-dependent K^+ channel and its dependence on the lipid membrane. *PNAS, 102*, 15441–15446.

Li, L., Mustafi, D., Fu, Q., Tereshko, V., Chen, D. L. L., Tice, J. D., et al. (2006). Nanoliter microfluidic hybrid method for simultaneous screening and optimization validated with crystallization of membrane proteins. *PNAS, 103*, 19243–19248.

Liu, Z., Yan, H., Wang, K., Kuang, T., Zhang, J., Gui, L., et al. (2004). Crystal structure of spinach major light-harvesting complex at 2.72 A resolution. *Nature, 428*, 287–292.

Lounaci, M., Rigolet, P., Casquillas, G. V., Huang, H. W., & Chen, Y. (2006). Toward a comparative study of protein crystallization in microfluidic chambers using vapor diffusion and batch techniques. *Microelectronics Engineering, 83*, 1673–1676.

Lunin, V. V., Dobrovetsky, E., Khutoreskaya, G., Zhang, R., Joachimiak, A., Doyle, D. A., et al. (2006). Crystal structure of the CorA Mg^{2+} transporter. *Nature, 440*, 833–837.

Magnani, F., Shibata, Y., Serrano-Vega, M. J., & Tate, C. G. (2008). Co-evolving stability and conformational homogeneity of the human adenosine A2a receptor. *PNAS, 105*, 10744–10749.

McLuskey, K., Gabrielsen, M., Kroner, F., Black, I., Cogdell, R. J., & Isaacs, N. W. (2008). A protocol for high throughput methods for the expression and purification of inner membrane proteins. *Molecular Membrane Biology, 25*, 599–608.

Michel, H. (1982). Three-dimensional crystals of a membrane protein complex. The photosynthetic reaction centre from *Rhodopseudomonas viridis*. *Journal of Molecular Biology, 158*, 567–572.

Miller, K. R. (1991). Two-dimensional crystals of the *Rhodopseudomonas viridis* photosynthetic reaction center. In H. Michel, (Ed.), *Crystallization of membrane proteins* (pp. 183–196). Boca Raton: CRC Press.

Newstead, S., Ferrandon, S., & Iwata, S. (2008). Rationalizing alpha-helical membrane protein crystallization. *Protein Science, 17*, 466–472.

Noble, M. E., Zeelen, J. P., Wierenga, R. K., Mainfroid, V., Goraj, K., Gohimont, A. C., et al. (1993). Structure of triosephosphate isomerase from *Escherichia coli* determined at 2.6 A resolution. *Acta Crystallographica. Section D, 49*, 403–417.

Nollert, P. (2004). Lipidic cubic phases as matrices for membrane protein crystallization. *Methods, 34*, 348–353.

Patro, S. Y., & Przybycien, T. M. (1996). Self-interaction chromatography: A tool for the study of protein-protein interactions in bioprocessing environments. *Biotechnology and Bioengineering, 52*, 193–203.

Prive, G. G. (2007). Detergents for the stabilization and crystallization of membrane proteins. *Methods, 41*, 388–397.

Pylypenko, O., Ignatev, A., Lundmark, R., Rasmuson, E., Carlsson, S. R., & Rak, A. (2008). A combinatorial approach to crystallization of PX-BAR unit of the human Sorting Nexin 9. *Journal of Structural Biology, 162*, 356–360.

Raman, P., Cherezov, V., & Caffrey, M. (2006). The membrane protein data bank. *Cellular and Molecular Life Sciences, 63*, 36–51.

Ringler, P., Heymann, B., & Engel, A. (2000). Two-dimensional crystallization of membrane proteins. In S. A. Baldwin, (Ed.), *Membrane transport* (pp. 229–268). Oxford: Oxford University Press.

Rosenbaum, D. M., Cherezov, V., Hanson, M. A., Rasmussen, S. G., Thian, F. S., Kobilka, T. S., et al. (2007). GPCR engineering yields high-resolution structural insights into beta2-adrenergic receptor function. *Science, 318*, 1266–1273.

Serrano-Vega, M. J., Magnani, F., Shibata, Y., & Tate, C. G. (2008). Conformational thermostabilization of the beta1-adrenergic receptor in a detergent-resistant form. *PNAS, 105*, 877–882.

Wadsten, P., Wohri, A. B., Snijder, A., Katona, G., Gardiner, A. T., Cogdell, R. J., et al. (2006). Lipidic sponge phase crystallization of membrane proteins. *Journal of Molecular Biology, 364*, 44–53.

Walter, T. S., Mancini, E. J., Kadlec, J., Graham, S. C., Assenberg, R., Ren, J., et al. (2008). Semi-automated microseeding of nanolitre crystallization experiments. *Acta Crystallographica. Section F, 64*, 14–18.

Witt, H. T., Krauss, N., Hinrichs, W., Witt, I., Fromme, P., & Saenge, W. (1992). Three-dimensional crystals of Photosystem I from *Synechococcus* sp. and X-ray structure analysis at 6 Å resolution. In *Proceeding of the 9th International Conference of Photosynthesis*, (pp. 521–528). Nagoya: Photosynthetic Research.

Wohri, A. B., Johansson, L. C., Wadsten-Hindrichsen, P., Wahlgren, W. Y., Fischer, G., Horsefield, R., et al. (2008). A lipidic sponge phase screen for membrane protein crystallization. *Structure (London, England: 1993), 16*, 1003–1009.

Yurkova, E. V., Demin, V. V., & Abdulaev, N. G. (1990). Crystallisation of membrane proteins: Bovine rhodopsin. *Biomedical Science, 1*, 585–590.

Zhang, H., Kurisu, G., Smith, J. L., & Cramer, W. A. (2003). A defined protein-detergent-lipid complex for crystallization of integral membrane proteins: The cytochrome b6f complex of oxygenic photosynthesis. *PNAS, 100*, 5160–5163.

CHAPTER 7

Tools to Enhance Membrane Protein Crystallization

W. William Wilson,* Gregg Whited,[†] Robert W. Payne,[‡] Charles Henry,[‡] David H. Johnson,[§] and Larry DeLucas[§]

*Department of Chemistry, Mississippi State University, Mississippi State, Mississippi 39762
[†]Genencor International, Inc., Palo Alto, California 94304
[‡]Department of Chemistry, Colorado State University, Fort Collins, Colorado 80523-1872
[§]Center for Biophysical Sciences and Engineering, University of Alabama at Birmingham, Birmingham, Alabama 35294-4400

Current Topics in Membranes, Volume 63 1063-5823/09 $35.00

DOI: 10.1016/S1063-5823(09)63007-6

I. OVERVIEW

This chapter describes a novel technique, self-interaction chromatography (SIC), to rapidly measure the second virial coefficient, *B*, (a thermodynamic term that provides semiquantitative information regarding protein-protein interactions) of proteins in a variety of cosolvents. SIC provides a rapid diagnostic for determining the optimum solution conditions that promote increased protein stability and minimize unwanted nonspecific aggregation. In addition, SIC allows the selection of specific solution conditions (including detergent type and concentration) for subsequent optimization of protein solubility, homogeneity, and crystallization. Balanced solubility and crystallization screens have been designed such that *B* values are experimentally measured via SIC for a relatively small number of conditions, providing a balanced representation of the total number of possible combinations of chemical variables. The screen conditions plus corresponding *B* values are then used to "train" a predictive algorithm (neural net) that subsequently produces an *in silico* screen, predicting the *B* values for all possible combinations of the chemical variables. The high-positive *B* values indicate solution conditions that should provide increased protein solubility and consequently, minimization of nonspecific protein aggregation. Slightly negative *B* values indicate solution conditions that have a higher probability of yielding crystals. The combined use of these innovative technologies provides a powerful approach to address several of the critical impediments (i.e., protein production, stability, purification, and crystallization) preventing membrane structure determinations.

II. INTRODUCTION

The preceding chapters discuss a variety of approaches to increase success rates for membrane protein expression, solubilization, purification, and crystallization. There have been a number of advances made in each of these areas during the past decade, particularly for prokaryotic membrane proteins. However, in spite of these improvements the crystallization of membrane proteins continues to present a major challenge. Crystallization success rates remain extremely low for membrane proteins when compared to aqueous proteins. Current state-of-the-art approaches for protein crystallization employ extreme brute force, using high-throughput systems that automatically prepare 500-2000 initial conditions per protein for both detergent-based and lipidic-based approaches. Yet, although the structural genomics era has led to the development of a number of high-throughput crystallization robotic systems (Krupka et al., 2002; Luft et al., 2001; Stevens, 2000), it is clear that robotics alone does not dramatically improve success rates for aqueous or membrane proteins. The ability to determine initial crystallization conditions or optimize existing crystallization conditions to provide diffraction quality crystals remains a major impediment. Low-crystallization success rates are a consequence of the challenge to find the correct combinations of a large number of relevant parameters, some of which include protein polydispersity, protein purity/homogeneity, precipitating agent, buffer/pH, temperature, protein solubility, concentration, flexibility of the protein itself, protein stability, and even the propensity of amino acids on the protein surface to form good protein-protein contacts. Additional factors that influence crystallization include lipid type and relative amounts (in the case of mixed lipid systems), detergent type and concentration, additives (polar, apolar, and amphipathic), protein instability, and limited protein quantity.

What is needed is a fundamental understanding of the chemical and molecular conditions necessary to produce soluble, nonaggregated, stable protein as well as conditions that induce crystallization versus precipitation or nonspecific aggregation. Trained crystallographers can be overwhelmed by the large volume of information produced by thousands of crystal screening experiments, and thereby miss subtle interrelationships between crystallization results and the numerous variables associated with crystallization experiments. To cope with this information overload, this chapter will describe a diagnostic high-throughput, knowledge-based approach that provides a mechanism to tease out the correct combination(s) of variables that play key roles for protein solubility (and therefore, indirectly, protein expression), protein polydispersity, protein stability, and protein crystallization.

III. PROTEIN-PROTEIN INTERACTIONS AND CRYSTALLIZATION

The process of converting individual protein molecules in solution into a macroscopic crystal suitable for diffraction studies so that a three-dimensional protein structure can be determined is often referred to as "protein crystallogenesis." The conversion process can generally be thought of as occurring in a sequence of stages described as prenucleation aggregation, nucleation, and postnucleation growth. In short, prenucleation aggregation is the association (aggregation) of protein monomers to form lower order (dimers, trimers, *n*-mers) protein aggregates. It is a dynamic process whereby association and dissociation of the aggregates are occurring simultaneously. Nucleation occurs when a protein aggregate has reached a critical size so that its growth will be self-sustained. Postnucleation growth is just the continued growth of the nucleus to create a microphase which continues to grow to form a macrophase. It is important to note that the ultimate outcome of the experiment regarding diffraction quality of the macrophase is inextricably linked to the prenucleation aggregation stage. Specifically, if the molecular arrangement of the aggregate formed is ordered, then the resulting nucleus will be ordered leading eventually to an ordered macrophase called a protein crystal. On the other hand, if the aggregate has a disordered molecular arrangement, then the nucleus will be disordered and the subsequent macrophase is amorphous and called protein precipitate. Thus, it is the thermodynamics of the protein-protein interactions in the specified solvent condition that ultimately dictates the nature of the process' final result.

Examples of relatively weak, noncovalent protein-protein interactions discussed here include excluded volume contributions as well as charge-charge, dipole-dipole, dipole-induced dipole, and hydrophobic interactions. In a general sense, these interactions all contribute to the nonideality of a protein solution, and the magnitude of the nonideal behavior can be measured by a thermodynamic solution parameter called the osmotic second virial coefficient, B. Negative values of B indicate net attractive interactions between protein molecules whereas positive values of B indicate net repulsive interactions. B values of zero indicate that attractive and repulsive interactions cancel, and the solution is referred to as an ideal solution.

The role of protein-protein interactions regarding crystallization as estimated by B is well established both experimentally and theoretically (Bonnete, Finet, & Tardieu, 1999; Ducruix, Guilloteau, Kautt, & Tardieu, 1996; George & Wilson, 1994; George et al., 1997; Malfois, Bonnete, Belloni, & Tardieu, 1996; Neal, Asthagiri, & Lenhoff, 1998; Neal, Asthagiri, Velov, Lenhoff, & Kaler, 1999; Rosenbaum, Zamora, & Zukoski, 1996). The important result is that a fairly narrow range of B values is correlated to solution conditions that are favorable to crystallization. In fact, a "crystallization slot" has been defined and corresponds to B values in the range of about -1×10^{-4} to -8×10^{-4} (mol ml/g^2). This range of B values indicates protein-protein interactions that are slightly or moderately attractive. If crystallization experiments are conducted in solution conditions at more negative B values, then there is a greater risk of forming amorphous solid phase because of stronger protein-protein attractions. Alternatively, experiments at more positive B values, where protein-protein repulsions dominate, are generally unsuccessful since protein concentrations that are impractically high are required to cause phase separation of any kind. Thus, the crystallization slot can be used as an effective guide for crystallization experiments by directing changes in a particular solution parameter (temperature, pH, ionic strength, etc.) that results in a more favorable B value. It is important to note that working at conditions within the crystallization slot does not guarantee a successful outcome, but experiments conducted outside the slot most assuredly reduces the probability of obtaining desirable crystals.

Another important aspect of protein-protein interactions is the influence on protein solubility, s. Second virial coefficient values are typically measured in protein solutions that are well undersaturated, that is, concentrations in the 1-5 mg/ml range. Experimental data correlating B with s values for several proteins (lysozyme, ovalbumin, equine serum albumin) in various solution conditions have been reported (Demoruelle et al., 2002; George et al., 1997; Gripon et al., 1997; Guo et al., 1999). The empirical correlation between B and s is intuitive at best since B is a dilute-solution parameter and s is a phase-transition parameter. Nevertheless, the experimental data suggested that a definite relationship between B and s must exist and subsequent publications (Haas, Drenth, & Wilson, 1999; Ruppert, Sandler, & Lenhoff, 2001) provided the theoretical basis for the direct link between B and s.

From a practical standpoint, the correlation between B and s offers distinct advantages for determining the solubility behavior of proteins. For example, for any rational design of crystallization trials, it is important to know if and to what extent a protein's solubility depends on solution parameters such as temperature, pH, ionic strength, type and concentration of additives, etc. Since it is impractical to determine absolute values for s by equilibrium studies as a function of so many solution variables, the measurement of B provides an attractive alternative for obtaining solubility behavior.

The concept of a crystallization slot has been experimentally verified for many water-soluble proteins and a limited number of membrane proteins. For example, trigonal and tetragonal crystals of outer membrane protein F (OmpF) porin, a bacterial outer membrane protein, formed within a narrow range of B values that was within the crystallization slot (Hitscherich, Kaplan, Allaman, Wiencek, & Loll, 2000). The implication of this work is that efforts to crystallize membrane proteins should emphasize placing the protein-detergent complex (PDC) in solutions that have B values inside the crystallization slot.

Although the framework relating B to the protein-protein interactions so significant for crystallization has been well established, its utilization as a primary-screening parameter for crystallization has been limited. To realize this goal, a way must be found to measure B that is relatively fast (minutes per B determination), requires small amounts of protein (microgram or submicrogram quantities per B determination) and lends itself to a robotic platform that is consistent with a high-throughput system. Traditional biophysical techniques used to measure B include static light scattering (SLS; Kratochvil, 1987), osmotic pressure (Tombs & Peacocke, 1974), and sedimentation equilibrium (Fujita, 1975), none of which currently meet the stated requirements for a high-throughput screen. In what follows, we describe an alternative method, a scale-down version of SIC that does meet the requirements for a high-throughput screen.

IV. SELF-INTERACTION CHROMATOGRAPHY

SIC was first introduced (Patro & Przybycien, 1996) as a method to measure protein-protein interactions in a given formulation condition. The technique is fundamentally an application of affinity chromatography with differences in the interpretation of protein elution and the preparation of column media. Although the materials and data analysis of SIC have been refined several times since 1995, the same basic steps are utilized:

(1) Covalently bind protein of interest to chromatography media (stationary phase) and load protein-bound media into a column.
(2) Flow formulation of interest over the column using high-performance liquid chromatography hardware.
(3) Inject soluble protein (mobile phase) into formulation flow and measure volume required to elute soluble protein (retention volume) as it interacts with the same protein covalently bound to column media.
(4) Compare retention volume measured in Step 3 with retention volume of the soluble protein injected into a column of inert media (dead column).

The basis of SIC is that increased attraction between the injected mobile-phase protein and covalently bound protein results in increased volume to elute the protein from the column. The comparison of retention volumes in the presence and absence of immobilized protein provides a normalized retention factor. This retention factor can then be used to compare the magnitude of protein self-interaction in the presence of different formulation conditions.

Two problems arose from the use of a simple retention factor to compare protein-protein interactions in various formulations. First, the thermodynamic model of the SIC experiment was not immediately apparent. This initially prevented SIC measurements from being independently verified by previously established thermodynamic measures of protein-protein interactions. Additionally, the control comparing retention volume of the protein over a protein-bound column with the same protein over inert media failed to account for excluded volume effects—the same effects utilized by size-exclusion chromatography to separate protein dimers and trimers from single species. Protein molecules with strong self-interactions could form oligomers which had access to less volume in the column and would therefore elute with less volume. Reduced elution volume of strongly interacting molecules in the mobile phase is counter to the SIC premise that mobile-phase molecules should interact with the stationary phase resulting in increased elution volume.

In 2002, Tessier applied an appropriate thermodynamic model and incorporated a noninteracting molecule to correct for excluded volumes (Tessier et al., 2002a). In addition to protein injections over the inert and protein-bound column media, acetone was also injected. The retention volume ratio between protein and acetone accounts for excluded volume contributions by the protein. As acetone is a noninteracting marker, the measure of acetone retention volume are required in Steps 3 and 4 only periodically to ensure that the live and dead columns have not changed physical configuration due to media compression or degradation of covalently bound protein. The measurements are then used to calculate the osmotic second virial coefficient (B value).

Protein-molecules in solution interact through a variety of forces including electrostatics, dipole-dipole, and van der Waals forces. MacMillan-Meyer solution theory expands ideal solution theories to account for molecular interactions. The molecular interaction parameter, B value, can be calculated from SIC measurements by the following equations (Tessier et al., 2002b):

$$V_0 = V_a \frac{V_r'}{V_a'} \tag{1}$$

$$k' = \frac{V_r - V_0}{V_0} \tag{2}$$

$$B = \frac{N_A}{MW^2}\left(B_{HS} - \frac{k'}{\phi\rho}\right) \tag{3}$$

Equation (1) is the void volume, V_o, adjusted for excluded volume of the protein. V_a is the eluted volume of the acetone injection over the protein-bound column and V_r'/V_a' is the ratio of the protein elution volume to acetone elution volume over the inert column. This ratio accounts for the protein's contribution to excluded volume. With this correction the retention factor, k' (Eq. 2), is calculated using the measured retention volume, V_r, of the protein over the protein-bound column. The k′ value is entered directly into the equation for the B value (Eq. 3). N_A, MW, B_{HS}, φ, and ρ are all based on the protein and media independent of protein-protein interactions. N_A and MW are Avagadro's number and the molecular weight of the protein, respectively. B_{HS} is the excluded volume contribution by the protein as a hard sphere and is calculated based on the molecular weight of the protein. The phase ratio, φ, is the ratio of available surface area to available volume and is an established value dependent on the media (DePhillips & Lenhoff, 2000). The number of protein molecules per unit area, ρ, is measured by a Pierce BCA assay of the protein-bound beads themselves. The retention factor then becomes the formulation-dependent variable resulting in a *B* value for the protein in a given formulation.

Use of the second virial coefficient as a measure of protein-protein interaction has two benefits: It can be independently verified with alternate measurement techniques, and it is directly correlated to protein solubility and ability to crystallize. *B* values measured by static light scattering (SLS) agree with SIC. The ability to use SLS as a reference has allowed validation of efforts to increase throughput and miniaturize SIC. The original SIC column introduced by Patro and Przybycien in 1996 utilized 28 mg of protein to prepare a 1.6 ml column of sepharose gel. Increased chromatographic accuracy and improved binding chemistry (Tessier et al., 2002a) reduced the quantity of required protein to 6.5 mg for a 1-ml column of Toyopearl AF-Formyl particles (Tosoh Bioscience). Current *B* value measurements are made on a 0.1-ml column requiring as little as 0.5 mg of protein for binding. After the column is prepared each *B* value measurement requires only 1 μg of protein (1 μl injection of 1 mg/ml protein solution) and can be measured in less than 10 min.

V. VALIDATION STUDIES: SIC ACCURATELY QUANTIFIES *B*

There are multiple examples where the SIC *B* values were demonstrated to directly correlate with *B* values via SLS for aqueous proteins (Garcia, Hadley, Wilson, & Henry, 2003; Tessier et al., 2002, 2003). The excellent correlation between *B* values obtained by SIC and SLS indicates that the immobilization chemistry used in SIC provides a set of random orientations mimicking the behavior of free proteins in solution. We have evaluated several proteins using SIC, and additional aqueous and membrane proteins have been evaluated by several other investigators (Berger, Gendron, Lenhoff, & Kaler, 2006; Berger, Gendron, Robinson, Kaler, & Lenhoff, 2005; Dumetz, Snellinger-O'Brien, Kaler, & Lenhoff, 2007; Garcia et al., 2003; Patro & Przybycien, 1996; Pjura, Lenhoff, Leonard, & Gittis, 2000; Tessier, Sandler, & Lenhoff, 2004; Tessier et al., 2002b, 2003; Valente, Kusum, Manning, Wilson, & Henry, 2005). Figure 1 shows *B* as a function of temperature and $(NH_4)_2SO_4$ concentration for concanavalin A. Concanavalin A is well known to form dimers at low pH (below 7) and tetramers at high pH (above 7). The data shown in Fig. 1 represent the dimer system (verified by size-exclusion chromatography) which exhibits retrograde solubility of concanavalin A. This demonstrates the ability to extend SIC to the measurement of complex multimeric proteins as well as accurately profiling retrograde solubility.

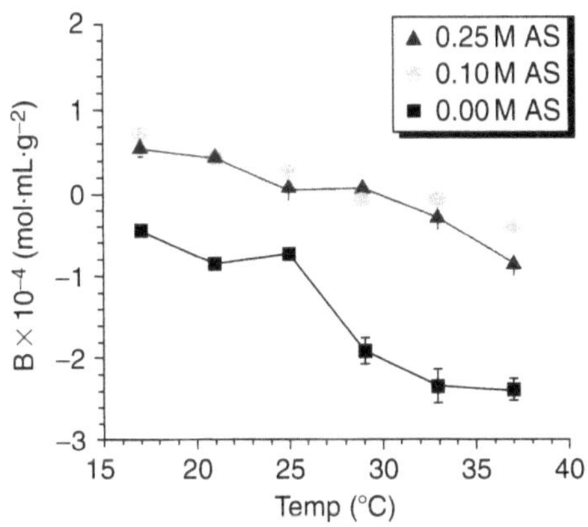

FIGURE 1 *B* as a function of temperature and $(NH_4)_2SO_4$ concentration for concanavalin A.

SIC data is virtually unaffected as the protein concentration is increased (provided you do not induce protein precipitation). At identical solution conditions the second virial coefficient value for human serum albumin was unaffected at concentrations as low as 1.0 mg/ml and as high as 120 mg/ml. The osmotic second virial coefficient is a dilute-solution property, thus conditions in each precipitating condition can be run with low-protein concentrations (avoiding precipitation on the column) yet the results are predictive of protein-protein interactions for higher protein concentrations. Thus, if protein aggregates at high concentrations, the *B* value measured at low concentrations (where it is well behaved) will accurately predict the strong attractions at high concentration. The measurement of *B* values, guided by the range found via the crystallization slot, can be used to improve crystallization conditions. In Fig. 2, crystals of thaumatin are grown by batch method at different *B*-value conditions (all residing within the crystallization slot for thaumatin). As *B* values become more negative, protein-protein attractive forces increase, producing more rapid crystal nucleation and growth. The result is more crystals but smaller crystals (and possibly more poorly ordered crystals) are produced at successively larger negative *B* values within the slot. Generally, lower negative *B* values (indicating more gentle protein attractive forces) produce fewer but larger crystals that typically grow more slowly. It has also been observed that *B* values for crystallization conditions of larger proteins (i.e., molecular weight greater than 60 kDa) generally fall in the −0.5 to −3.0 range of the crystallization slot, implying weaker protein attraction.

These and previous *B* value experiments performed on more than 60 different aqueous proteins clearly demonstrate the potential of *B* values used as a diagnostic for initial crystallization conditions as well as improving

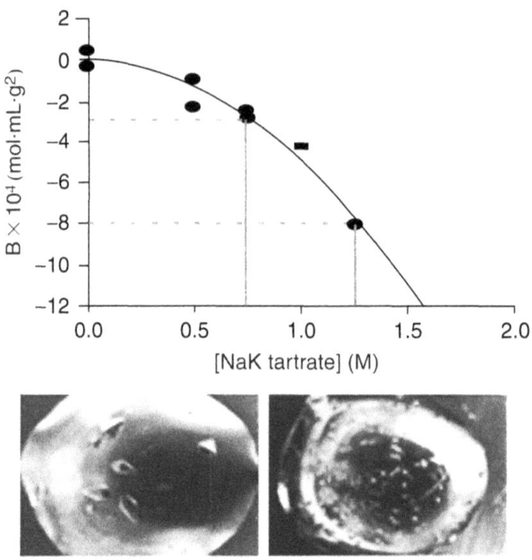

FIGURE 2 Thaumatin crystals grown at disparate *B* values within the crystallization slot.

upon existing conditions (Berger et al., 2005, 2006; Dumetz et al., 2007; Garcia et al., 2003; Patro & Przybycien, 1996; Pjura et al., 2000; Tessier et al., 2002a, 2003, 2004; Valente et al., 2005).

A. Excipients

Since validating SIC versus SLS for simple salt solutions, we have begun to characterize protein self-interaction in more complex cosolvent systems, commencing with sugars and polyols. Pharmaceutical companies have used such molecules to enhance protein stability for protein therapeutic formulations. This research has found that protein aggregation is controlled by both conformational stability and colloidal stability, with either becoming the rate-limiting effect, depending on the solution conditions (Tsumoto, Ejima, Kita, & Arakawa, 2005). Companies manipulate solution conditions during preformulation studies to maximize *B*, thereby promoting the development of useful formulations that exhibit long-term storage stability. There are two general classes of excipients, those that block hydrophobic patches (i.e., arginine, glutamine) (Tsumoto et al.) on the surface of proteins thereby decreasing nonspecific aggregation and those that stabilize protein conformation (i.e., trehalose, sorbitol, mannitol) by their effect on the structure and properties of solvent water (Arakawa & Timasheff, 1985; Bhat & Timasheff,

1992; Chi, Krishnan, Randolph, & Carpenter, 2003; Kaushik & Bhat, 2003; Kendrick et al., 1997; Lins, Pereira, & Hunenberger, 2004; Timasheff, 1998; Tsumoto et al.; Valente et al., 2005). Crystallization approaches used by crystallographers include inducing the protein to assume a more compact shape (usually by truncation of the N-terminus, a loop, or binding of a substrate), avoiding nonspecific aggregation (usually by decreasing protein concentration or adjusting ionic strength, or by addition of a single additive such as arginine), providing more suitable lattice contacts (usually by site-directed mutation of surface residues or cocrystallization with another protein or peptide). Yet, minimal effort has been made by the crystallographic community to routinely use known excipients that affect these same aspects of a protein's conformation and self-interaction. Studies indicate that increases in protein stability and solubility scale with the concentration of sucrose, mannitol, or trehalose used as a cosolvent (Arakawa & Timasheff; Bhat & Timasheff; Kaushik & Bhat; Kendrick et al.; Lins et al.; Timasheff; Tsumoto et al.). Furthermore, it is well documented that specific combinations of these different excipient classes often provide an even greater effect, and that SIC is an effective tool for elucidating complex interactions between cosolvent additives (Arakawa & Timasheff; Chi et al.; Kaushik & Bhat; Lins et al.; Timasheff; Tsumoto et al.). Figure 3 shows B as a function of concentration for sucrose, trehalose, and mannitol using lysozyme as the model protein. The mobile phase contained either 2% or 5% NaCl. Experiments were also run in the absence of NaCl but behaved qualitatively similar to the 2% studies (no change with increasing cosolvent concentration). Several important findings were noted in these SIC experiments. First, as the concentration of cosolvent increased, B became more positive, confirming the ability of SIC to follow the

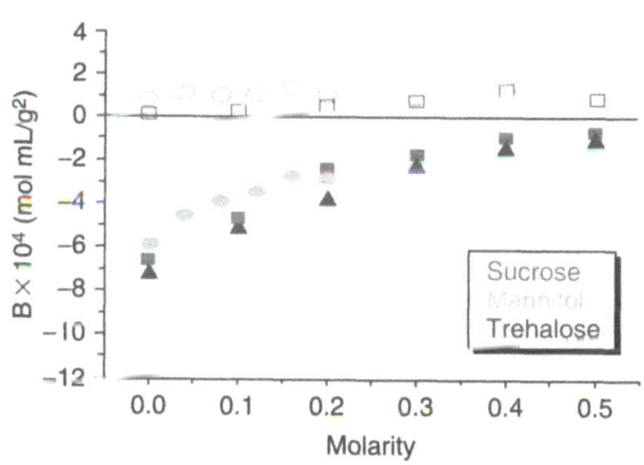

FIGURE 3 B as a function of sugar and polyol concentration for lysozyme. Open symbols are for 2% NaCl mobile phase. Closed symbols are for 5% NaCl. Squares = sucrose, circles = mannitol, and triangles = trehalose.

impact of sugars and polyols on protein self-interaction. Second, *B* values were most impacted in the presence of 5% NaCl. This makes sense from a biophysical interaction perspective. Electrostatic interactions are long range and dominate protein interactions at low-ionic strength. As ionic strength increases, so does charge shielding and electrostatic repulsion becomes less dominant. Under these conditions, short-range attractive hydrophobic interactions dominate. It is these hydrophobic interactions that are mediated by the sugars and polyols. This work showed that the second virial coefficient (measured by SIC) is an effective tool for elucidating complex interactions between cosolvent additives such as arginine, glutamic acid, sucrose, trehalose, mannitol, and glycine (Valente et al.). It was found that for this particular protein, the combination of arginine and glutamic acid is more effective at reducing intermolecular attraction than any single cosolvent in the study. This was interpreted that the charged and liphatic portions of these amino acids were favorably interacting with oppositely charged and hydrophobic portions of the protein's surface, respectively (Fig. 4).

Thus, different excipient mixtures can be explored via the proposed high-throughput SIC system in an effort to find combinations that produce a more stable protein with *B* values that lie within the crystallization slot. Alternatively, these studies can be used to find suitable conditions to reduce membrane protein aggregation during extraction and purification procedures.

B. Surface Mutations

SIC is able to discriminate site-directed mutations of a protein's surface amino acids provided that they do, in fact, affect protein-protein interactions, that is, solubility. SIC can provide sensitive, quantitative data to guide molecular engineering to support membrane protein solubility and stability (to identify surface changes that reduce unwanted aggregation) and crystallization (by finding substitutions that place proteins within the crystallization slot for specific solvent conditions).

C. Membrane Proteins

In addition to our work with soluble proteins and peptides, we and others have also evaluated the ability of SIC to measure *B* for membrane proteins. It should be noted that SLS is a very difficult method for measuring *B* of membrane proteins due to interference of detergent micelles. The following present examples of how membrane protein *B* values (experimentally obtained via SIC) can be used to provide valuable information regarding the proper choice and concentration of detergent(s), additives, and precipitants to optimize conditions for crystallization.

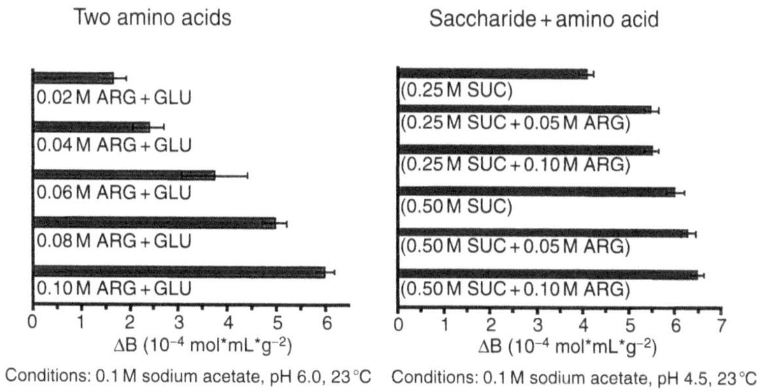

FIGURE 4 Co-solvent effect of different excipient concentrations.

D. Effects of Polyethylene Glycol (PEG) on Outer Membrane Protein-X (OMPX)

PEG is a common cosolvent used to crystallize soluble and membrane proteins. The effects of different PEG concentrations (0-30%) at pH 8.5, 0.5% C_8E_4, 20 mM Tris-HCl, 20 mM NaCl mixed with a dilute solution of OMPX are shown in Fig. 5. At 0% PEG, *B* was -4.0×10^{-4} mol ml/g^2, but when PEG was increased from 0 to 30%, *B* decreased to -14×10^{-4} mol ml/g^2. The screening experiment was performed to determine if the instrumental setup was sensitive enough to detect changes in *B*. The decrease in *B* with increasing PEG concentration indicates that PDC interactions are favored, in agreement with previous results for PDC (Beja, Spudich, Spudich, Leclerc, & DeLong, 2001). Other accounts of PEG versus *B* have shown similar trends, as the concentration of PEG increases *B* has decreased (Garavito & Ferguson-Miller, 2001; Loll, Hitscherich, Aseyev, Allaman, & Wiencek, 2002; Niegowski, Hedren, Nordlund, & Eshaghi, 2006).

E. Conformational Stability of Proteorhodopsin

As mentioned in previous chapters, in some cases mixed detergent systems can enhance protein stability and crystallization. The following provides an example of how SIC can be used to investigate and optimize different combinations and concentrations for specific membrane proteins. Proteorhodopsin (pR), a member of the rhodopsin protein family, was studied as a function of pH, surfactant concentration, and surfactant type. Like other

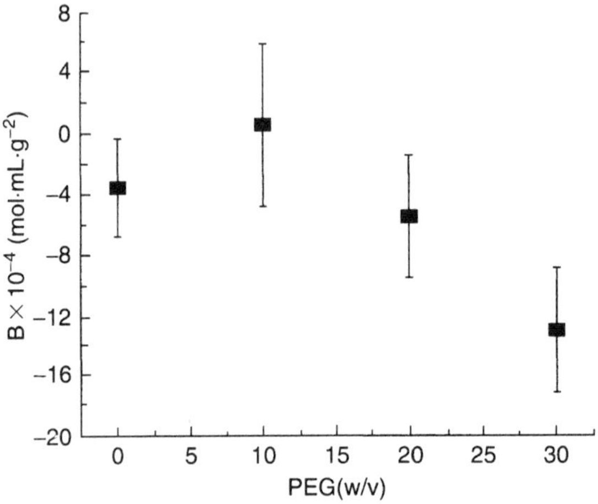

FIGURE 5 The effects of different PEG concentrations (0-30%, w/v) on the second virial coefficient at pH 8.5. 0.5% C_8E_4, 20 mM Tris-HCl, 20 mM NaCl. Experimental condition: flow rate 25 μL/min, injection volume 0.5 μL, wavelength 280 nm.

members of the rhodopsin family, pR undergoes light cycling at alkaline pH. Furthermore, this light cycling is very dependent on the conformation of the protein. Changes in the native structure of pR were monitored by measuring the absorbance of the retinal at 531 and 381 nm (Lam & Packer, 1983). Data collected for the absorbance at 531 nm for each screening condition is presented as a contour plot (Fig. 6). Absorbance at 531 nm was measured for concentrations of Tween-20 and dodecylmaltoside (DDM) ranging from 0.08 to 0.34 mM and 0.00 to 0.09 mM, respectively. Two lines that represent the critical micelle concentration (CMC) of the detergents divide the contour plot; the horizontal line is the CMC of 0.06 mM PEG sorbitan monolaurate-20 (Tween-20) and the vertical line is the CMC of 0.17 mM DDM. The contour plot shows three areas of significant change at 531 nm. Absorbance at 531 nm is at a minimum in the lower left corner of the contour plot where the concentration of the DDM is below the CMC. The second minimum at 531 nM is on the right-hand side of the contour plot when DDM is above the CMC and Tween-20 is below the CMC. Lastly, there is a maximum when Tween-20 is above the CMC at all concentrations of DDM. The conformation of pR was measured in the presence of increasing concentration of CHAPS, Tween-20, and DDM as shown in Fig. 6B and C. No significant change in the absorbance at 531 nm was measured for either system. The absorbance of the retinal fluctuated between 0.07 and 0.09 AU (Fig. 6B and C).

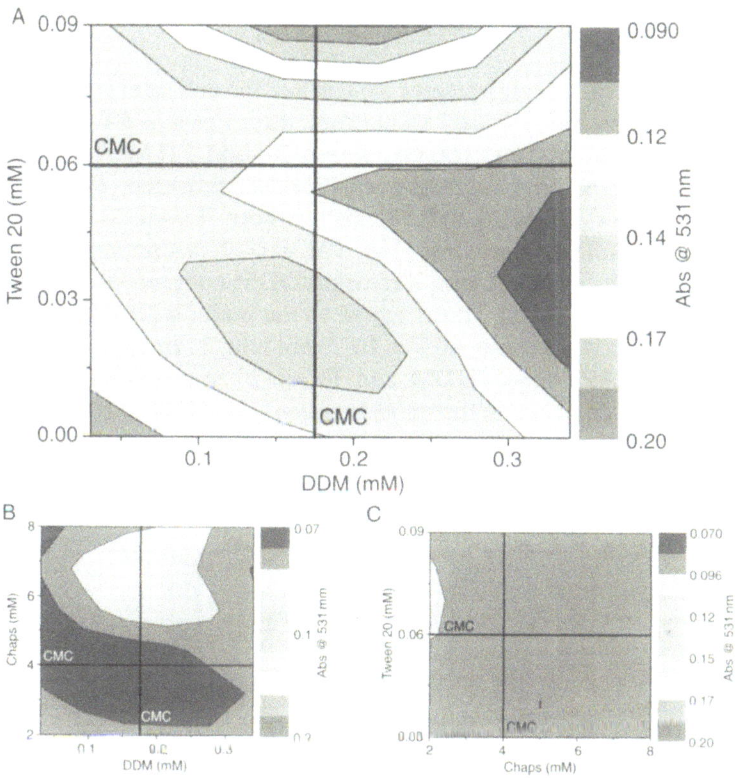

FIGURE 6 (A-C) Conformational stability of pR measured by changes of the retinal absorbance at 531 nm: (A) DDM versus Tween-20, (B) DDM versus CHAPS, and (C) CHAPS versus Tween-20, in pH 7 and 20 mM MES. (See Color Insert.)

Both the detergent type and concentration can affect conformational stability of pR. When two nonionic detergents are mixed, the conformational stability is dependent on detergent concentration. When the concentration of Tween-20 is below the CMC, stability of the pR is dependent on the concentration of DDM. If concentration of DDM is below the CMC, not enough detergent molecules are available to stabilize the membrane protein by shielding the hydrophobic area from the solvent. The data shows when DDM is above the CMC and Tween-20 is below the CMC, the membrane protein is not completely stabilized. This might be due to Tween-20 preventing the DDM from completely covering the hydrophobic area of the membrane protein by steric effects. For zwitterionic versus nonionic systems, the zwitterionic detergent dominates the conformational stability of pR.

F. Measuring B for Mixed Detergent Systems

The effects of a mixed detergent system on the colloidal stability of pR were evaluated by SIC. Figure 7 shows the contour plot of *B* for pR in three different detergent mixtures (DDM, Tween-20, and CHAPS) at pH 7 and 20 mM MES. The first screening experiment measured *B* at different concentrations of DDM (0.30-0.37 mM) without Tween-20 and CHAPS present in the buffer system (Fig. 7A). The data shows an improvement in the colloidal stability of pR with increasing DDM concentrations from 0.08 to 0.37 mM. When Tween-20 was added to the buffer with DDM, *B* values fluctuated between -3.0 and $+0.7 \times 10^{-4}$ mol ml/g^{-2}. In Fig. 7B and C, the addition of CHAPS versus DDM and Tween-20 resulted in no significant change in *B*, which stayed in the range between -3 and 0.7×10^{-4} mol ml/g^{2}.

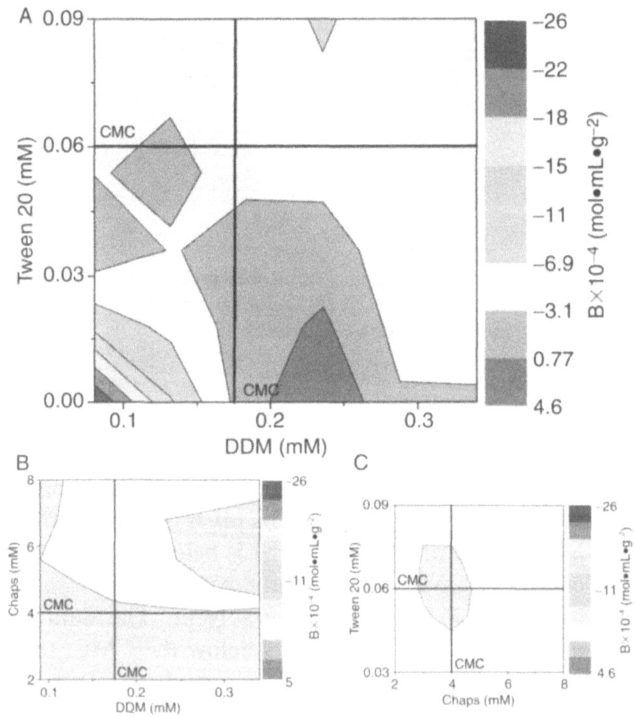

FIGURE 7 (A-C) Colloidal stability of pR measured by changes in *B*: (A) DDM versus Tween-20, (B) DDM versus CHAPS, and (C) CHAPS versus Tween-20, in pH 7 and 20 mM MES. (See Color Insert.)

When pR is in the presence of two different detergents, B does not significantly change when the concentration of either detergent is below the CMC.

G. Assessing Additive Effect on Bacteriorhodopsin

Investigations with SIC were performed to assess the effect of mixed micelles and an additive (1,3-heptane triol) at the reported crystallization conditions for bacteriorhodopsin (10 mM $NaPO_4$, 250 mM NaCl, 6% PEG 3350, 1% *N*-octyl-β-D-glucopyranoside (BOG). The same condition was prepared but with varying percentages of DDM at 0.1% (1.2 × CMC) as shown in Table I (all experiments were performed in triplicate and averaged). The mixture of 75% BOG + 25% DDM makes the PDC more attractive, moving to the far negative area of the crystallization slot at -8×10^{-4} mol ml/g^2. This effect is less in the 50/50 mixture of OBG and DDM. Increasing percentages of DDM cause shorter retention times (less negative B values) until at 100% DDM the solubilization increases dramatically and the measured B value is positive (net repulsion). These experiments imply that the solubilizing nature of DDM is not additive. Heptane-triol is used to improve crystallization conditions by altering the micelle structure in the PDC. As can be seen, in both the BOG and mixed micelle cases, the heptane-triol increases the protein self-attraction (Table I).

TABLE I
SIC B Values for Bacteriorhodopsin

Live column data	$B \times 10^{-4}$ (mol × mL/g^2)	S.D.
Buffer 1: 100% BOG	−2.7	1.5
Buffer 2: 75% BOG, 25% DDM	−8.1	0.9
Buffer 3: 50% BOG, 50% DDM	−3.2	1.0
Buffer 4: 25% BOG, 75% DDM	−0.8	2.1
Buffer 5: 100% DDM	5.1	0.9
Buffer 6: 100% BOG + 80 mM HT	−3.6	0.5
Buffer 7: 75% BOG, 25% DDM + 80 mM HT	−9.5	0.1

H. Other Studies

A compelling example of the use of B values for membrane protein crystallization was reported in two publications (Hitscherich et al., 2000; Loll et al., 2002). The solution behavior of bacterial outer membrane protein, OmpF porin, was studied by SLS in a variety of crystallization solutions. B was demonstrated to be a clear predictor of the crystallization behavior of porin. Both tetragonal and trigonal porin crystals were found to form in a narrow range of B values which were only located within the "crystallization slot" as defined by Wilson (George & Wilson, 1994; George et al., 1997; Guo et al., 1999) for aqueous proteins. B values were also used to study the effect of precipitants such as PEG on micelle size and interacting forces between micelles (Berger et al., 2006). B values for detergent micelles (free of protein) at the identical crystallization conditions exhibited similar B values to the PDCs. Thus, based on this one example where detergent appears to dominate the interactions between PDCs, the authors suggest that for any given detergent, membrane protein crystallization screens may be designed by simply manipulating the detergent-solution properties until B measured values lie within the crystallization slot, thereby minimizing the amount of protein required for crystallization screening and improving productivity (Hitscherich et al., 2000; Loll et al., 2002). In studies performed by Berger the interactions leading to the crystallization of bacteriorhodopsin solubilized in N-octyl-β-D-glucoside (bR-$C_8\beta G_1$ PDCs) were investigated using SIC (Berger et al., 2005). B values for solubilized bacteriorhodopsin were determined in the presence of varying concentrations of ammonium sulfate, sodium formate, and sodium malonate. Attractive PDC interactions were observed as cloud point temperature was approached for the various salts, suggesting that surfactant interactions may play an important role for membrane protein crystallization. Furthermore, these studies suggest that the interaction trends are strongly influenced by micelle structure as well as surfactant-phase behavior, both of which are sensitive to salt and surfactant concentration. This study demonstrates the value of using SIC (with other biophysical techniques) to elucidate the mechanisms that determine PDC interactions for a particular protein. Their data exhibit a strong correlation between the measured B value of the bR-$C_8\beta G_1$ PDCs and the corresponding cloud-point temperature of $C_8\beta G_1$. Crystallization trials resulted in a subset of these conditions where crystallization occurred in the "crystallization slot" for aqueous proteins. Over the range of salt concentrations used in bacteriorhodopsin crystallization, the CMC of β-Octyl-D-glucoside ($C_8\beta G_1$) decreases by at least an order of magnitude. Lorber, DeLucas, and Bishop (1991) observed similar results previously.

Berger also used SIC and other biophysical techniques to investigate effects of additives on surfactant-phase behavior for bacteriorhodopsin crystallization (Berger et al., 2006). It was determined for bacteriorhodopsin that additives (i.e., heptane-triol and 1,6-hexane-diol) that raise the cloud-point temperature do so by affecting free micelles in solution rather than PDCs to promote crystallization. A possible explanation offered is that these additives are weakly surface active, without an apparent CMC, acting as cosolutes rather than cosurfactants. As cosolutes the additives were suspected to provide additional hydrophobic surfaces with which the surfactant monomers can interact. However, 1,2-heptane-diol was found to act differently, displaying a CMC of 0.81 M, lowering the cloud-point temperature and requiring lower salt concentrations for the onset of PDC interactions. A possible explanation offered by Berger is that 1,2-heptane-diol may partition into the surfactant portion of the PDC (as well as into the detergent micelles), thereby influencing PDC interactions (Berger et al.). It was suggested that differences in phase behavior with additive structure may be influenced by the amphiphilicity of the additive. This concept is similar to the small amphiphile concept (Michel & Oesterhelt, 1980). In contrast, heptane-triol and 1,6-heptane-diol display only weak surface activity and act indirectly through decreased micelle interactions and size. Berger's papers clearly demonstrate that the second virial coefficient combined with other techniques can provide insight into the complex nature of PDC interactions, information that is important to developing rational approaches to membrane protein crystallization. It should be noted that these conclusions were base on experiments performed in the absence of PEG, unlike our preliminary investigations with bacteriorhodopsin described above.

I. SIC Studies for Cystic Fibrosis Transmembrane Regulatory (CFTR) Protein

CFTR protein plays a key role in chloride conductance in normal airway cells. Single point mutations of CFTR can cause improper processing of the protein within epithelial cells or unresponsiveness to other interacting molecules important for CFTR activity. Progress toward a structure determination of full-length CFTR has been difficult due to its low abundance in typical expression systems as well as its limited solubility and stability (Riordan, 1993). CFTR produced via overexpression systems typically exhibits moderate to significant aggregation (dimers and a small amount of higher aggregates) following purification over a metal chelating column followed by a Superdex 200, size-exclusion chromatography. Slightly aggregated protein was solubilized in DDM detergent (protein concentration at

0.1 mg/ml) in a variety of buffered solutions, one of which replicated the reported two-dimensional crystallization conditions (Rosenberg, Kamis, Aleksandrov, Ford, & Riordan, 2004). The crystallization conditions for CFTR are 50 mM Tris (pH 8), 5% (w/v) glycerol, 0.1 mg/ml dodecyl maltoside, 110 mM ammonium sulfate plus 11% (w/v) PEG6000. DDM-solubilized protein (~1.2× CMC) in buffer is mixed with the crystallization buffer and equilibrated by vapor diffusion against a dehydrating reservoir, leading to two-dimensional crystals (Riordan). To probe the crystallization conditions, we applied CFTR to a SIC column equilibrated with the reported crystallization buffers but with varying concentrations of PEG 6000. All conditions were run in triplicate, from which the average *B* value was calculated (Table II shows the averaged *B* values and standard deviations). Solution conditions begin with a moderately attractive *B* values, and become slightly but measurably more attractive up to about 5-10% PEG. At 15% PEG, an early peak not previously seen on the chromatogram may indicate aggregated protein. As shown in Table I, all conditions were either at or close to the reported two-dimensional crystallization condition and as such, exhibited *B* values within the defined "crystallization slot" for aqueous proteins. Representative SIC chromatograms for this highly sensitive/labile protein are shown in Fig. 8. The peaks are well behaved and reproducible (all measurements were taken in triplicate) indicating that SIC is indeed a viable method for "difficult-to-work-with-proteins." The slight shoulder that elutes prior to the predominant peak is suspected to be aggregated CFTR (Rosenberg et al.). If this is true, SIC may also represent an alternative method to separate aggregated CFTR from monomeric CFTR, once the column length and solution conditions are optimized for peak separation. This additional capability of SIC was reported previously (Tessier et al., 2004).

All of the results presented plus others not mentioned are obviously preliminary insofar as they are based on less than 10 membrane proteins. However, combined with more than 70 aqueous proteins where the second

TABLE II

B Values for CFTR in Different Concentrations of PEG 6000

B values-DDM	$B \times 10^{-4}$ (mol × mL/g^2)	S.D.
Buffer 1: no PEG Black	−1.3	0.1
Buffer 2: 1% PEG6K Green	−1.4	0.1
Buffer 3: 2% PEG6K Blue	−1.5	0.1
Buffer 4: 5% PEG6K Red	−1.6	0.1
Buffer 5: 10% PEG6K Purple	−1.5	0.2

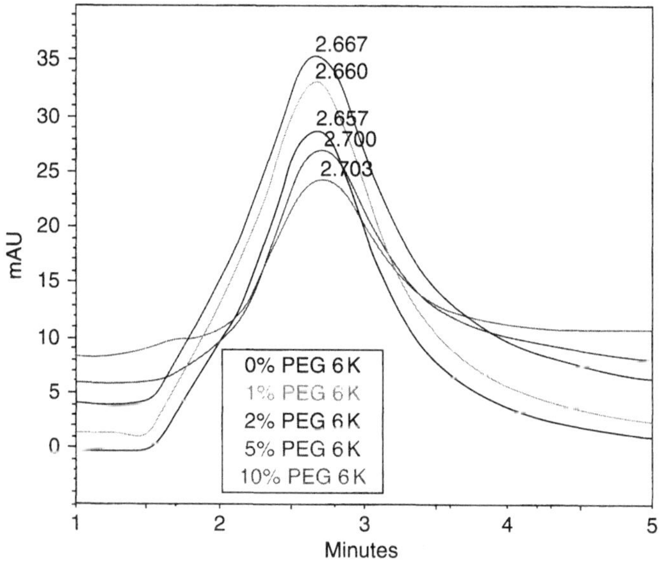

FIGURE 8 CFTR-SIC chromatograms.

virial coefficient was demonstrated to be an accurate predictor of solubility and crystallization conditions these initial results suggest that *B* values may also represent useful diagnostic information for membrane protein solubility, stability and crystallization.

J. Development of a High-Throughput, Knowledge-Based Approach to Membrane Crystallization

The use of *B* values for aqueous and membrane proteins can provide valuable information to improve/optimize aqueous or membrane protein solubility, stability, and crystallization. However, there are a large number of different variables and variable concentrations that must be explored. A sufficient set of conditions must be investigated so as not to miss important combinations/concentrations that might improve a protein's solubility, homogeneity, stability, and ability to produce high-quality crystals. At least 25 factors (McPherson, 1990) can influence the formation of protein crystals including pH, temperature, ionic strength, additives, and precipitants. Within the different factors are individual variables such as pH which can take any number of values as well as categories of additives. The category of "additives" can include hundreds of different compounds each with its own concentration variable. An exhaustive (factorial) search of all possible variable combinations is infeasible. A common tool in the search for formulation conditions amenable to producing protein crystals is the screening kit. These kits generally incorporate formulation conditions which have produced protein crystals along with perturbations around those conditions. Unfortunately, proteins with an expected novel fold or membrane association are interesting targets, but do not have a widely established base of previous crystallization conditions. Therefore, when crystal screens produce little to no positive results (only precipitate or clear drops) there are few leads toward potential crystallization conditions. Also, in the event of a negative result (clear drop) there is no information available about why a particular crystallization condition failed and how formulation parameters influence protein-protein interaction. *B* value measurement by SIC provides protein-protein interaction information in every formulation condition tested.

Even with a measure of protein-protein interaction at every formulation condition there remains the problem of dimensionality of an exhaustive search of parameters. The artificial neural network (ANN) is a technique that has been successfully employed for pattern recognition in high-dimensional parameter space (Bishop, 1994). An ANN, like traditional linear regression models, is a method of mapping input parameters (formulation variables) to an output variable (*B* value). In the case of linear regression the equation, $Y = a_i X_i + b$, is set up where X_i is each input variable (pH, additive concentration, etc.) and Y is the variable which is attempted to be modeled (*B* value). The linear regression is a mathematical solution which solves for a_i and b to give the minimum error over all known values of Y and X_i. Cross terms (interactions, $X_i X_j$) and polynomial terms (X_i^2) may be added to form a more robust generalized linear model (GLM). In the case of the ANN, output parameters are mapped from input parameters via a more complex, nonlinear model. The fundamental mapping unit is a weighted sum of inputs, $Z = \sum w_i X_i$, which is the input to a neuron activation function (Fig. 9A). The initial parameter variables (X_i) are input to an initial set of neurons (input layer) and the output of their activation functions serve as inputs to the next set of neurons (hidden layer). The output of the hidden layer is then used as input to a final set of neurons (output layer). In the case of *B* value modeling, the output layer consists of a single neuron (Fig. 9B). The output of that neuron is the *B* value calculated by the neural network model given a formulation's parameters as input. Like the GLM, the goal is to determine weight values which produce minimal error between model output and observed *B* values. Unlike the GLM there is not a mathematical regression function to determine optimum weight values. Instead each weight's contribution to the error between output *B* value and observed *B* value is assigned (back-propagation of error). Then the weights are incrementally updated to

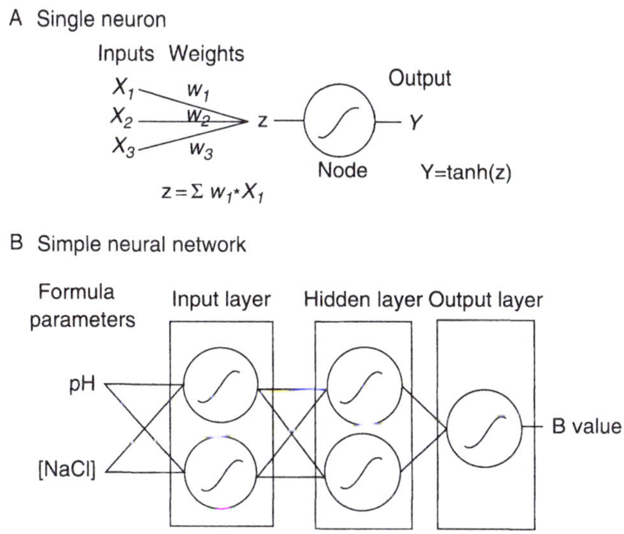

FIGURE 9 A simple neural network with only two variable inputs, pH and NaCl concentration.

reduce root mean square error (RMSE) between the model output and observed output (learning). The ability of the model to generalize to formulations not already used for training is maintained by excluding a subset of the observed *B* value measurements from the training process (validation set). After each incremental update to the ANN's weights the updated model is used to predict the *B* value of each formulation in the validation set. After several thousand iterations of training, optimal weights are chosen as the point of minimum RMSE between the ANN's prediction and observed *B* values over the validation set. After a model is determined it can then be used to computationally predict *B* values for all possible combinations of formulation parameters.

The primary benefit of an ANN model over the GLM is that the choice of which parameter interactions to model is not explicitly defined in advance for the ANN model. The ANN learning process updates weights dynamically to include complex parameter interactions. These complex parameter interactions would likely be impossible for a human to identify by simply looking at a table of parameters and resulting *B* values. The use of ANNs for screen analysis has proven effective in crystallization (DeLucas et al., 2005). An ANN model of crystallization success as a function of condition parameters was able to predict novel crystallization conditions which produced improved crystals for a variety of proteins.

The application of ANN to protein-protein interaction screens is perhaps more promising. It was shown (Johnson, Parupudi, Wilson, & DeLucas, 2009) that *B* values can be predicted with a high level of accuracy by an ANN model trained on a small set of *B* value measurements. Only 81 measured formulation conditions were used to train an ANN model. The model was then used to predict *B* values for 12,626 formulations. Twenty of the formulations were verified experimentally and the predictions were found to have an error rate somewhat larger than the error inherent to SIC *B* value measurements, but the results were clearly semiquantitative in predicting *B* values for unmeasured formulations (Fig. 10). The combination of a

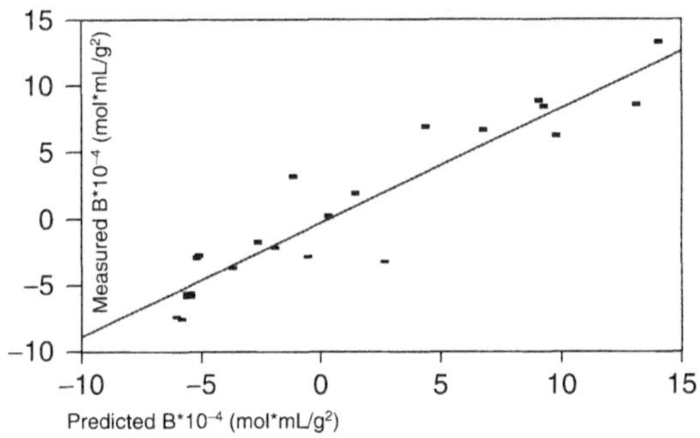

FIGURE 10 Predicted versus observed *B* values for excipient screen ($N = 20$, RMSE = 2.50).

FIGURE 11 High-throughput SIC system.

high-throughput incomplete factorial screen of 96 formulation conditions with immediate neural network analysis could produce, in a single day, a model of how formulation parameters affect protein-protein interaction. Predictions of the ANN could be used to improve protein solubility by determining formulations which reduce protein-protein interactions. The predictions could also direct crystallization screens by eliminating formulations which cause proteins to associate in a nonideal way (outside the crystallization slot) improving the rate of successful crystallization.

Excluding the overhead of creating multiple columns a parallel system can divide the time required per *B* value measurement by the number of parallel columns. The prototype system shown in Fig. 11 can measure the *B* value of a protein in six different formulation conditions simultaneously. With the addition of fully automated formulation handling, this would be capable of making 96 (a standard plate) *B* value measurements in triplicate in a single day. The ability to rapidly measure *B* values will enable use of performing incomplete factorial screens followed by neural net analysis. It is expected that this capability will help guide crystallographers to arrive at optimum formulation conditions to improve protein solubility, stability, and crystallization.

References

Arakawa, T., & Timasheff, S. N. (1985). The stabilization of proteins by osmolytes. *Biophysics Journal, 47*, 411–414.

Beja, O., Spudich, E. N., Spudich, J. L., Leclerc, M., & DeLong, E. F. (2001). Proteorhodopsin phototrophy in the ocean. *Nature, 411*(6839), 786–789.

Berger, B. W., Gendron, C. M., Lenhoff, A. M., & Kaler, E. W. (2006). Effects of additives on surfactant phase behavior relevant to bacteriorhodopsin crystallization. *Protein Science, 15*, 2682–2696.

Berger, B. W., Gendron, C. M., Robinson, C. R., Kaler, E. W., & Lenhoff, A. M. (2005). The role of protein and surfactant interactions in membrane-protein crystallization. *Acta Crystallographica D, D61*, 724–730.

Bhat, R., & Timasheff, S. N. (1992). Steric exclusion is the principal source of the preferential hydration of proteins in the presence of polyethylene glycols. *Protein Science, 1*, 1133–1143.

Bishop, C. M. (1994). Neural networks and their applications. *The Review of Scientific Instruments, 65*, 1803–1832.

Bonnete, F., Finet, S., & Tardieu, A. (1999). Second virial coefficient: Variations with lysozyme crystallization conditions. *Journal of Crystal Growth, 196*, 403–413.

Chi, E. Y., Krishnan, S., Randolph, T. W., & Carpenter, J. F. (2003). Physical stability of proteins in aqueous solution: Mechanism and driving forces in nonnative protein aggregation. *Pharmaceutical Research, 20*, 1325–1336.

DeLucas, L. J., Hamrick, D., Cosenza, L., Nagy, L., McCombs, D., Bray, T., et al. (2005). Protein crystallization: Virtual screening and optimization. *Progress in Biophysics and Molecular Biology, 88*, 285–309.

Demoruelle, K., Guo, B., Kao, S., McDonald, H., Nikic, D., Holman, S., et al. (2002). Correlation between the osmotic second virial coefficient and solubility for equine serum albumin and ovalbumin. *Acta Crystallographica, D58*, 1544–1548.

DePhillips, P., & Lenhoff, A. M. (2000). Pore size distributions of cation-exchange adsorbents determined by inverse size-exclusion chromatography. *Journal of Chromatography, 883*, 39–54.

Ducruix, A., Guilloteau, J., Kautt, M., & Tardieu, A. (1996). Protein interactions as seen by solution X-ray scattering prior to crystallogenesis. *Journal of Crystal Growth, 168*, 28–39.

Dumetz, A. C., Snellinger-O'Brien, A. M., Kaler, E. W., & Lenhoff, A. M. (2007). Patterns of protein-protein interactions in salt solutions and implications for protein crystallization. *Protein Science, 16*, 1867–1877.

Fujita, H. (1975). *Foundations of ultracentrifugal analysis.* New York: Wiley.

Garavito, R. M., & Ferguson-Miller, S. (2001). Detergents as tools in membrane biochemistry. *Journal of Biological Chemistry, 276*, 32403–32406.

Garcia, C. D., Hadley, D. J., Wilson, W. W., & Henry, C. S. (2003). Measuring protein interactions by microchip self-interaction chromatography. *Biotechnology Progress, 19*, 1006–1010.

George, A., Chiang, Y., Guo, B., Arabshahi, A., Cai, Z., & Wilson, W. (1997). Second virial coefficient as a predictor in protein crystal growth. *Methods in Enzymology, 276*, 100–110.

George, A., & Wilson, W. (1994). Predicting protein crystallization from a dilute solution property. *Acta Crystallographica, D50*, 361–365.

Gripon, C., Legrand, L., Rosenman, I., Vidal, O., Robert, M., & Boue, F. (1997). Lysozyme-lysozyme interactions in under-and supersaturated solutions: A simple relation between the second virial coefficients in H_2O and D_2O. *Journal of Crystal Growth, 178*, 575–584.

Guo, B., Kao, S., McDonald, H., Asanov, A., Combs, L., & Wilson, W. (1999). Correlation of second virial coefficients and solubilities useful in protein crystal growth. *Journal of Crystal Growth, 196*, 424–433.

Haas, C., Drenth, J., & Wilson, W. (1999). Relation between the solubility of proteins in aqueous solutions and the second virial coefficient of the solution. *The Journal of Physical Chemistry B, 103*, 2808–2811.

Hitscherich, C., Kaplan, J., Allaman, M., Wiencek, J., & Loll, P. (2000). Static light scattering studies of OmpF porin: Implications for integral membrane protein crystallization. *Protein Science, 9*, 1559–1566.

Johnson, D., Parupudi, A., Wilson, W., & DeLucas, L. (2009). High-throughput self-interaction chromatography: Applications in protein formulation prediction. *Pharmaceutical Research, 26*, 296–305.

Kaushik, J. K., & Bhat, R. (2003). "Why is trehalose an exceptional protein stabilizer?" *The Journal of Biological Chemistry, 278*, 26458–26465.

Kendrick, B. S., Chang, B. S., Arakawa, T., Peterson, B., Randolph, T. W., Manning, M. C., et al. (1997). Preferential exclusion of sucrose from recombinant interleukin-1 receptor antagonist: Role in restricted conformational mobility and compaction of native state. *Proceedings of the National Academy of Sciences of the United States of America, 94*, 11917–11922.

Kratochvil, P. (1987). *Classical light scattering from polymer solutions.* Amsterdam: Elsevier.

Krupka, H. I., Rupp, B., Segelke, B. W., Lekin, T. P., Wright, D., Wu, H.-C., et al. (2002). The highspeed Hydra-Plus-One system for automated high-throughput protein crystallography. *Acta Crystallographica, D58*, 1523–1526.

Lam, E., & Packer, L. (1983). Non-ionic detergent effects on spectroscopic characteristics and the photocycle of bacteriorhodopsin in purple membranes. *Archives of Biochemistry and Biophysics, 221*, 557–564.

Lins, R. D., Pereira, C. S., & Hunenberger, P. H. (2004). Trehalose-protein interaction in aqueous solution. *Proteins: Structure, Function and Bioinformatics, 55*, 177–186.

Loll, P. J., Hitscherich, C., Aseyev, V., Allaman, M., & Wiencek, J. (2002). Assessing micellar interaction and growth in detergent solutions used to crystallize integral membrane proteins. *Crystal Growth and Design, 2*(6), 533–539.

Lorber, B., DeLucas, L., & Bishop, B. (1991). Changes in the physico-chemical properties of the detergent octyl glucoside during membrane protein crystallization using a salt as the precipitant. *Journal of Crystal Growth, 110*, 103–113.

Luft, J. R., Wolfley, J., Jurisica, I., Glasgow, J., Fortier, S., & DeTitta, G. T. (2001). Macromolecular crystallization in a high throughput laboratory—The search phase. *Journal of Crystal Growth, 232*, 591–595.

Malfois, M., Bonnete, F., Belloni, L., & Tardieu, A. (1996). A model of attractive interactions to account for fluid-fluid phase separation of protein solutions. *The Journal of Chemical Physics, 105*, 3290–3300.

McPherson, A. (1990). Current approaches to macromolecular crystallization. *European Journal of Biochemistry, 189*, 1–23.

Michel, H., & Oesterhelt, D. (1980). Three-dimensional crystals of membrane proteins: Bacteriorhodopsin. *Proceedings of the National Academy of Sciences of the United States of America, 77*, 1283–1285.

Neal, B., Asthagiri, D., & Lenhoff, A. (1998). Molecular origins of osmotic second virial coefficients of proteins. *Biophysical Journal, 75*, 2469–2477.

Neal, B., Asthagiri, D., Velov, O., Lenhoff, A., & Kaler, E. (1999). Why is the osmotic second virial coefficient related to protein crystallization? *Journal of Crystal Growth, 196*, 377–387.

Niegowski, D., Hedren, M., Nordlund, P., & Eshaghi, S. (2006). A simple strategy towards membrane protein purification and crystallization. *International Journal of Biological Macromolecules, 39*, 83–87.

Patro, S. Y., & Przybycien, T. M. (1996). Self-interaction chromatography: A tool for the study of protein-protein interactions in bioprocessing environments. *Biotechnology and Bioengineering, 52*, 193–203.

Pjura, P. E., Lenhoff, A. M., Leonard, S. A., & Gittis, A. G. (2000). Protein crystallization by design: Chymotrypsinogen without precipitants. *Journal of Molecular Biology, 300*, 235–239.

Riordan, J. R. (1993). The cystic fibrosis transmembrane conductance regulator. *Annual Review of Physiology, 55*, 609–630.

Rosenbaum, D., Zamora, P., & Zukoski, C. (1996). Phase behavior of small attractive colloidal particles. *Physical Review Letters, 1*, 150–153.

Rosenberg, M. F., Kamis, A. B., Aleksandrov, L. A., Ford, R. C., & Riordan, J. R. (2004). Purification and crystallization of the cystic fibrosis transmembrane conductance regulator (CFTR). *The Journal of Biological Chemistry, 279*, 39051–39057.

Ruppert, S., Sandler, S., & Lenhoff, A. (2001). Correlation between the osmotic second virial coefficient and the solubility of proteins. *Biotechnology Progress, 17*, 182–187.

Stevens, R. C. (2000). High throughput protein crystallization. *Current Opinion in Structural Biology, 10*, 558–563.

Tessier, P. M., Johnson, H. R., Pazhianur, R., Berger, B. W., Prentice, J. L., Bahnson, B. J., et al. (2003). Predictive crystallization of ribonuclease A via rapid screening of osmotic second virial coefficients. *Proteins: Structure, Function and Genetics, 50*, 303–311.

Tessier, P. M., Lenhoff, A. M., & Sandler, S. I. (2002b). Rapid measurement of protein osmotic second virial coefficients by self-interaction chromatography. *Biophysical Journal, 82*, 1620–1631.

Tessier, P. M., Sandler, S. I., & Lenhoff, A. M. (2004). Direct measurement of protein osmotic second virial cross coefficients by cross-interaction chromatography. *Protein Science, 13*, 1379–1390.

Tessier, P. M., Vandrey, S. D., Berger, B. W., Pazhianur, R., Sandler, S. I., & Lenhoff, A. M., (2002a). Self-interaction Chromatography: A novel screening method for rational Protein crystallization. *Acta Crystallographica* D58, 1531–1535.

Timasheff, S. N. (1998). Control of protein stability and reactions by weakly interacting cosolvents: The simplicity of the complicated. *Advances in Protein Chemistry, 51*, 355–432.

Tombs, M., & Peacocke, A. (1974). *The osmotic pressure of biological macromolecules.* Oxford: Clarendon Press.

Tsumoto, K., Ejima, D., Kita, Y., & Arakawa, T. (2005). Review: Why is arginine effective in suppressing aggregation? *Protein and Peptide Letters, 12*, 613–619.

Valente, J. J., Kusum, S. V., Manning, M. C., Wilson, W. W., & Henry, C. S. (2005). Second virial coefficient studies of cosolvent-induced protein self interaction. *Biophysical Journal, 89*, 4211–4218.

CHAPTER 8

Advances in Microfluidic Membrane Protein Crystallization Techniques

Cory Gerdts* and Peter Nollert†
*deCODE biostructures, woodridge, Illinois 60517
†Emerald BioSystems, Inc., Bainbridge Island, Washington 98110

I. OVERVIEW

Microfluidic devices for protein crystallization have been designed to consume small amounts of protein, be inexpensive and amenable to use in high-throughput protein crystallization efforts. Although a relative new addition to the arsenal of tools available to membrane protein crystallographers, recent reports on membrane protein crystallizations in microfluidic devices are promising. Consuming 5-100 times less sample volume as compared to traditional pipetting technologies, they provide unique crystallization regimes for (i) free interface diffusion (FID), (ii) gradients, and (iii) microbatch experiments in plugs. Here we review those microfluidic crystallization methodologies that have proven their utility for membrane protein crystallization. As the peculiar advantages and shortcomings of the individual systems become better understood, further integration and their specific significance for membrane proteins will aid in guiding the future development of these technologies.

Current Topics in Membranes, Volume 63 1063-5823/09 $35.00

DOI: 10.1016/S1063-5823(09)63008-8

II. MEMBRANE PROTEIN CRYSTALLIZATION

The pursuit of crystallizing membrane proteins suffers from a variety of ailments such as low protein expression yields, difficult purification of active protein species, and awkward fluid handling properties that are due to the protein's requirement to be solubilized in an amphiphile environment. Membrane protein crystallization poses a substantial challenge, the reason of which is that transmembrane proteins remain functional and folded only when a proper microenvironment is supplied that allows interaction with a high and a low dielectric medium with the proteins' aqueous and membraneous moieties, respectively. Crystal growth from a solution requires solvent modifications that provide an energy gain for proteins to aggregate in a well-organized manner. Since the physical chemistry of aqueous and oily solutions is fundamentally different, providing the required microenvironment while changing the solvent bulk properties to induce crystallization is rather difficult to establish. To increase screening efficiency of the somewhat incompatible multidimensional phase space, rational, and pragmatic solutions have been devised (Wiener, 2004) as well as practical recipes for detergent formulations and crystallization screening reagents, the latter of which have been developed by trial and error (Newstead, Ferrandon, & Iwata, 2008). Hence, reagent formulations are now available that allow an approach to crystallization of membrane proteins in a manner similar to that of soluble proteins.

III. MICROFLUIDICS

The field of microfluidics aims to provide relief to some of the problems encountered in small volume liquid handling and specifically to protein crystallography. Microfluidics inherently limits protein sample consumption (much less than 100 nl per experiment), an essential advantage when protein supply is limited and many experiments are required to identify a productive crystallization condition. New advances in protein sample handling and environmental control offered by microfluidic FID and plug-based crystallization provide intriguing new options to protein crystallization researchers.

Crystallization in microfluidic devices is fundamentally different from traditional crystallization in small volumes. While crystallization experiments carried out in very small volumes using conventional techniques such as vapor diffusion or batch under oil crystallization can be rationalized as scaled-down versions of their larger counterparts (Chayen, Shaw, & Blow, 1992; Yeh, 2003), the equilibration kinetics and crystallization times are shortened (Stevens, 2000). When volumes much lower than ca. 100 nl are used and crystallization experiments ensue in the confined space of a microfluidic channel, the flow regime changes from turbulent to laminar flow (defined by the dimensionless *Reynolds number*), resulting in a crystallization regime that may be very different from that in scaled-up experiments. A comprehensive assessment of microfluidic versus normal flows is given in (Squires & Quake, 2005). For the experimenter these unique properties can be used to explore otherwise inaccessible areas in the multidimensional protein crystallization phase space. On the other hand, crystals may not grow to sufficient size in small volumes, making it impossible for X-ray experiments to be conducted. In the latter case, the so-called "scale-up problem" must be overcome in order to grow crystals amenable to X-ray crystallographic structure determination. Furthermore, the physical recovery of crystals from the confined space of the crystallization chamber is desired for cryopreservation and X-ray diffraction. Regardless of such pitfalls, access to low volume crystallization screening can be a tremendous economic advantage when working with precious membrane protein samples. Microfluidic technology can provide useful crystalline material and information because a larger number of different crystallization experiments can be assayed.

In this context it is important to note that in membrane protein samples, subjected to microfluidic crystallization experiments, the presence of amphiphiles reduces the surface energy of sample solutions. In microfluidic crystallization experiments, surface effects dominate viscous forces and fluid behavior is affected according to the *Capillary number* (*Ca*) ($Ca = \mu U/\lambda$, where μ = shear viscosity, U = flow velocity, and λ = surface tension in air) (Squires & Quake, 2005). Because volumetric flow rates are small, microfluidic systems are generally operated with a small *Ca*. The surface tension of the fluid, which is greatly affected by the lipid/detergents in the membrane protein sample, affects the *Ca* and becomes an important factor to consider in the design of a microfluidic system (Tice, Song, Lyon, & Ismagilov, 2003). As a practical matter, however, these features need to be considered mainly when designing microfluidic crystallization devices.

Finally, due to the large surface-to-volume ratio of small liquid volume samples, the interaction of components of the crystallization cocktail, such as the diffusion of solvents and amphiphiles into the channel walls, need to be minimized if possible.

In the following we review the current state of microfluidic membrane protein crystallization. While several innovative microfluidic devices for protein crystallization have been generated, limited experimental results have been reported in the literature thus far. Nevertheless, we can draw several conclusions from this first glimpse on what may one day become a standard crystallization methodology.

IV. CRYSTALLIZATION BY FID IN POLYDIMETHYLSILOXANE (PDMS) DEVICES

FID within microfluidic devices was introduced as a method for protein crystallization in 2002 by Quake and coworkers (Hansen, Skordalakes, Berger, & Quake, 2002). Using an integrated set of microfluidic valves and microchannels, fluids are segmented and combined by establishing a small interface region (Thorsen, Maerkl, & Quake, 2002). Mixing between the desired fluids is achieved by free diffusion through the established interface without convective flow. In the microfluidic FID protein crystallization system, the interface that is established is between the protein solution and the precipitant. Due to the unique layering process in manufacturing and the PDMS material properties, microvalves and microchannels are easily replicated many times in a small space due to the photolithographic production process.

The FID method aids in combining a single protein sample with a wide variety of different chemical cocktails for sparse matrix screening. Its compartmentalized experiments are created using an integrated network of microvalves and microchannels that prevent cross-contamination between experiments. The microvalves operate in a mechanical manner, meaning that fluids of all types and properties can be used in the system. In protein crystallography—and especially membrane protein crystallography—where a wide variety of fluid properties (such as viscosity, surface tension, and pH) is prevalent, this kind of flexibility is important.

Shortly after the technology was introduced, it was released as a commercial product by Fluidigm Corporation and became available to the structural biology community. The TOPAZ protein crystallization system from Fluidigm offers FID crystallization chips that allow screening of ca. 1 μl protein sample against 96 precipitants (1.96 version), 4 proteins against 96 precipitants (4.96 version), or 8 proteins against 96 precipitants (8.96). Although to date there are no reports demonstrating the utility of this device for membrane protein crystallizations, it may be just a matter of time for such reports to appear in the published literature.

With volumes well below 10 nl, crystal sizes are usually small and crystals cannot be readily retrieved from the PDMS chip. As a result, any crystals that are grown must be reproduced in a scaled-up experiment for analysis by X-ray diffraction. The success of the scale-up process seems to vary from laboratory to laboratory and from protein to protein. This may be due to several factors including the "translation" from FID-style crystallization conditions to vapor diffusion- or microbatch-style crystallization performed in the scale-up experiment.

An intriguing feature of PDMS-based microfluidic devices are their tendency to absorb many small molecule compounds, including amphiphiles such as detergents (Seddon, Curnow, & Booth, 2004). Since the detergent concentration is a critical parameter in membrane protein crystallization (Wiener, 2004), this feature may dramatically affect the crystallization outcome in a negative or positive manner. While reducing the detergent concentration to the point where membrane protein-lipid-detergent complexes become the dominant species is desired for crystal nucleation and growth, reduction of the detergent concentration may lead to depletion of the membrane protein-lipid-detergent complex and thus favor disordered aggregation. Thus, the interplay of surface to volume ratio, partition coefficient, and kinetics of detergent depletion determine whether a favorable crystallization environment is provided for a particular membrane protein.

Employing PDMS-based microfluidic crystallization chambers in a unique arrangement of channels, chambers, and valves, Anderson, Hansen, and Quake (2006) report the crystallization of the membrane proteins cytochrome Cbb3 from *Rhodobacter sphaeroides* and wild-type bacteriorhodopsin from *Halobacterium* (bR) as well as its mutant form bR D85S in PDMS-based devices that employ the FID method (Fig. 1). In a first round of experiments, these microfluidic devices were used to create rational phase diagram-based crystallization experiments. The solubility information was subsequently used to design and perform customized crystallization trials. For all proteins tested, this strategy doubled the crystallization success rate over common crystallization methods. Crystals of the membrane proteins tested were reproduced in microfluidic devices with larger experiment volumes, resulting in 6.7 Å resolution X-ray diffraction data for bacteriorhodopsin (bR D85S) crystals, 16 Å for wild-type crystals, and 14.5 Å for cytochrome Cbb3 crystals.

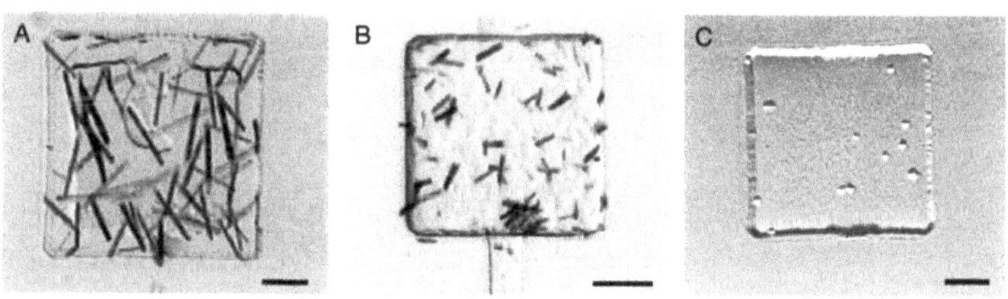

FIGURE 1 Crystallization of (A) wild-type bacteriorhodopsin from *Halobacterium* (bR), (B) mutant form D85S bacteriorhodopsin from *Halobacterium*, and (C) cytochrome Cbb3 from *Rhodobacter sphaeroides* in PDMS-based microfluidic crystallization devices employing the free interface diffusion methodology. Scale bars: 100 μm. (Reproduced with permission of PNAS from Anderson et al., 2006.)

A more sophisticated microfluidic formulation device was used by Sommer and Larsen (2005) for the generation of a solubility phase diagram of the integral membrane protein SERCA. Utilization of this data for scaled-up using traditional crystallization methods yielded crystals of this calcium pump.

V. MEMBRANE PROTEIN CRYSTALLIZATION IN GRADIENTS ESTABLISHED IN MICROCHANNELS

Crystallization by capillary counterdiffusion are well understood experiments that have been used to identify the optimal crystal growth conditions within a single experiment (Ng, Gavira, & Garcia-Ruiz, 2003). In an effort to miniaturize this crystallization methodology, Ng, Clark, Stevens, and Kuhn, (2008) employed plastic microchannels for the crystallization of the membrane protein bacteriorhodopsin and obtained crystals that diffracted to 6 Å resolution (Fig. 2).

The devices were microchannel plates molded with a cyclic olefin copolymer and with channel widths ranging from 0.1 to 0.3 mm, lengths of 20-26 mm, and a height of 0.1 mm, providing microchannel volumes of 0.2-1.96 μl. Although this volume is rather large in the context of microfluidic crystallization, each counterdiffusion crystallization experiment allows one to effectively scan a range of concentration and diffusion values for a single formulation. During an experiment, detergent concentrations vary within the length of the microfluidic channel owing to the fact that different diffusion rates of free detergent and detergent in membrane protein-lipid-detergent complexes differ starkly.

A similar microfluidic device called Crystal Former has 16 individual capillary channels and is commercially available through Microlytic ApS.

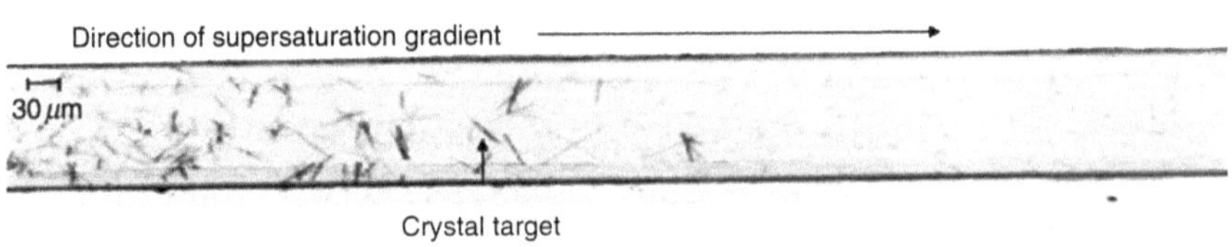

FIGURE 2 Crystallization of bacteriorhodopsin in plastic microchannels employing counterdiffusion. (Reprinted with permission of IUCr from Ng et al., 2008.)

VI. PLUG-BASED MEMBRANE PROTEIN CRYSTALLIZATION IN MICROCAPILLARIES

Plug-based protein crystallization was introduced in 2003 by Ismagilov and coworkers (Zheng, Spencer, & Ismagilov, 2003). In such plug-based protein crystallization experiments, the individual aqueous solutions used to formulate the crystallization experiments enter the microchip in separate streams until they reach an area where they are combined and spontaneously break up into a series of drops or "plugs." The channel dimensions and materials used as well as the content and flow rates of the aqueous and oil liquids determine the size of formed plugs. During the plug formation process the aqueous crystallization solution becomes surrounded by the inert and immiscible carrier fluid and is subsequently transported within a microfluidic channel with a diameter of 200 μm (Song, Tice, & Ismagilov, 2003). While new plugs are formed, those already created are pushed further into a microfluidic storage channel. Usually three aqueous streams, protein solution, buffer, and a crystallization formulation enter the plug formation area where they break up into plugs. As plugs are formed the aqueous contents mix within seconds by internal recirculation.

Each plug represents a separate and distinct microbatch-style crystallization experiment that is incubated for crystal growth. The size and hence volume of individual plugs can be set by adjusting the flow rates, and for protein crystallization experiments volumes in the 10 nl range are generally employed. Furthermore, the composition of the plugs is controlled by varying the relative flow rates of the streams that are entering the microchip. This simple concept allows for on-chip formulation of carefully controlled concentrations of the components in each plug. Very fine gradients can be generated over a series of plugs to carefully interrogate crystallization phase space (Fig. 3). Sparse matrix screening is accomplished by generating a preformed cartridge of precipitants and merging them with a stream of protein. Crystals of membrane proteins, photosynthetic reaction center from *Rhodopseudomonas viridis* and Porin from *Rhodobacter capsulatus*, have been grown in plugs using this approach (Fig. 4). Such crystals were harvested by flowing the plug containing the crystals out of the microfluidic channel, subjected to cryopreservation and X-ray diffraction experiments, yielding X-ray diffraction to 1.95 and 1.9 Å resolution for Porin and photosynthetic reaction center, respectively. Datasets collected were used to for structure determination by molecular replacement (Li et al., 2006).

Alternatively, crystals in plugs can be diffracted by X-rays without removing them from the plug or the microfluidic channel. For the soluble test protein Thaumatin a merged X-ray diffraction dataset with up to 1.86 Å resolution was obtained and used for structure determination (Yadav et al., 2005).

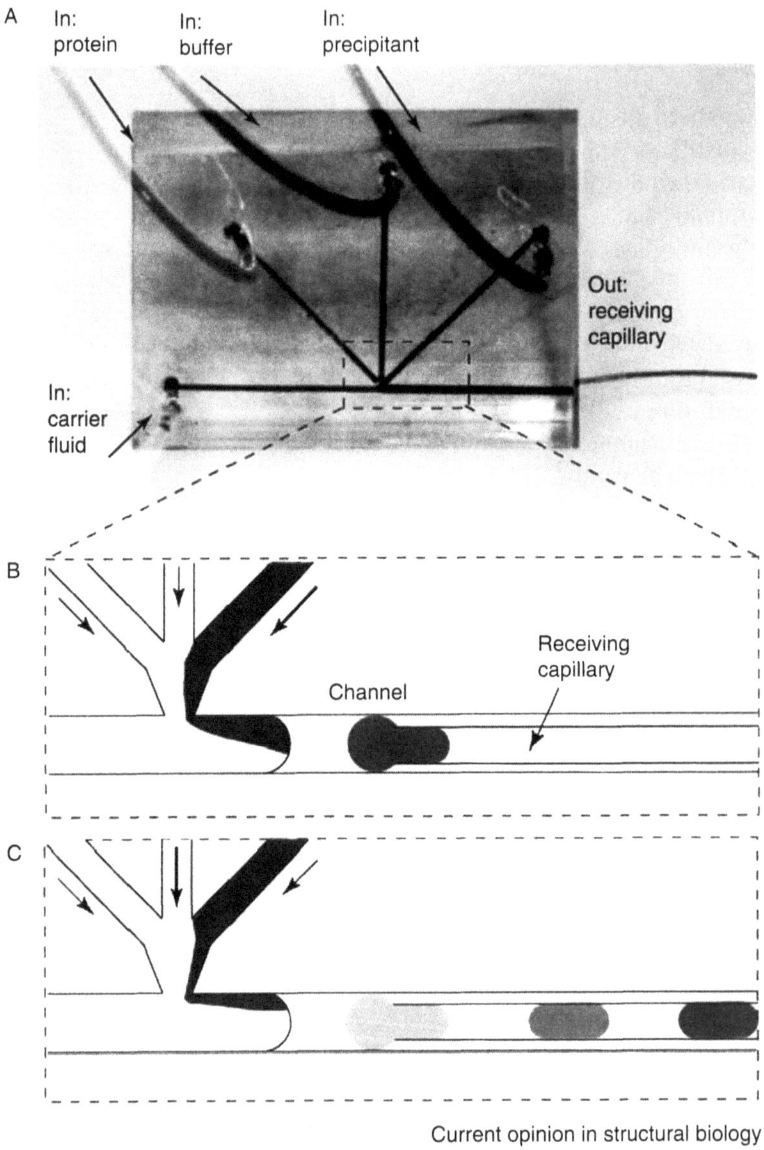

FIGURE 3 Devices and concept for plug-based protein crystallization in microfluidic devices. Left panel: optimization of protein crystallization experiments with the plug-crystallization approach. (A) Microfluidic device for generating aqueous plugs. Three inlets for aqueous liquids (protein, buffer, and precipitant) and the carrier fluid are combined in this PDMS-based microfluidic device. Optimization screens are produced with varying flow rates

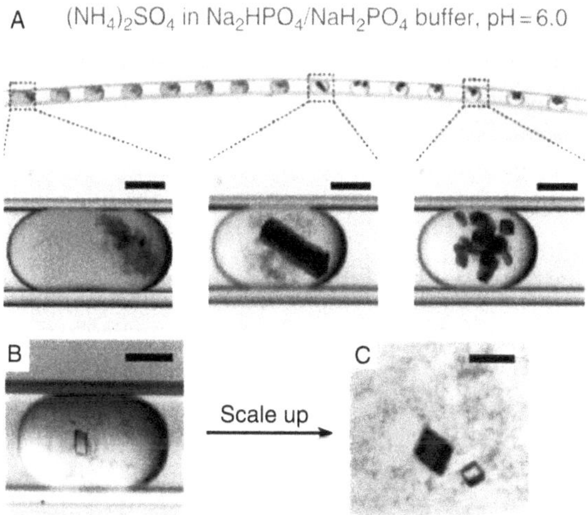

FIGURE 4 Crystallization of photosynthetic reaction center from *R. viridis* and Porin from *R. capsulatus* in plugs. Shown are results of sparse matrix screening in plugs using the hybrid methodology. (A) Crystals of photosynthetic reaction center from *R. viridis* grown in plugs within a Teflon capillary. A transition from precipitation to single crystals is shown. Plug volume is ca. 10-15 nl. (B) Crystals of *R. viridis* and Porin from *R. capsulatus* formed in a plug. (C) Larger crystals of Porin obtained by scaling up trials into 600 μm diameter glass capillaries. Scale bars: 100 μm. (Reprinted with permission of PNAS from Li et al., 2006.)

It appears that such an *in situ* X-ray diffraction approach should work for membrane proteins as well, however at this point in time this has not been reported.

Microfluidic devices with the appropriate mixing chamber and integrated storage capillaries have been commercialized by Emerald BioSystems, Inc. and are called CrystalCards. A single CrystalCard holds two separate sets of experiments with each channel system capable of producing ca. 400 crystallization plugs.

VII. CONCLUSION

Since their recent introduction, the use of microfluidic devices for membrane protein crystallization has been described in the scientific literature primarily for test proteins rather than actual new target proteins.

of the aqueous streams. (B) With high precipitant (dark) flow rate, the concentration of the precipitant in the resulting plug is high and it becomes low (C) as the flow rate of the buffer is increased and the flow rate of the precipitation is decreased. Plug volume is in the range of 10 nl. (Reprinted with permission of Elsevier from Zheng, Gerdts, & Ismagilov, 2005).

Nevertheless, the successful generation of crystals in such devices bodes well for their future application and their general adoption, especially if their application is cost effective. Here we have briefly reviewed the use of different microfluidic devices that allow nanoliter crystallization by (i) FID, (ii) by gradient crystallization, and (iii) as microbatch experiments in plugs. It is difficult to compare their crystallization success or the quality of the generated crystals without proper side-by-side comparison experiments. Thus far the best crystals have been grown employing the plug-based approach in microchannels where crystals grow in a way that makes them readily amenable to cryopreservation and subsequent X-ray crystallographic characterization.

From a practical point of view, the spatial confinement of the crystallization experiment generally poses new challenges for handling soluble and for membrane protein samples alike. Furthermore, little is known about the interaction of organic solvents and amphiphiles, specifically detergents with the walls of microfluidic crystallization containers. The partitioning of such small molecules may be utilized to one's advantage, effectively employing detergent concentrations lower than those in the starting sample. In special cases the detergent concentration may approach the critical micellar concentration, providing a crystallization regime that is deemed useful to membrane protein crystallization (Wiener, 2004). Finally, special microfluidic devices have proven their utility when applied to generate solubility phase diagrams from a large number of experiments, even when crystals are not obtained or are too small to be diffracted. The resulting data can then be used to guide scaled-up traditional protein crystallization experiments.

Opportunities to further develop microfluidic devices specifically for membrane protein crystallizations hence lie in areas where (i) detergent concentrations can be modulated within the experimental setup, and (ii) the preparation of crystallization experiments is further integrated with upstream processes such as purification, concentration, and formulation as well as with downstream processes such as *in situ* X-ray diffraction. *In situ* X-ray diffraction presents an opportunity for microfluidics since such devices may help prevent damage of the often fragile membrane protein crystals during handling. Eventually, crystal growth, X-ray data collection, and phase determination may become feasible without any crystal manipulation.

Acknowledgments

We thank ATCG3D funded by the NIGMS and NCRR under the PSI-2 Specialized Center program (U54 GM074961) for support. P. Nollert was supported by the NIH Roadmap grant PO1 GM075913. Both authors are affiliated with Emerald BioSystems, Inc.

References

Anderson, M. J., Hansen, C. L., & Quake, S. R. (2006). Phase knowledge enables rational screens for protein crystallization. *PNAS, 103*(45), 16746–16751.

Chayen, N. E., Shaw, S. P. D., & Blow, D. M. (1992). Microbatch crystallization under oil—a new technique allowing many small-volume crystallization trials. *Journal of Crystal Growth, 122*, 176–180.

Hansen, C. L., Skordalakes, E., Berger, J. M., & Quake, S. R. (2002). A robust and scalable microfluidic metering method that allows protein crystal growth by free interface diffusion. *Proceedings of the National Academy of Sciences of the United States of America, 99*(26), 16531–16536.

Li, L., Mustafi, D., Fu, Q., Tereshko, V., Chen, D. L., Tice, J. D., et al. (2006). Nanoliter microfluidic hybrid method for simultaneous screening and optimization validated with crystallization of membrane proteins. *PNAS, 103*, 19243–19248.

Newstead, S., Ferrandon, S., & Iwata, S. (2008). Rationalizing alpha-helical membrane protein crystallization. *Protein Science: A publication of the Protein Society, 17*(3), 466–472.

Ng, J. D., Clark, P. J., Stevens, R. C., & Kuhn, P. (2008). *In situ* X-ray analysis of protein crystals in low-birefringent and X-ray transmissive plastic microchannels. *Acta Crystallographica, D64*, 189–197.

Ng, J. D., Gavira, J. A., & Garcia-Ruiz, J. M. (2003). Protein crystallization by capillary counterdiffusion for applied crystallographic structure determination. *Journal of Structural Biology, 142*(1), 218–231.

Sommer, M. O. A., & Larsen, S. (2005). Crystallizing proteins on the basis of their precipitation diagram determined using a microfluidic formulator. *Journal of Synchrotron Radiation, 12*, 779–785.

Song, H., Tice, J. D., & Ismagilov, R. F. (2003). A microfluidic system for controlling reaction networks in time. *Angewandte Chemie International Edition, 42*(7), 768–772.

Squires, T. M., & Quake, S. R. (2005). Microfluidics: Fluid physics at the nanoliter scale. *Reviews of Modern Physics, 77*(3), 977–1026.

Seddon, A. M., Curnow, P., & Booth, P. J. (2004). Membrane proteins, lipids and detergents: Not just a soap opera. *Biochimica et biophysica Acta, 1666*, 105–117.

Stevens, R. (2000). High-throughput protein crystallization. *Current Opinion in Structural Biology, 10*(1), 558–563.

Tice, J. D., Song, H., Lyon, A. D., & Ismagilov, R. F. (2003). Formation of droplets and mixing in multiphase microfluidics at low values of the reynolds and the capillary numbers. *Langmuir: The ACS Journal of Surfaces and Colloids, 19*(22), 9127–9133.

Thorsen, T., Maerkl, S. J., & Quake, S. R. (2002). Microfluidic large-scale integration. *Science, 298*(5593), 580–584.

Wiener, M. C. (2004). A pedestrian guide to membrane protein crystallization. *Methods, 34*, 364–372.

Yadav, M. K., Gerdts, C. J., Sanishvili, R., Smith, W. W., Roach, L. R., Ismagilov, R. F., et al. (2005). *In situ* data collection and structure refinement from microcapillary protein crystallization. *Journal of Applied Crystallography, 38*, 900–905.

Yeh, J. I. (2003). A manual nanoscale method for protein crystallization. *Acta Crystallographica. Section D, 59*, 1408–1413.

Zheng, B., Gerdts, J. C., & Ismagilov, R. R. (2005). Using nanoliter plugs in microfluidics to facilitate and understand protein crystallization. *Current Opinion in Structural Biology, 15*, 548–555.

Zheng, B., Spencer, L. R., & Ismagilov, R. F. (2003). Screening of protein crystallization conditions on a microfluidic chip using nanoliter-size droplets. *Journal of the American Chemical Society, 125*(37), 11170–11171.

CHAPTER 9

Crystallization of Photosynthetic Membrane Proteins

Petra Fromme and Ingo Grotjohann

Department of Chemistry and Biochemistry, Arizona State University, Tempe, Arizona 85287-1604

Current Topics in Membranes, Volume 63 1063-5823/09 $35.00

DOI: 10.1016/S1063-5823(09)63009-X

I. OVERVIEW

Photosynthesis is one of the most important processes on earth as it converts light energy into chemical energy. The first and most important steps of this process are catalyzed by large membrane protein complexes that are involved in light capturing, energy transfer, and electron transfer. In the past 15 years, great progress has been made towards the crystallization and structure determination of the antenna complexes and photoreaction centers, which are the most complex membrane proteins that have been crystallized to date. This chapter summarizes the important factors that have been critical for the successful crystallization of the large protein-cofactor complexes, with a special focus on the lessons learned from the crystallization and structure determination of Photosystems I and II (PSI and PSII, respectively).

II. OVERVIEW OF CRYSTALLIZATION OF PHOTOSYNTHETIC MEMBRANE PROTEINS

Photosynthetic membrane proteins have long been an active focus of research. This was partially due to the fact that these proteins are responsible for the production of all oxygen in our atmosphere. Photosynthetic organisms are the primary producers of nearly all the biomass and bioenergy produced on Earth. From a practical point of view, they are abundant in their source organisms, and their pigment content facilitates tracking them in their native form in experiments. Because of its high importance, it comes as no surprise that the first membrane protein crystallized was a photosynthetic protein complex, the Photosynthetic Reaction Center of *Rhodopseudomonas viridis* (Deisenhofer, Epp, Miki, Huber, & Michel, 1985; Deisenhofer, Epp, Sinning, & Michel, 1995).

In the last 10 years more structures of photosynthetic membrane proteins have been determined by X-ray crystallography, and each of these structures was based on a result of long-time efforts of the groups that worked often for more than 10 years on these projects. The structures are summarized in Table I. The table shows that closely related proteins often (but not always) can be crystallized with similar techniques. The bacteriorhodopsin from the

TABLE I

Overview of the So Far Crystallized Photosynthetic Membrane Proteins and Their Crystallization Conditions

Protein complex	Organism	Crystallization technique	Detergents	Precipitants	Additives	PDB code
Photosynthetic Reaction Center	*Rhodopseudomonas viridis*	Vapor diffusion (Hanging Drop)	*N,N*-Dimethyl-dodecylamine-N-oxide (LDAO) or *N*-Octyl-*β*-d-glucopyranoside (OG)	Ammonium sulfate	1,2,3-Heptanetriol	1prc
	Rhodobacter sphaeroides	Vapor diffusion		PEG 4000, NaCl	1,2,3-Heptanetriol, benzamidine	1pss, 2rcr, 1aij
		Lipidic cubic phase/batch		Jeffamine M-600, ammonium sulfate	EDTA	1ogv
	Thermochromatium tepidum	Vapor diffusion (sitting drop)		PEG 4000, NaCl		1eys
Photosynthetic Reaction Center/ LHC Lh1	*Rhodopseudomonas palustris*	Vapor diffusion (sitting drop)	Sucrose monocholate	PEG-MME 2000	Spermidine, magnesium chloride	1pyh
LHC Lh-2	*Rhodopseudomonas acidophila*	Vapor diffusion (sitting drop)	*N*-Octyl-*β*-D-glucopyranoside (OG)	Potassium phosphate	Benzamidine	1nkz
	Rhodospirillum molischianum	Vapor diffusion (sitting drop)	*N,N*-Dimethyl-dodecylamine-N-oxide (LDAO)	Ammonium sulfate	1,2,3-heptanetriol	1lgh
LHCII	*Spinacia oleracea*	Vapor diffusion (sitting drop)	*N*-Nonyl-*β*-D-glucopyranoside	Sodium citrate	Deoxy-BigCHAP, DGDG	1rwt
	Pisum sativum	Vapor diffusion (hanging drop)	*N*-Nonyl-*β*-D-glucopyranoside	Sodium chloride, glycerol, PEG-MME 350		2bhw

(*Continued*)

TABLE I (*Continued*)

Protein complex	Organism	Crystallization technique	Detergents	Precipitants	Additives	PDB code
PSI	*Thermosynechococcus elongatus*	Dialysis, micro-/macroseeding	*N*-Dodecyl-β-D-malto-pyranoside (DDM)	(Magnesium sulfate)		2pps, 1jb0
	Pisum sativum	Vapor diffusion (sitting drop)	*N*-Dodecyl-β-D-thiomaltopyranoside	PEG 6000, ammonium citrate		1qzv, 2o01
PSII	*TS elongatus*	Batch	*N*-Dodecyl-β-D-malto-pyranoside (DDM)	PEG 2000	Calcium chloride	1fe1, 2axt
		Vapor diffusion (hanging drop)	*N*-Dodecyl-β-D-malto-pyranoside (DDM), Dodecyloctaoxyethy-lene (C12E8)	PEG 4000	Glycerol	1s5l
	Thermosynechococcus vulcanus	Dialysis	*N*-Dodecyl-β-D-malto-pyranoside (DDM)	PEG 1450	Calcium chloride	1izl
Bacteriorhodopsin	*Halobacterium salinarium*	Lipidic cubic phase/batch	*N*-Octyl-β-D-glucopyra-noside (OG)	Sodium/potassi-um phosphate	2-Methyl pentanediol	1ap9, 1qhj, 1brx, 1c3w
		Vapor diffusion (hanging drop)	*N*-Octyl-β-D-glucopyra-noside (OG)	Sodium phosphate	Benzamidine	1brr
Cytochrome b_6f	*Mastigocladus laminosus*	Vapor diffusion	*N*-Undecyl-β-D-maltopyranoside	PEG 400, mag-nesium chloride	1,2-Dioleyl-*sn*-gly-cero-3-phosphocholine	1vf5, 2d2c, 2e74
	Chlamydomonas rheinhardtii	Vapor diffusion (hanging drop)	*N*-Dodecyl-β-D-malto-pyranoside (DDM)	PEG-MME 350	Glycerol	1q90

The PDB codes in the last column can be used to find the associated publications in the PDB database (http://www.rcsb.org/pdb/home/home.do).

archean *Halobacterium salinarium* from the Dead Sea and other saltwater bodies is one of the most thoroughly studied light-driven proton pumps so far (Belrhali et al., 1999; Essen, Siegert, Lehmann, & Oesterhelt, 1998; Luecke, Richter, & Lanyi, 1998; Luecke, Schobert, Richter, Cartailler, & Lanyi, 1999; Pebay-Peyroula, Rummel, Rosenbusch, & Landau, 1997). As far as photosynthesis is concerned, our insight into this group of proteins is already well established.

For anoxygenic photosynthesis, detailed structures from the purple bacterial reaction centers (pbRCs) from purple nonsulfur (Nogi, Fathir, Kobayashi, Nozawa, & Miki, 2000) and purple sulfur bacteria (Chang, el-Kabbani, Tiede, Norris, & Schiffer, 1991; Katona, Andreasson, Landau, Andreasson, & Neutze, 2003; Roszak et al., 2003; Stowell et al., 1997; Yeates, Komiya, Rees, Allen, & Feher, 1987) as well as their antenna complexes (Koepke, Hu, Muenke, Schulten, & Michel, 1996; Papiz, Prince, Howard, Cogdell, & Isaacs, 2003; Roszak et al., 2003) have been determined, which allow a detailed view on the structure-function relationship of these complexes. The structures of anoxygenic-type I bacterial reaction centers from green sulfur bacteria and heliobacteria still wait to be solved, among these some of the probably most ancient types of reaction centers.

The most complex photosynthetic apparatus can be found in cyanobacteria, algae, and plants, which perform oxygenic photosynthesis and contain two Photosystems. All three major protein complexes, PSI (Amunts, Drory, & Nelson, 2007; Jordan et al., 2001; Schubert et al., 1997; Witt et al., 1992), PSII (Ferreira, Iverson, Maghlaoui, Barber, & Iwata, 2004; Kamiya & Shen, 2003; Loll, Kern, Saenger, Zouni, & Biesiadka, 2005; Zouni et al., 2001), and the cytochrome b_6f complex (Kurisu, Zhang, Smith, & Cramer, 2003; Stroebel, Choquet, Popot, & Picot, 2003; Yamashita, Zhang, & Cramer, 2007; Yan, Kurisu, & Cramer, 2006) have been crystallized and their structures have been determined. The nearly complete set of protein structures is, so far, only known for thermophilic cyanobacteria, and we will discuss why this might be so later in this chapter. The structure of cytochrome b_6f complex from *Chlamydomonas rheinhardtii* (Stroebel et al., 2003) is the only structure of a photosynthetic membrane protein from algae. PSI (Amunts et al., 2007; Ben-Shem, Frolow, & Nelson, 2003) and the main plant antenna complexes (Liu et al., 2004; Standfuss, Terwisscha van Scheltinga, Lamborghini, & Kuhlbrandt, 2005) are the only photosynthetic membrane protein complexes that have been structurally solved from higher plants, so far.

It has been an exciting time of discovery, and much of our insight into photosynthetic processes has been tightly coupled to pioneering work in the crystallization techniques for membrane proteins. The remainder of this chapter will mainly focus on the experience the authors have with the crystallization of PSI and PSII.

III. BIOLOGICAL PARAMETERS THAT INFLUENCE CRYSTALLIZATION

A. Biological Source

Most of the photosynthetic membrane proteins that have been successfully crystallized to date are large multiprotein-cofactor complexes that have been isolated from the natural source, cyanobacteria, algae, or plants. Because of the complexity of the protein-cofactor interaction it was anticipated that these complex proteins cannot be heterologously expressed. However, there is one example of successful reconstitution of a heterologously overexpressed photosynthetic membrane protein: the light-harvesting complex II (LHCII) from spinach (Standfuss et al., 2005). The overexpressed LHCII proteins were folded in the presence of specific carotenoids, as well as chlorophylls and the structural and functional identity of the reconstituted protein with the native complex was proven by extensive spectroscopic studies (Pascal, Gastaldelli, Ceoldo, Bassi, & Robert, 2001). The structure of the LHCII was determined by Kuhlbrand and coworkers (Barros, Royant, Standfuss, Dreuw, & Kuhlbrandt, 2009; Standfuss et al., 2005) and very closely resembles the structure of the LHII isolated from pea determined by Wang and coworkers (Liu et al., 2004). The success of the crystallization of the LHCII proves that complex membrane proteins with cofactors can be reconstituted and crystallized. However, most of the other complexes (PSI, PSII, and the Cyt b_6f complex) are much more complex, undergoing posttranslational modification and requiring multiple proteins for assembly. As a result, thus far, these proteins have only been available via isolation from natural sources.

The use of thermophilic cyanobacteria provided the first breakthrough in the attempts to crystallize PSI and PSII (Fromme & Witt, 1998; Zouni, Jordan, Schlodder). Despite the fact that crystals of PSII from higher plants have been reported (Adir, 1999), the only structures of PSII that have been determined so far are from thermophilic cyanobacteria (*Thermosynechococcus (TS) elongatus* and *TS vulcanus*) (Ferreira et al., 2004; Kamiya & Shen, 2003; Loll et al., 2005; Zouni et al., 2001). The use of thermophilic proteins for crystallization is not limited to photosynthetic membrane proteins, but has also been successfully used for the crystallization of the soluble domain of complex I from the respiratory chain (Kohlstadt et al., 2008) and many soluble proteins (there are currently 222 structures of thermophilic proteins deposited in the Protein Data Base (PDB)). The use of a thermophilic model organism has the major advantage that the organism's proteins are more stable than their mesophilic counterparts. An additional positive aspect, which has often been overseen, is that the native proteases that are present in the thermophilic cells work very slowly at low temperatures. Thereby, proteolytic digestion of the proteins during cell disruption and membrane preparation can be minimized. The crystal structures of the cyanobacterial b_6f complex was determined from a thermophilic cyanobacterium (*Mastigocladus laminosus*) (Kurisu et al., 2003) and from the green algae *C. rheinhardtii* (Stroebel et al., 2003). Thus far, the structure of the b_6f complex from *Chlamydomonas* is the only membrane protein structure from any red or green algae, and the only membrane protein structure from an oxygenic photosynthetic organism that grows photoheterotrophically and for which a mutagenesis system has been established. A serious obstacle for the study of the structure-function relationship in photosynthetic membrane proteins is that no membrane protein crystal structures have been solved from those photosynthetic model organisms that are most commonly used in mutagenesis experiments, *Synechocystis PCC 6803* and *Arabidopsis*.

In a heroic effort, Nelson and coworkers have achieved the crystallization of the plant PSI-LHCI complex from pea and the first structure at 4.5 Å resolution (Ben-Shem et al., 2003) has been recently improved to 3.5 Å (Amunts et al., 2007). Spinach and pea are also the sources for the two crystal structures of the LHCII (Liu et al., 2004; Standfuss et al., 2005). These results show that successful isolation and crystallization of large membrane proteins from plants is possible.

Different purple bacteria have been widely used for the crystallization of the pbRC and the LH2 and LH1 complexes (Koepke et al., 1996; Papiz et al., 2003; Roszak et al., 2003). It is remarkable that no crystal structure has yet been determined from any anoxygenic bacterium that contains one of the ancient homodimeric-type I reaction centers. The major problems are the oxygen sensitivity of the organisms and isolated proteins as well as the instability of the protein complexes.

B. Physiological State of the Organism

While the choice of model organism is important for the success of crystallization experiments, the physiological status of the organisms is critical for isolation and crystallization of intact homogenous membrane protein complexes. Intact and homogenous proteins can only be isolated from healthy cells in the exponential growth phase. Of all the different parameters, light is the most critical factor for the crystallization as it triggers the formation and disassembly of PSI and PSII-antenna supercomplexes and the photodamage and repair of PSII. We discuss here in more detail two examples from our experience with the crystallization of PSI and PSII from the thermophilic cyanobacterium, *TS elongatus*. PSI trimers can only be successfully isolated and crystallized from low-light adapted cells. Under these conditions the cells contain a PSI:PSII ratio of 6:1 and all PSI is trimeric and stable. The whole cells show an absorption maximum of 681 nm. When the light intensity is increased, the PSI trimers are degraded and the PSI:PSII ratio is decreased to (2–3):1. Under these conditions, most of the PSI is monomeric. The remaining PSI trimers show extensive proteolytic degradation of PsaD and PsaB and are useless for crystallization experiments. Cells grown under medium light contain more PSII/chlorophyll and, thereby, a higher overall yield of PSII can be achieved from these cells. However, due to higher light intensity, more of the PSII is in the process of repair and the cells contain significantly higher amounts of monomeric PSII in different stages of assembly. The best sources for active, dimeric homogenous PSII from *TS elongatus* are, therefore, also protein preparations from low-light adapted cells. Despite the fact that the oxygen evolution/chlorophyll is much lower in these cells due to the high PSI content, PSII preparations from these cells are superior to the PSII preparations from cells grown at higher light intensities.

C. Oligomeric State

The majority of large membrane protein complexes involved in photosynthesis form oligomers or supercomplexes, which are the active forms of the proteins *in vivo*. In the few cases where a monomer is the dominant oligomeric state in the membrane, the protein forms supercomplexes with antenna proteins. Examples include the plant PSI which forms a supercomplex with four LHCI proteins, or the PbRC which is located in the center of a large pbRC-LH1 complex, where 15 LH1 subunits surround one PbRC (Roszak et al., 2003).

It is very important to maintain the native oligomeric state during the isolation and crystallization, as only these native oligomeric states are stable and can be crystallized. PSII and the cytochrome b_6f complex are dimers in all photosynthetic organisms and there is no report of any crystals of the PSII or the b_6f monomer. The b_6f complex is an obligate functional dimer with a domain swap of the Rieske FeS protein and is only functional in dimeric form. In the case of PSII, monomers are still active and show oxygen evolving activity but are less stable and more heterogeneous. All attempts to crystallize any monomeric PSII have failed due to this heterogeneity and instability of the PSII monomers.

PSI and the LHCII complex are trimers and a monomerization can be observed under special environmental conditions. For PSI, crystals of monomers have been reported (Jekow, Fromme, Witt, & Saenger, 1995;

Jekow et al., 1996), but they never produced diffraction patterns better than 8 Å resolution. This is very likely due to the fact that the monomer is less stable than the trimer and subunits PsaL, I, and M are quickly lost, leading to heterogeneity of the monomer preparation. Plant PSI is a monomer *in vivo* and forms a supercomplex with four LHCI proteins. The stability of the complex and the growth conditions of the pea plants were critical for successful crystallization (Amunts et al., 2007; Ben-Shem et al., 2003). The LH2 complex from purple bacteria is remarkable as it shows different stoichiometry in different organisms (Cogdell et al., 2003; Koepke et al., 1996).

For all proteins which form native oligomers, it is most important for successful crystallization that the protein preparation be homogenous, containing only the desired oligomeric state and no monomers. Monomers of the same protein are the worst contamination that can be present in the protein preparation. They are much more harmful than a severe contamination with an unrelated protein. Foreign proteins can be relatively easily removed using crystallization as a last purification step, while monomers cannot be excluded. This is due to the fact that the monomers form the same contacts as the intact oligomer but crystal growth stops at the position where the monomer is incorporated. Monomers are not amenable for crystal growth, in cases where the oligomerization axis is a crystallographic axis. This occurs for PSI from *TS elongatus*, where the trimeric axis is also the threefold crystallographic axis. In this case, one monomer per 10,000 trimers hinders the successful crystallization of the PSI trimer.

Another important point that has to be accounted for is the change of the protein:detergent ratio in different oligomeric forms of a protein. The detergent surrounds the membrane protein like a swim ring and the ratio of detergent:protein decreases with an increase in the diameter, or size, of the oligomer or supercomplex. This leads to a decrease in solubility of the oligomer in comparison to monomer. In the case of PSI from *TS elongatus* the effect is dramatic—while the PSI trimer has a solubility of less than 1 μg/ml at low ionic strength (5 mM MES, pH 6.4, 0.02% β-dodecylmaltoside), the PSI monomer has a solubility of more than 100 mg/ml at the same ionic strength and therefore cannot be crystallized at low ionic strength.

D. Presence or Absence of Lipids

The reputation of lipids in the field of membrane protein crystallization has dramatically changed. While early protocols for membrane protein crystallization have recommended the complete delipidation of the protein, lipids have more recently been shown to be important for the structure and stability of most membrane proteins. Lipids have been identified in all structures of photosynthetic membrane proteins and they seem to play an important role in the function of the complexes. In PSI, PSII, and the b_6f complex, the lipids are essential parts of the protein and are located inside the protein complex, not at the detergent layer (Jordan et al., 2001; Kurisu et al., 2003; Loll et al., 2005). The addition of lipids was one of the critical elements in the crystallization of the b_6f complex from cyanobacteria (Zhang, Kurisu, Smith, & Cramer, 2003).

Lipids may have multiple functions, but in all complexes they have important structural roles, as the complexes fall apart in the absence of the lipids. In addition to their structural role, lipids have been proposed to play a role in the speed of the electron transfer along the two electron transfer branches in PSI (Jordan et al., 2001). In the b_6f complex, lipids may play a role in the Q-cycle and they may modulate the quinone binding sites (Kurisu et al., 2003). PSII is very special as it has, with 14 lipids, the highest lipid content of all membrane proteins for which structures have been determined (Loll et al., 2005). The lipids form clusters within the protein that are located at the monomer-monomer interface, the Q_B quinone binding site and at the interface of the D1 protein with CP43 (Loll et al., 2005). The high lipid content is very likely essential for the repair cycle of PSII, where the D1 protein is damaged and therefore, in the *in vivo* system it must be replaced every 30 min in bright sunlight (Aro, 1999).

A completely new type of membrane protein crystal (Type III) has been grown from the LHCII of spinach by the group of Wang and coworkers (Liu et al., 2004), where a lipid-protein vesicle forms the building block of the crystals. This new crystal type is very promising as it maintains the proteins in their native lipid environment. It will be very interesting to see if this method will become more popular for crystallization of membrane proteins.

Another unique method for membrane protein crystallization involving lipids is the crystallization in lipidic cubic phases, invented by Landau and Rosenbusch (Chiu et al., 2000). Bacteriorhodopsin and the bacterial reaction center are the photosynthetic membrane proteins that have been successfully crystallized using this method (Misquitta et al., 2004; Nollert, Navarro, & Landau, 2002). The reader is referred to Chapter 4 of this book for a detailed description of the crystallization in lipidic cubic phases.

E. Reduction or Oxidation

Reaction centers in photosynthesis catalyze light-driven electron transport across membranes. The reduction and oxidation state of the electron transport chain is important for the cocrystallization of the membrane protein complexes with their soluble electron carriers. The first cocrystallization of the PbRC with cytochrome *c* was reported by Adir, Okamura, and Feher, (1994) and the initial crystals were further improved to a resolution of 2.4 Å (Axelrod et al., 2002). Furthermore, PSI has been cocrystallized with ferredoxin (Fromme, Bottin, Krauss, & Setif, 2002). The crystals have recently been improved to 3.5 Å resolution and the publication of the structure is in progress (Yu et al., in preparation). The redox state of the two binding partners, the large membrane protein complex and the soluble electron transfer protein, are critical for successful cocrystallization. In the case of the PbRC-Cyt *c* crystals, complex formation is only observed between reduced cytochrome *c* and PbRC. In the case of PSI-ferredoxin cocrystals, the complex is only formed between oxidized ferredoxin and PSI.

The cocrystals of PSI with ferredoxin must be grown, mounted, and frozen under very dim green light as the crystals dissolve in less than 5 min upon illumination with white light. The instability of the crystals follows the natural function of ferredoxin as an electron carrier. PSI is fully functional in the crystals; thereby, illumination leads to electron transfer to the bound ferredoxin which is reduced and leaves the binding site. The crystal structure shows that ferredoxin is involved in crystal contacts. Therefore, dissociation of ferredoxin leads to dissolution of these crystals.

F. Posttranslational Modifications

Photosynthetic membrane proteins undergo extensive posttranslational modifications as cofactor binding, cleavage of signal peptides, and phosphorylation. We discuss here PSII as an example. PSII contains 19 proteins and more than 50 cofactors. The assembly of the oxygen evolving Mn_4Ca cluster in PSII involves photo-induced stepwise binding of the four Mn atoms and the Ca; the cleavage of the C-terminal end of the D1 protein is involved in the assembly process. Multiple proteins are involved in the process as assembly-disassembly of PSII during photoinhibition and repair. The PSII assembly-disassembly is highly dynamic and occurs in the cell at a high rate. As an estimate, even under low light, about 30% of PSII in cells is under reconstruction in different stages of assembly-disassembly, a situation that can be problematic for crystallization efforts. The solution to this dilemma involved use of crystallization as the last purification step, which was first developed for the crystallization of PSI (Fromme & Witt, 1998) and then adapted for the PSII crystallization. Only the active intact PSII dimer is precipitated, while all the partially assembled PSII complexes stay soluble. Threefold precipitation/crystallization led to large numbers of small very pure PSII crystals that were dissolved for the final crystallization experiments using the batch method (these initial crystals were also used as seed crystals using the microseeding technique described later in this chapter).

G. Sequence Heterogeneity

Sequence heterogeneity is a problem for the crystallization of several photosynthetic membrane proteins, especially PSII and the LHCII. All cyanobacteria contain multiple genes for the core protein of PSII (the D1 protein) which are expressed under different environmental conditions. When PSII is isolated from the natural organism, it therefore may contain a mixture of the different gene products. This might be one of the reasons for the limitation of the X-ray resolution of the current PSII crystals (~3 Å). Recently, Loll et al. (2008) have modeled the structures of the individual D1 proteins and found major differences. This problem may in the future be solved by silencing or deleting of all but one of the multiple gene copies.

In the case of the LHCII, several gene copies exist and it has been shown that the LHCII complex, when isolated from the natural sources, reflects these sequence heterogeneities.

This problem was one of the major reasons why the group of Kühlbrand worked so diligently on the heterologous expression, reconstitution and crystallization of the recombinant LHCII. Their recombinant expression approach eventually resulted in the first structural study of a mutant complex with altered subunit composition (Barros et al., 2009).

IV. PHYSICAL-CHEMICAL PARAMETERS THAT INFLUENCE CRYSTALLIZATION

A. Supersaturation

Now that we have looked at the biological parameters that are important to produce a membrane protein preparation suitable for crystallization experiments, we will examine the physical-chemical parameters that influence the solubility of the protein during the crystallization phase. Initially, the protein is in a defined oligomeric state in a monodisperse solution, and we want to gently force it out of solution into the ordered state of a crystal. The different states of protein solubility are best described by a phase diagram. Figure 1A shows an example of such a phase diagram as it can be found in the literature [for a recent example see (Chayen & Saridakis, 2008)]. The first-phase border (the saturation line), depicted on the left side of Fig. 1A, separates the unsaturated zone from the supersaturated zone.

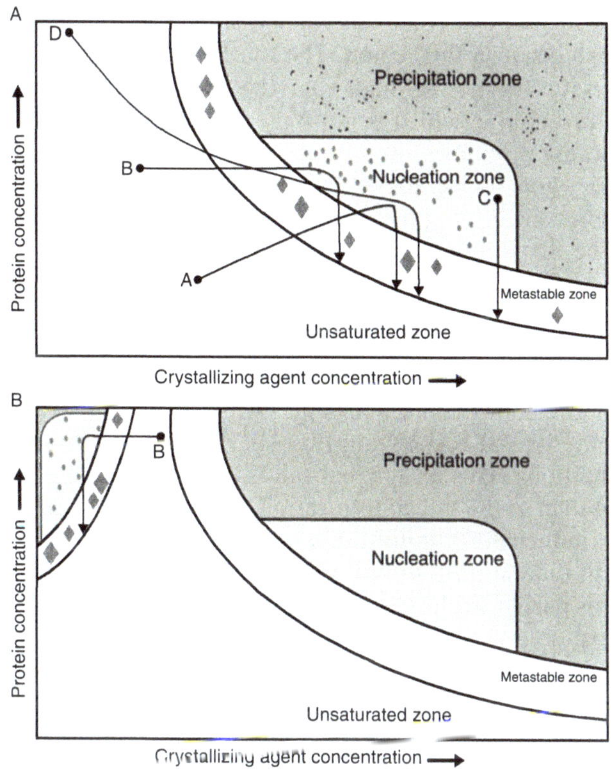

FIGURE 1 Phase diagrams of different crystallization techniques. (A) Idealized phase diagram that describes the crystallization behavior of most proteins (modified after Chayen & Saridakis, 2008). The unsaturated zone does not allow for crystal growth. The zone of supersaturation is subdivided into the metastable zone, which does not allow for spontaneous crystal growth but yields the best crystals; the nucleation zone, where spontaneous crystal growth originates; and the precipitation zone, which yields amorphous precipitate. The ways through the phase diagram is shown for four crystallization techniques: (A) vapor diffusion, (B) dialysis, (C) batch, and (D) free interphase diffusion. For details see Section IV. (B) The crystallization of PSI shows that phase diagrams look often more complicated than the idealized image in Fig. 1A. PSI crystallizes under low-salt conditions. Zones are color-matched to the descriptions on the right side. Part (B) shows what happens during crystal formation during PSI dialysis against low-salt buffers.

We must transition the protein solution into the supersaturated area to form or grow crystals. The supersaturated zone is subdivided into three further zones: The metastable zone allows for moderate slow growth of preformed crystals, but crystals will not spontaneously begin growing there because the protein molecules must overcome an activation energy barrier first and the supersaturation is not yet high enough to form stable crystal seeds which exceed the critical radius. In the nucleation zone, the protein supersaturation is sufficiently high to overcome this barrier, and spontaneous nucleation and crystal growth occur in this region. The third area is the precipitation zone, where supersaturation is so high that the protein leaves the solution as amorphous precipitate, which is not only unsuitable for crystallization but it may also cause protein denaturation.

The primary goal during the initial crystallization experiments is to reach the nucleation zone of the phase diagram thereby resulting in nucleation or growth of the first crystals. To produce improved crystals (higher quality and/or size), slow crystal growth in the metastable zone is desirable. This can be achieved by lowering the activation energy barrier by the use of seeding techniques, as discussed later in this chapter.

It is important to know that phase diagrams of membrane proteins can look much more complicated than Fig. 1A. As an example, a model of the phase diagram for PSI is shown in Fig. 1B. PSI crystallizes under low ionic strength conditions. This means that salt is required to retain the protein in solution and then as the salt concentration is lowered to the protein becomes less soluble, inducing crystallization. Most proteins are less soluble at low ionic strength than at medium salt concentration (see also Chapter 1 of this book) but this part of the phase diagram is rarely used for crystallization as it cannot be explored using vapor diffusion. With the presence of detergents, lipids, or other amphiphiles in a crystallization setup, phase diagrams can be more complicated (Nollert, 2005). In normal vapor diffusion setups of membrane proteins, a separation of detergent-rich phases from phases that are low in detergent is an often observed phenomenon (see Chapter 6 of this book for more details).

B. Temperature

The crystallization of membrane proteins is also a temperature-dependent process. The solubility of most proteins increases with temperature, although the opposite can be found with some of them. On the other hand, proteins might denature at higher temperatures, which means that a close observation of the temperature stability of the protein is advised.

The temperature dependency of protein crystallization is often not fully investigated, which means that the available data of successful crystallizations tend to be biased towards temperature ranges that are common in laboratories. If you look at the Membrane Protein Database (http://www.mpdb.ul.ie/) for proteins that crystallize at temperatures up to 30 °C, you find two clusters at normal refrigerator temperatures (0–-5 °C) and room temperature (21–25 °C), whereas the temperature ranges 6–15 °C or 26–30 °C sport hardly any entries. A large cluster of successful crystallizations can be found between 16 and 20 °C. This is the normal crystallization temperature for polyethylene glycols (PEGs), which tend to get too viscous at lower temperatures. This is also the temperature where PSII crystallizes best: its optimal crystallization temperature lies between 17 and 18 °C.

The temperature dependency of the PSI crystallization has been investigated in detail between 4 and 52 °C (Fromme, 2003). Small crystals of PSI were heated and the concentration of PSI in the supernatant was determined (for more details see section on phase diagrams below), and the protein was cooled down again in order to look for reversibility. These experiments showed that the solubility of PSI depends only weakly on temperature, but the protein started to show signs of denaturing above 43 °C. Comparison of crystal growth at different temperatures reveals that crystallization at 4 °C results in the most highly ordered crystals. One important point to keep in mind is that it is not necessarily only the protein that is responsible for the fast deterioration of protein crystals at higher temperatures, but that also the detergent might be unstable at higher temperatures. *N*-Dodecyl-*β*-D-maltopyranoside (DDM), the detergent used in PSI crystallization, tends to hydrolyze relatively quickly at elevated temperatures. The hydrolysis of the detergent is the main factor in the fast decrease of the diffraction quality of PSI crystals and occurs at a fast rate of approximately 1 Å/week upon storage of crystals.

C. *Velocity of Equilibrium*

If crystallization experiments lead to large crystals with visible defects and poor diffraction quality, one of the issues might be that they grew too fast. Slowing down the velocity of crystal growth may lead to crystals with improved order and diffraction quality. PSI can be used as an example for defects caused by fast crystal growth. Adding a single seed to a highly supersaturated solution of PSI, can induce the growth of 2 mm crystals of PSI at room temperature over a lunch break, which show indication of defects caused by too fast growth: they are hollow and look like rockets with multiple spikes at one end. These crystals have an extremely high mosaicity of 5–10° and cannot be used for X-ray structure analysis. The temperature experiments mentioned above showed that crystals grown at 4 °C in the course of 4 days diffracted best, and one possible explanation for this effect might be a slower and more ordered growth of the crystals.

Of course, we might face a conundrum here, as proteins are often not very stable, as it is the case for PSII. Most of the photosynthetic membrane proteins have been crystallized very fast, overnight or in less than a week, which is sometimes the only possible compromise between the needs of better crystal order and protein stability.

D. Diffusion and Convection

Diffusion and convection influence the crystallization of proteins in several ways. Crystals do not hover free in solution but often sink down to the bottom of the crystallization vessel. Thereby, the different faces of the crystals grow with different velocities due to the diffusion limitations and interactions with the walls of the crystallization vessel or the air-water interface (in the case hanging drop is used for crystallization). Convection is an additional major factor that cannot be easily excluded on earth. It leads to a constant mixing of the solution. Thereby, the crystal growth is not purely diffusion-controlled, but convection leads to a rapid mixing of the solution. Fresh protein is brought by convection and diffusion to the protein depletion zone surrounding the growing crystal. Crystallization in highly viscous solutions or in gels has been successfully used for smaller proteins to minimize these effects and allow a slower more controlled growth of crystals (Cudney, Patel, & Mcpherson, 1994). However, this method has not been successfully applied to the crystallization of large membrane protein complexes. The likely reason is that the diffusion constant for large proteins in gels or high-viscosity solutions is so slow that no nucleation occurs and the production of seeds is slowed down dramatically. We want to show here the example of the PSI crystallization in sucrose. PSI is the largest membrane protein crystallized to date with a MW of 1,056,000 Da. It has a MW of 1,300,000 Da, including the detergent micelle. The protein crystals have solvent a content of 78% and are frozen in 2 M sucrose, thereby the growth of PSI crystals in sucrose was highly desired to minimize mechanical stress during incubation of crystal in the cryoprotectant. PSI crystals have a density which is equal to 1.4 M sucrose, so they hover free in solution under these conditions and we have attempted to crystallize PSI by dialysis in 1.4 M sucrose. The results of the experiments showed that nucleation could be achieved only at sucrose concentrations below 0.8 M. Seeding crystals grow slowly in solution concentrations as high as 1.1 M sucrose, but above this limit no crystal growth is observed, even if the determination of the phase diagram has shown that the solution is highly supersaturated. The crystallization in 0.75 M sucrose in combination with micro- and macroseeding was very successful and finally led to well-ordered crystals that diffracted X-rays to 2.5 Å resolution (Jordan et al., 2001). The use of sucrose provided an advantage in that the velocity of the equilibrium of the salt in the dialysis chambers (measured by conductivity measurements) was reduced by a factor of 3, and the diffusion rate of PSI was decreased, leading to slower growth of better ordered single crystals. Furthermore, the crystals suffered less from stepwise incubation in the cryo-solution, as they already contained 0.75 M sucrose. Because of these improvements, they showed less increase in mosaicity during freezing and diffracted to higher resolution than the crystals grown in the absence of sucrose.

E. Crystallization Under Microgravity

The improvements of crystals that are grown at higher viscosity or in gels are impressive; however, the optimal conditions for crystal growth can be established under microgravity. There are great benefits from crystallization under microgravity: the crystal growth is purely diffusion controlled, the nucleation rate is lowered, sedimentation does not interfere with crystal growth and the absence of convection allows for the undisturbed growth of the single crystals. However, microgravity crystallization experiments are controversial and have been criticized by the public and the scientific community; with the major points of criticism being the large costs and the difficulties to prove that breakthroughs in the diffraction quality can be achieved using this high-tech effort. A detailed discussion of these points is beyond the scope of this book chapter, but we want to briefly summarize promising results that have been achieved for the crystallization of PSI under microgravity. PSI was crystallized under microgravity using the Advanced Protein Crystallization Facility (APCF) during three missions: USML-2, LMS, and STS107. PSI crystals were successfully grown during the USML-2 and the LMS mission with very promising results. At the time of the USML 2 mission in 1995, the best crystals grown on earth were 0.5 mm long, 0.2 mm in diameter, and diffracted to a maximal resolution of 5 Å. This was also the quality and size of the ground control crystals of the mission, grown on earth. The crystals grown under microgravity were up to 20 times larger than all crystals that had been grown on earth previously (Fig. 2). The largest crystal was 4 mm long and 1.5 mm in diameter. It diffracted to 3.2 Å and we were able to collect a complete native data set at 4 °C at the Synchrotron DESY in Hamburg on this crystal. The PSI crystals decay very rapidly in the X ray beam, but we were able to shift this crystal after each image during data collection and collect 200 images on this single crystal at 4 °C. The data set could be evaluated to 3.4 Å resolution and is still the only complete PSI data set that have ever been collected on one single nonfrozen crystal of PSI. This native data set from a microgravity grown crystal formed the basis for the improved crystal structure of PSI at 4 Å resolution (Fromme 1998; Klukas et al., 1999a,b).

At the LMS mission, we explored the combination of microgravity with growth of crystals in sucrose and determined that the nucleation and growth rates are decreased by a factor of 10 under microgravity. The last mission in 2002 was the STS107 mission, in which we had planned to combine all the

FIGURE 2 Large single crystal of PSI, grown in the space shuttle Columbia during the USML-2 Mission, the crystals was 4 mm in length and 1.5 mm in diameter, it is mounted in an X-ray capillary; the bottom a ruler (one bar = 1 cm) is shown. This crystal was used to collect the only complete native data set at room temperature from a single PSI crystals at DESY. The data was evaluated to 3.4 Å; at that time best earth-grown crystals diffracted to 5 Å (Fromme, 2003).

experience of PSI crystallization on earth, including macro- and microseeding techniques, with the optimal growth environment under microgravity with the aim to grow crystal that may diffract better than 2 Å. Everything went well until the reentry of the space shuttle, were it exploded over Texas and the life of all astronauts on board and all experiments on the shuttle were lost. We have developed in the meantime new crystallization chambers which would allow the automated use of seeding and we would like to combine these chambers with the online surveillance of the experiments on earth. The new science laboratories at the International Space Station (ISS) would be perfectly suited for new sets of crystallization experiments; however, the current manpower of two astronauts at the ISS does not allow them to conduct any scientific experiments. The future of crystallization under microgravity is therefore limited by factors other than science and crystallization efforts must be currently focused on improvement of crystallization techniques and strategies for rational design crystallization on earth.

F. Mode of Nucleation

The mode of nucleation also strongly influences the crystallization of proteins. When no seeds are present, crystals are only formed when the solution is in the nucleation zone. For large molecules, often very high supersaturation levels are needed to reach the nucleation zone, often leading to a shower of small crystals or the growth of large crystals with severe growth defects, which are hollow and look like teeth. A solution to this problem is the use of seeding techniques. Regarding seeding techniques, one can discriminate between two different modes: heterogeneous nucleation and homogenous nucleation. Heterogeneous nucleation includes the induction of nucleation by addition of seeds that do not consist of the protein to be crystallized. With PSI, we were able to induce nucleation by hair (human hair, cat-whiskers are not necessary), wool, and dodecanol fibers. PSI crystallization can also be successfully induced in a homogenous electric field, with all crystals grown on the positive electrode. The best crystals, however, were grown using homogenous nucleation using small well defined PSI crystals as seeds using the micro- and macroseeding techniques described in the following section.

V. CRYSTALLIZATION TECHNIQUES

This chapter only briefly mentions the main crystallization techniques; Chapter 7 of this book describes the individual techniques in more detail.

Figure 1A shows the pathways of the different crystallization regimes in the phase diagram for the first four methods mentioned here.

A. Vapor Diffusion

The technique of vapor diffusion is the most commonly used method for protein crystallization (trace A in Fig. 1A). A droplet of a protein solution is mixed with a droplet of precipitant solution from a reservoir and placed in a tightly closed compartment with the reservoir solution. As the reservoir solution typically has a higher osmotic pressure than the protein containing droplet, water evaporates from the protein containing droplets until equilibrium is reached, that is, until the concentration of solutes in the protein droplet and the reservoir is equal. Crystals may grow, if the protein solution reaches the nucleation zone. Table I shows that vapor diffusion has been established as a highly successful technique for crystallization of photosynthetic membrane proteins. This method has also been used to grow crystals of the majority of membrane protein structures determined thus far.

The popularity of the method might have to do with the fact that it is easy to use and can be set up even in a laboratory that is not specialized in crystallization. The use of the sitting or hanging drop is mostly based on user preference as the two vapor diffusion methods are fundamentally equivalent. The sitting drop is slightly easier to handle, as the protein (crystal) does not hang from an inverted coverslip. Furthermore, sitting drop experiments accommodate crystallization robotics more easily than do hanging drop experiments, albeit one robotic system, the Mosquito-crystallization robot, can also be used to prepare hanging drops.

A slight disadvantage of the sitting drop technique for membrane protein crystallization is that the crystals often are so tightly attached to the plastic surfaces that they cannot be harvested. We have grown beautiful large crystals of the PSI-ferredoxin complex using the sitting drop technique, but all crystals were so tightly attached to the plastic surface that we were not able to harvest and mount any crystals from these experiments. Hanging drop experiments have the advantage that the majority of the crystals grow at bottom of the drop, which corresponds to the water-air interface, and can therefore be easily recovered. It should be noted that siliconized glass surfaces have a much lower "sticking"-tendency for membrane protein crystals than do plastic surfaces.

B. Dialysis

The dialysis method of crystallization, like vapor diffusion, involves the exchange of precipitant during the time course of the experiment. However, the pathway through the phase diagram differs between the two methods (Fig. 1). The protein solution and the reservoir are separated by a semipermeable membrane, which allows free diffusion of molecules that are smaller than the exclusion limit of the dialysis membrane. Thus, the protein and detergent concentrations remain constant during the time course of the experiment. The exception is the use of nonpermeable precipitants (e.g., higher molecular weight PEG) (see Chapter 7 for more details on the limits and exceptions of the use of PEG for dialysis experiments).

In a dialysis experiment, equilibrium is reached and the precipitant concentration changed without noticeable influence of the sample volume (traces B in both, Fig. 1A and B). In the case of PSI, which crystallizes under low ionic strength conditions, dialysis is the only possible technique for crystallization, as this prevents dilution of the high protein concentration required for crystallization (Witt et al., 1992).

The only commercially available crystallization reactors for dialysis are dialysis cells called "buttons." These buttons are difficult to handle and are often prone to leak protein solution at the button membrane. These difficulties might be one of the reasons why dialysis is not commonly used for crystallization as better and easier crystal handling is desirable.

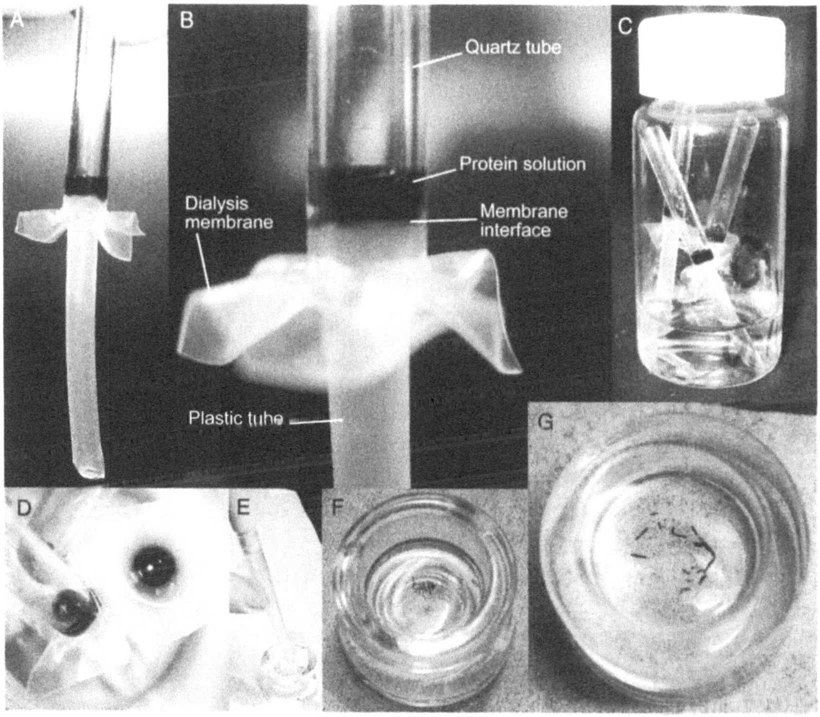

FIGURE 3 The dialysis crystallization setup. (A, B) Crystallization vessel for the dialysis method. A plastic tube with 3 mm outer diameter is inserted into a quartz tube with 3 mm inner diameter. Sandwiched in between, a dialysis membrane with 16 kDa exclusion size keeps the 4 μl protein in the quartz tube. (C) The plastic tube is filled with buffer and connects to the 4 ml reservoir in a scintillation vial. (D) Micro- and macroseeding take place through the upper opening of the quartz tube, which can be seen here. (E) Crystals are handled with 1-ml pipette tips. (F, G) Crystals for macroseeding in glass vessels. The crystals are up to 2.5 mm in length.

We will therefore describe here in more detail the setup for the crystallization of PSI by dialysis, which is routinely used in our lab, and which is simple, cheap, reliable, and easy to handle and assemble. The dialysis reactors that are used for crystallization of PSI are shown in Fig. 3A and B.

The top parts of the reactors, which harbor the protein solution, consist of quartz tubes with an inner diameter of 3 mm and a length of 3 cm. We normally purchase several long quartz tubes with an inner diameter of 3 mm in bulk. They are cut into pieces of 3 cm in length and are cautiously fire-polished at the ends. They can be easily cleaned, sterilized and reused for several years. The bottom part of the reactor consists of polyethylene tubes with an outer diameter of 3 mm. These are standard "hard" tubes that are used for FLPC, Fast Protein Liquid Chromatography. (A note: "standard" refers to the European metric system; it can be sometimes difficult in the US to purchase metric tubes; however, the only important point is that the outer diameter of the plastic tubes matches the inner diameter of the quartz tube, thereby the setup can also be made on an inch-based system.) The polyethylene tubes are cut into pieces of 2 cm length; one end is blunt, while the other end is beveled. The reactors are assembled by sandwiching a 1 × 1 cm piece of dialysis membrane (12–14 kDa size exclusion limits) between the quartz tube and the blunt end of the polyethylene tube. The polyethylene tube is positioned 1–2 mm into the quartz tube, thereby sealing the setup. The plastic tube is completely filled with buffer that resembles the sample buffer of the protein by use of a Hamilton syringe with blunt ends. The tip of the syringe directly touches the membrane during filling; thereby, avoiding the creation of air bubbles. Twenty microliters of the same buffer is delivered into the quartz tube to keep the membrane moist. The tightness of the setup can be easily checked: a flexible plastic tube with a 5 ml plastic syringe is attached to the top part of the quartz tube and gentle under-pressure is applied. A small leak is immediately visible as the solution moves up into the quartz tube. The assembled reactors are placed in 20-ml scintillation vials containing 4 ml of the sample buffer (Fig. 3C), and are ready to use or can be stored for weeks. When large numbers of reactors are assembled for storage and use with various proteins, they are assembled with water, which is replaced by the appropriate buffer by use of the Hamilton syringe before initiation of individual dialysis experiments. After equilibration with dialysis buffer, the quartz tube is filled with 4 μl of protein solution, as seen in Fig. 3A and B using gel-loading pipette tips. The dialysis is started by placing the reactors into the reaction vessel with 4 ml of the desired precipitant solution. It is important that the solution in the plastic tube corresponds to the protein buffer and does not already contain the final concentration of the precipitant. It functions as a "buffer" reservoir that allows slow diffusion of the precipitant into the protein chamber. This buffer reservoir is entirely missing when standard dialysis buttons are used. The equilibration of the solution requires 1–3 days, depending on the viscosity of the solution. Seeding techniques can be applied easily and with only minimal disturbance of the equilibration of solutions. Seeding takes place with a seeding loop made from hair, which is inserted from the top into the quartz tube. The loop is made from human, clean hair that is glued into a Pasteur pipette. Figure 3f and g show seeding crystals that have been grown by dialysis with microseeding.

C. Batch

The batch method is the simplest procedure for crystallization but each experiment corresponds to just one point in the phase diagram (trace C in Fig. 1A). The protein is mixed with the precipitant solution and placed in an airtight vessel. The easiest way to perform batch experiments, especially with robots, is the "microbatch" method, which uses standard 96-well plates sealed with paraffin oil (Chayen, Stewart, Maeder, & Blow, 1990). As the conditions do not change from those of the starting point, except the eventual decrease in protein concentration due to crystallization or precipitation, the method requires at least a rough estimate of potential crystallization conditions and is therefore less flexible than the vapor diffusion method. However, the advantages are that microbatch experiments are easy to prepare in large numbers and each experiment requires only small amounts of protein. Nevertheless, batch crystallization is less commonly used for screening than vapor diffusion, in spite of the fact that it is one of the superior methods for well-defined growth of large single crystals (after the best crystallization conditions have been identified by examination of the phase diagram). In contrast to dialysis, there are no limitations to the type precipitant used. PSII is an example where the batch method is used in combination with seeding to grow large well-ordered single crystals. All currently published crystals of PSII from *TS elongatus* use variations of the batch method for crystallization of PSII, as batch grown crystals of PSII are usually better than those that were made with vapor diffusion. There are two reasons why batch crystallization leads to better crystals of PSII than vapor diffusion: (a) PSII is not very stable and must be crystallized fast. A vapor diffusion experiment starts in the nonsaturated phase, somewhere off the ideal conditions, and PSII seems to deteriorate during the time needed for equilibration. Second, PSII does not tolerate high detergent concentration and the increase of detergent in the protein drop during the vapor diffusion experiment destabilizes PSII. Both of these negative effects can be avoided, by the use of batch experiments. The protein is directly mixed with the precipitant, leading to a point in the metastable phase of the phase diagram. One small seed crystal is added and the solution is placed in capillary and sealed. PSII crystals are grown under batch conditions with PEG 2000 and calcium chloride in the precipitant solution in capillaries (Kern et al., 2005; Zouni et al., 2000). Ferreira et al. (2004) used the hanging drop method for PSII crystallization, but the published conditions are essentially equivalent and close to batch conditions. Figure 4 shows batch crystallization experiments in quartz tubes with slight variations in the precipitant. Figure 4D shows

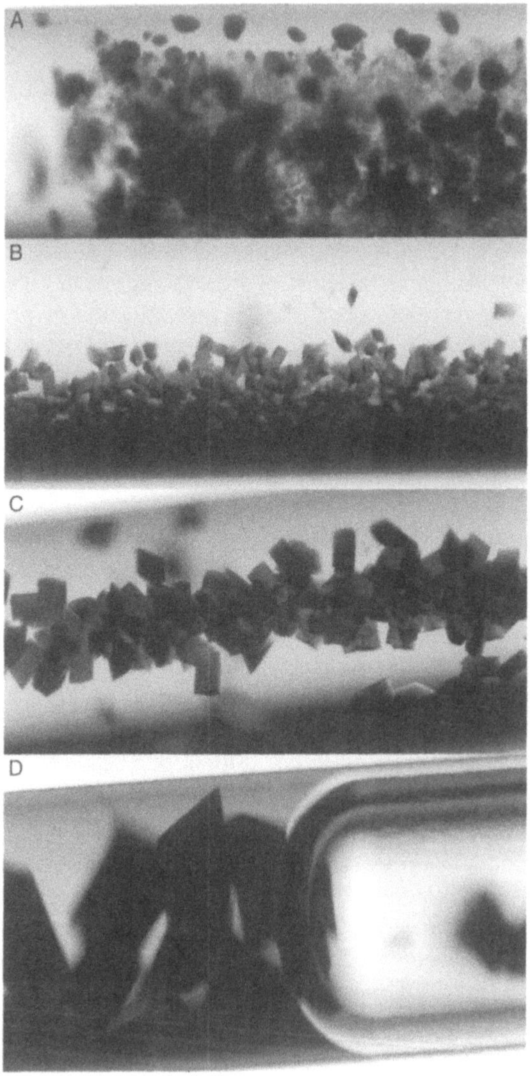

FIGURE 4 Batch crystallization of PSII. In a batch crystallization setup, here in wax-sealed quartz tubes, the ideal crystallization conditions have to be exactly matched from the very beginning on, (A) a suboptimal mixture leading to rounded crystals and some amorphous precipitate in the background, (B) too many seeds or a too high precipitant concentration lead to many small crystals, (C) well-formed crystals with sharp edges, but still too small, (D) the lowest precipitant concentration yielded larger crystals with sharp edges. Note that air bubbles provide an additional phase border in this setup, and sometimes the largest crystals can be found right at their borders.

an optimal result where few, relatively large crystals have been grown with sharp edges, which showed good diffraction quality to about 3 Å resolution.

D. Free Interface Diffusion

Free interface diffusion is a method that has only rarely be used for crystallization of membrane proteins, very likely because the current commercially available reactors (Granada crystallization box) (Garcia-Ruiz, Gonzalez-Ramirez, Gavira, & Otalora, 2002; Kuta Smatanova, Gavira, Rezacova, Vacha, & Garcia-Ruiz, 2006) require a rather high amount of protein. The protein solution is filled into a small capillary and the end of the capillary is "closed" with a thin gel. Subsequently, the capillary is placed in the box that contains the precipitant solution. The protein and the precipitant can diffuse freely through the gel interface, thereby a broad gradient of protein and precipitant is established along the axis of the capillary (see trace D in Fig. 1A). The advantage of this method is the large range of precipitant/protein concentrations that can be covered in one experiment. If combined with interferometry this method may also be used to determine the optimal growth conditions for the crystals. For more details, the reader is referred to the excellent review of Ng, Gavira, and Garcia-Ruiz (2003).

E. Lipidic Cubic Phase

Crystallization in lipidic cubic phases, first invented by Landau and Rosenbusch, has been successfully used to crystallize several membrane proteins including the photosynthetic membrane protein bacteriorhodopsin (Chiu et al., 2000; Misquitta et al., 2004) and the pbRC (Katona et al., 2003). The lipidic cubic phase is a continuous phase of lipid bilayers formed by monooleine with water-filled channels. The membrane proteins diffuse in the lipid bilayer and supersaturation is induced by diffusion of the precipitant into the water-filled channels. It has been postulated that a lamellar phase may be involved in the nucleation event (Nollert, Qiu, Caffrey, Rosenbusch, & Landau, 2001). Recently, modification of the lipid has led to an increase of the water-filled cavity size which allows even larger proteins to be crystallized in the lipidic cubic phases (Caffrey, 2008). The sponge phase has also been shown to be very suitable for membrane protein crystallization (Wadsten et al., 2006). For more information about this very useful new technique, the reader is referred to Chapter 4 for a detailed description of the lipidic cubic phase crystallization method.

VI. DETERMINATION OF PHASE DIAGRAMS

A. Reason for the Determination of a Phase Diagram

The determination of phase diagrams is very useful for the crystallization of membrane proteins and has been successfully applied for the rational design crystallization of PSI and PSII. The major question frequently asked is how a phase diagram is constructed for individual proteins. The phase diagram for a protein can be divided into the unsaturated phase and the supersaturated phase (see Fig. 1A). The supersaturated phase splits into three zones: the metastable zone, the nucleation zone, and the precipitation zone. The metastable is the zone at lowest supersaturation, where preformed nuclei can grow but the supersaturation is too low for the formation of nuclei that are stable. These are nuclei that are larger than the critical radius r_c. When the supersaturation reaches the nucleation zone, crystals can be formed. This phase must be reached to observe crystals, when no seeding is involved. At very high supersaturation—in the precipitation zone—the protein precipitates out of the solution in form of nonordered aggregates. This is the simplest phase diagram (it can be more complex for membrane proteins, involving two nonaqueous phases or lipid phases). We will focus here on the main methods for the determination of the general phase diagram, which can be extended for the use of more complex phase diagrams.

B. How to Determine the Solubility Curve?

The first and most important step in developing phase diagrams is measurement of the solubility curve, that is, the determination of the supersaturation borderline. This supersaturation borderline is difficult to determine in the direction of increasing supersaturation, except when seeding crystals are available (see below). The best and most accurate method to determine the supersaturation curve is by dissolving precipitate or microcrystals. We will describe here the setup that has been used for the phase diagrams of the PSI solubility, which can be easily applied for many other membrane proteins. The experiments start with precipitate or microcrystals that can be reversibly precipitated. The whole protein batch is transferred and stored in form of the precipitate/microcrystals. Ten microliters of the protein suspension in the precipitant solution is placed in a small reaction vessel and stirred with a microstir bar for 20 min. The reaction vessel is centrifuged to sediment the precipitate and a 0.5 μl sample of the supernatant is taken out using a 0.5 μl microcapillary (these capillaries fill due to capillary pressure and are very accurate). The precipitant concentration is lowered by addition of 0.5 μl buffer without precipitant and the solution is stirred for another 20 min. The protein concentration in the 0.5 μl samples can be very precisely measured using a Hellma microcuvette by absorption at 280 nm (for colored proteins any other suitable wavelength can be used). These cuvettes allow the accurate measurement of absorption spectra in drops of 0.5–5 μl volume and fit into any commercial photometer. The procedure is repeated until nearly all of the precipitate/microcrystals have been dissolved. It is highly recommended to test the reversibility of the process by adding precipitant to the reaction vessel corresponding to the initial precipitant concentration. Thereby it can be proven that the solubility is the same as initially determined. This is especially important for the determination of the solubility as a function of the temperature. This experiment is easily performed in a thermo cycler. In the case of PSI from the thermophilic cyanobacterium *TS elongatus*, the dissolution of crystals was fully reversible up to 43 °C. These results indicate that the protein in detergent micelles is fairly stable but not at higher temperatures which represent the optimal growth conditions for *TS elongatus* at 56 °C.

The solubility curve can also be used to examine the quality of different protein batches. A common problem in protein crystallization is that the quality of protein batches vary slightly, which can lead to problems in reproducing optimal crystallization conditions. We routinely determine the solubility at three well-defined precipitant concentrations (in case of PSI these are different low ionic strength salt concentrations) as a measure of the quality of the protein preparation. The lower the solubility (in a well-defined range), the higher the quality of the protein preparation. When the solubility is too high (indicative of impurities or heterogeneity in the sample) recrystallization is used to further purify the protein until the desired solubility (and thereby protein quality) is reached.

C. A Fast Run Through the Phase Diagram

Another test of the protein quality of PSI preparations is a fast run through the phase diagram under the microscope as shown in Fig. 5. PSI (1 μl) at 80 mg/ml (in a buffer containing 50 mM $MgSO_4$) and a 1 μl drop of precipitant (buffer without salt) are placed under the microscope and brought in contact. At the interface, first amorphous precipitate is formed, but soon darker droplets occur which transform into fast growing crystals within 1–2 min. As the drop slowly evaporates, the salt concentration increases. At first, the amorphous precipitate dissolves while the crystals are still growing, but toward the end the crystals also dissolve. This test can also be repeated with one or two seed crystals in the precipitant drop. In this

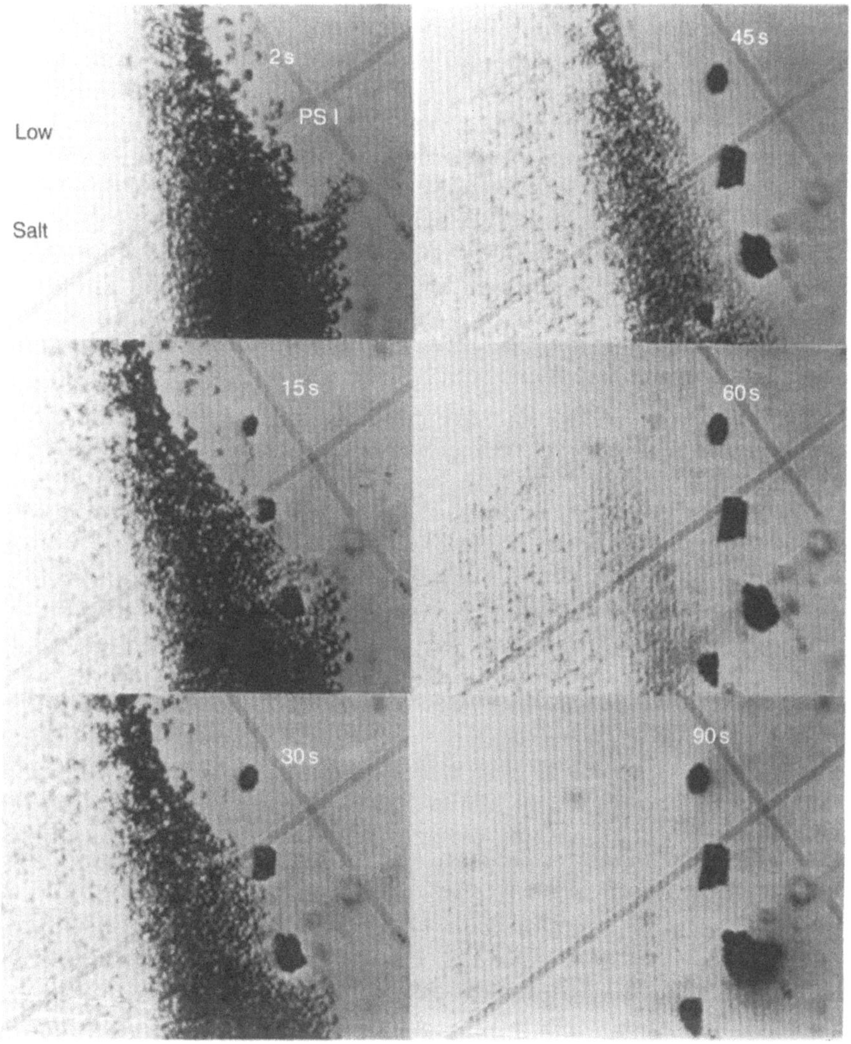

FIGURE 5 The "fast crystallization test" with PSI. The "fast crystallization test" allows for a quick quality assessment of a solution of PSI. One microliter of highly concentrated protein solution in a high salt buffer is put next to 1 μl of distilled water under the microscope. At the phase border, precipitation takes place, with "oily" droplets of phase separation in the direction of the protein solution. After 15 s, the first crystals can be seen growing in the protein solution in the vicinity of the phase separation droplets. The crystals continue to grow and use up the precipitate in this process, until larger crystals are seen after 90 s. Subsequently, these crystals disappear again (not shown), because the resulting salt concentration after complete mixture of both liquids lies still in the unsaturated part of the phase diagram.

case—within seconds—thousands of small crystals begin to grow. This test corresponds to a fast run through the phase diagram and is very useful to test the quality of the protein preparation. Minor proteolytic digestion of the protein, which cannot be detected by SDS gel electrophoresis, results in the failure of the drop test. Under these conditions only amorphous precipitate and drops that appear like cauliflower are formed but no crystal growth is observed, even when seed crystals are added.

Another fast and elegant method for solubility determination is the recently developed counter-diffusion method (Gerdts et al., 2008). This new method is very exciting and a commercial microfluidic instrument has recently been developed by Peter Nollert and colleagues. The reader is referred to Chapter 8 of this book for a detailed description of this system.

D. Determination of the Borderline Between the Metastable and Nucleation Zone

1. Using Dialysis

An accurate and rapid method to determine the borderline between metastable and nucleation zone uses dialysis reactors (see section on seeding) for determination of the complete phase diagram. Only 4 μl of the protein solution is used for each experiment. We screen the phase diagram for 10 different protein concentrations between 10 and 80 mg/ml. The reactors are set up so that all reactors are in the nonsupersaturated zone at the beginning of the experiment (in case of PSI, we start with a salt concentration of 50 mM $MgSO_4$; under these conditions, the solubility of PSI is very high (120 mg/l)). The reactors are assembled and filled, so that the precipitant concentration is equal in all parts of the reactor (the protein, the plastic tube below the membrane, and the 4-ml reservoir). The reactors are transferred into a new glass-vial with higher precipitant concentration (in case of PSI lower ionic strength) and equilibrated for 1 day. It should be noted that the time for equilibrium may vary depending on the viscosity of the solution and the temperature; we determine via conductivity measurements that the equilibration is completed after one day at 4 °C at low viscosity and after 3 days when the equilibrium is performed in 0.75 M sucrose (i.e., a higher viscosity solution). Subsequently, either a small crystal or a suspension of precipitate/microcrystals is added using a seeding loop. As the reactors are open on the top, the seeds can be added without disturbing the experiment. The fate of the seeds is followed under the microscope. When the seeds are dissolved, the solution is still nonsaturated and the reactor is placed into the glass-vial with increasing precipitant concentration, equilibrated and the procedure repeated until the seeds are stable. At this point, the metastable phase is reached.

Close to the supersaturation borderline the seeds are stable but do not grow, at higher supersaturation crystals increase in size. The first crystals often grow during the phase diagram determination, for a new protein, where only precipitate had been observed previously. When the supersaturation is further increased and the nucleation zone reached, hundreds of small crystals are formed. The seeding crystals are often covered with small crystals, resembling a porcupine. The advantage of the use of this dialysis-based method is that screening can be performed in one run with the same protein solution and subsequent seeding steps. It works very well when the precipitant is salt (either low or high ionic strength) or any other small molecule, which permeates freely through the dialysis membranes. Unfortunately, this is not the case for higher molecular weight PEGs, which are very frequently used for crystallization of membrane proteins. In this case, the determination of the phase diagrams must be performed using one of the batch-or counter-diffusion methods.

2. Batch Technique

The determination of a complete phase diagram using batch methods is easy but protein consuming, and is therefore best performed using the microbatch method under oil (Chayen, 2003). A small drop of protein (as low as 50 nl), which contains small seeds, is placed under oil to hinder evaporation. The fate of the seeds is followed under the microscope and evaluated as described above for the microdialysis experiments. One drop is needed for each data point, so experiments are generally prepared in the 96-well formats. The major challenge is accurately placing the seeds into the protein solution. It is preferable to apply microseeding using the strike-through technique; however, an evaporation problem exists due to the small drop size. Manual preparation requires two people: one holds two filled pipettes (one with the protein and one with the oil) while the second person strikes the hair with the seeds through the drop immediately after the protein drop is deployed. The first person then adds the oil on top of the protein solution. With good teamwork this takes less than 10 s. There is a recent report on a robotic system that can add seeds to a protein solution (D'Arcy, Villard, & Marsh, 2007), which would make this process easy although we do not have experience with this system.

3. Batch-Vapor Diffusion Experiment

Most researchers like to use vapor diffusion as the crystallization technique, but this method is less precise for phase diagram determinations because protein, detergent, and precipitant are increased simultaneously, making it difficult to determine the exact point of the equilibrium. However, the phase diagram can also be determined using vapor diffusion with the

Nextal crystallization plates. They contain screw caps which allow the easy addition of seeds to the protein drop and solution to the reservoir. The experiment starts with a batch experiment in hanging drop. The concentration of all solutes in the drop and reservoir are calculated to make sure that the concentration of all solutes is exactly the same in the drop and reservoir (corresponding thereby to a batch experiment). The seeds are added by microseeding using strike-through seeding or a single crystal is added using a seeding loop. When the seeds are completely dissolved, the solution is unsaturated. The concentration of precipitant can now be stepwise increased by either adding precipitant to the reservoir, or transfer of the screw cap to the next higher precipitant reservoir. Fresh seeding crystals are added after 2–3 days of equilibration and the fate of the crystals is determined as described above. The procedure is repeated until the nucleation zone is reached.

VII. SEEDING TECHNIQUES

We have discussed the importance of controlling and slowing down the crystallization process as this affects crystals and the limitation of the number of seeds, in order to achieve the growth of large, well-ordered single crystals. In order to achieve this goal, crystal growth is optimally achieved at a lower level of supersaturation, in the metastable zone (Fig. 1A) using seeding techniques.

A. Heterogeneous Seeding

Crystallization can also be induced by heterogeneous seeds that contaminate the solution and lead to the induction of crystal growth. A single piece of dust in the experiment might turn out to be the focus of extensive crystal growth. This can occur from hydrolyzed detergent resulting in dodecanol crystals that act as nucleants causing a shower of small membrane protein crystals. We therefore try to avoid dust and unwanted particles in the solution. However, it is also possible to harness this seeding power and use it to one's advantage. We have tested multiple heterogeneous seeds for the induction of crystal growth of PSI, PSII, and other membrane proteins and found that hair, wool, or other protein-based fibers can very often be successfully used as seeds, while cotton and other carbohydrate-based fibers had no effect on the nucleation rate. When hair is used, detergents should be removed from the hair, as these might interfere with the detergent in the crystallization buffer. Table I illustrates one example of heterogeneous seeding, where bacteriorhodopsin crystals are grown in a vapor diffusion setup using benzamidine crystals as seeds (Essen et al., 1998). Furthermore, all crystals used for the 4 Å crystal structure of PSI published in 1996

(Krauss et al., 1996) were grown using human hair as seeds. If the reader is further interested in the theory behind heterogeneous seeding, the articles by Chayen, Saridakis, and Sear (2006) and Stolyarova, Saridakis, Chayen, and Nemirovsky (2006) provide further insights into these techniques.

B. Homogenous Seeding

Homogenous seeding requires previously grown small microcrystals of the protein of interest. Often, small crystals are already observed via broad screening of crystallization conditions. These crystals might consist of multiple crystals or crystals that are much too small for X-ray structure analysis, but they may be very useful for seeding experiments. For microseeding, crystals or crystal fragments of a size smaller than is visible by the naked eye are sufficient. A suspension of microcrystals can be added to the drop or an existing larger crystal (this can even be poorly formed crystals) is crushed into small fragments.

The crystal is crushed with the tip of a hair, the hair is subsequently swiped through the protein droplet and the reaction vessel is closed immediately. We found that the seeds should not be visible, tiny crystal particles that are attached to the hair are best suited for "strike-through" seeding. We have used this microseeding technique to grow PSII using the batch method (Fig. 4). When using PEG as precipitant, seeding with larger, intact crystals is often counterproductive. The reason is that membrane protein crystals grown in PEG are often surrounded by a "skin" consisting of PEG and detergent. As seeding requires a crystal growth-inducing surface exposed, nucleation does not occur if the crystal is covered by a detergent-PEG skin. Intact microcrystals can only be used as seeds when they are freshly grown and do not have a skin. Furthermore, the crystallization using intact seeds may include conditions, where the seed crystal is partially dissolved.

The most sophisticated seeding technique sequence, which we routinely apply in our laboratory, was developed for improvement of the structure of PSI to 2.5 Å (Jordan et al., 2001). The method involves three discrete crystallization steps with microseeding in the second crystallization, followed by macroseeding in the third and final step. In the course of this process, the protein solution of PSI is dialyzed against low ionic strength, leading to a decrease in solubility of PSI by 2 orders of magnitude.

The first crystallization step is part of the protein preparation. It is used as the last purification step of PSI, where the complete protein preparation is crystallized under low ionic strength conditions. PSI is purified by ion-exchange chromatography and eluted from the column in 140 mM $MgSO_4$. The protein is concentrated by ultrafiltration and subsequently diluted with buffer without salt, leading to a concentration of 6 mM $MgSO_4$. The crystals grow on the ultrafiltration membrane in the next concentration step. Using ultrafiltration, the protein concentration is increased without changing any other solvent parameter. At one point (40 mg/ml), the nucleation zone is reached, and nearly all the PSI precipitates as a shower of tiny crystals. PSI is stored in the form of these small crystals at 4 °C. They can be dissolved to obtain the high protein concentration for the crystallization experiments or alternatively serve (nondissolved) as seeds for the next microseeding step.

1. Microseeding of PSI

The dialysis reactors described above (Fig. 3) are assembled and filled with 4 μl of PSI solution. The protein solution has a concentration of 80 mg/ml in a buffer containing 50 mM $MgSO_4$.in 5mM MES pH 6.4, 0.02% β-dodecyl-maltoside. The protein solution in the microdialysis reactors is dialyzed at 4 °C for 17 h against the buffer containing 9, 10, and 11 mM $MgSO_4$. The protein solution has not yet reached the equilibrium; it is in the metastable zone. The tiny crystals from the original protein preparation are diluted in 10 mM $MgSO_4$, leading to a suspension that contains 10–25 tiny crystals in 1 μl solution. These seeds are transferred with a hair-loop (1.5 mm) into the protein solution in the dialysis reactors. After 1 day, medium size seeding crystals can be harvested from the dialysis reactors (Fig. 3f and g).

2. Macroseeding of PSI

The setup for the macroseeding experiments is similar to the one described above, but the buffer in the reservoir and the plastic tube of the reactor contains 30 mM $MgSO_4$. The solution is still unsaturated but close to the solubility borderline. After 1 day of equilibration at 4 °C, one freshly grown medium size seeding crystal (already visible by eye, length approximately 0.1 mm, diameter 0.2 mm) is transferred into the protein solution using the seeding loops described above. At this point, the salt concentration is further reduced by dialysis against lower ionic strength buffer (11–9 mM $MgSO_4$). The seeding crystal begins to dissolve and the solution must reach the metastable phase before the crystal is completely dissolved. Therefore, the size of the seed is critical for the success of the macroseeding experiment. Under optimized conditions, the seeding crystal grows in the metastable zone to one large, well ordered, single crystal. Crystals up to 3 mm size can be grown using this procedure, with crystal size limited by the diameter of the quartz tube. Quartz tubes of 5-mm diameter have been used to grow larger crystals of up to 5 mm in size to support neutron diffraction experiments. This example shows that large, well-diffracting crystals of large membrane protein complexes (PSI has a MW of 1,056,000 Da) can be achieved using a multistep seeding procedure. For membrane protein crystallization, seeding is a very powerful technique that is worth routine investigation.

References

Adir, N. (1999). Crystallization of the oxygen-evolving reaction centre of photosystem II in nine different detergent mixtures. *Acta Crystallographica. Section D, Biological Crystallography, 55*, 891–894.

Adir, N., Okamura, M. Y., & Feher, G. (1994). Co-crystallization and Preliminary Structure Determination of the Photosynthetic Reaction Center and Cytochrome C2 complex from Rb. Sphaeroides. *Biophysical Journal, 66*, A127.

Amunts, A., Drory, O., & Nelson, N. (2007). The structure of a plant photosystem I supercomplex at 3.4 Å resolution. *Nature, 447*, 58–63.

Aro, E. M. (1999). Photodamage and repair of photosystem 11. *Photochemistry and Photobiology, 69*, 70s.

Axelrod, H. L., Abresch, E. C., Okamura, M. Y., Yeh, A. P., Rees, D. C., & Feher, G. (2002). X-ray Structure Determination of the Cytochrome c2: Reaction Center Electron Transfer Complex from Rhodobacter sphaeroides. *Journal of Molecular Biology, 319*, 501–515.

Barros, T., Royant, A., Standfuss, J., Dreuw, A., & Kuhlbrandt, W. (2009). Crystal structure of plant light-harvesting complex shows the active, energy-transmitting state. *EMBO Journal, 28*, 298–306.

Belrhali, H., Nollert, P., Royant, A., Menzel, C., Rosenbusch, J. P., Landau, E. M., et al. (1999). Protein, lipid and water organization in bacteriorhodopsin crystals: A molecular view of the purple membrane at 1.9 Å resolution. *Structure, 7*, 909–917.

Ben-Shem, A., Frolow, F., & Nelson, N. (2003). Crystal structure of plant photosystem I. *Nature, 426*, 630–635.

Caffrey, M. (2008). *Annual Review of Biophysics.*

Chang, C. H., el-Kabbani, O., Tiede, D., Norris, J., & Schiffer, M. (1991). Structure of the membrane-bound protein photosynthetic reaction center from Rhodobacter sphaeroides. *Biochemistry, 30*, 5352–5360.

Chayen, N. E. (2003). Protein crystallization for genomics: Throughput versus output. *Journal of Structural and Functional Genomics, 4*, 115–120.

Chayen, N. E., & Saridakis, E. (2008). *Nature Methods, 5*, 147–153.

Chayen, N. E., Saridakis, E., & Sear, R. P. (2006). Experiment and theory for heterogeneous nucleation of protein crystals in a porous medium. *Proceedings of the National Academy of Sciences of the United States of America, 103*, 597–601.

Chayen, N. E., Stewart, P. D. S., Maeder, D. L., & Blow, D. M. (1990). An automated system for micro-batch protein crystallization and screening. *Journal of Applied Crystallography, 23*, 297–302.

Chiu, M. L., Nollert, P., Loewen, M. C., Belrhali, H., Pebay-Peyroula, E., Rosenbusch, J. P., et al. (2000). Crystallization in cubo: General applicability to membrane proteins. *Acta Crystallographica. Section D, Biological Crystallography, 56*, 781–784.

Cogdell, R. J., Isaacs, N. W., Freer, A. A., Howard, T. D., Gardiner, A. T., Prince, S. M., et al. (2003). The structural basis of light-harvesting in purple bacteria. *FEBS Letters, 555*, 35–39.

Cudney, B., Patel, S., & Mcpherson, A. (1994). Crystallization of macromolecules in silica gels. *Acta Crystallographica. Section D, Biological Crystallography, 50*, 479–483.

D'Arcy, A., Villard, F., & Marsh, M. (2007). An automated microseed matrix-screening method for protein crystallization. *Acta Crystallographica. Section D, Biological Crystallography, 63*, 550–554.

Deisenhofer, J., Epp, O., Miki, K., Huber, R., & Michel, H. (1985). Structure of the protein subunits in the photosynthetic reaction centre of Rhodopseudomonas viridis at 3 Å resolution. *Nature, 318*, 618–624.

Deisenhofer, J., Epp, O., Sinning, I., & Michel, H. (1995). Crystallographic refinement at 2.3 Å Resolution and Refined Model of the Photosynthetic Reaction Centre from Rhodopseudomonas viridis. *Journal of Molecular Biology, 246*, 429–457.

Essen, L. O., Siegert, R., Lehmann, W. D., & Oesterhelt, D. (1998). Lipid patches in membrane protein oligomers: Crystal structure of the bacteriorhodopsin-lipid complex. *Proceedings of the National Academy of Sciences of the United States of America, 95*, 11673–11678.

Ferreira, K. N., Iverson, T. M., Maghlaoui, K., Barber, J., & Iwata, S. (2004). Architecture of the Photosynthetic Oxygen-Evolving Center. *Science, 303*, 1831–1838.

Fromme, P. (1998). Kristallisation und strukturelle Untersuchungen am Photosystem I. *In* Physical chemistry p. 165. Technical University Berlin, Berlin.

Fromme, P. (2003). Crystallization of photosystem I.In S. Iwata, (Ed.), *Methods and results in crystallization of membrane proteins* p. 355. International University Line, La Jolla, CA.

Fromme, P., Bottin, H., Krauss, N., & Setif, P. (2002). Crystallization and Electron Paramagnetic Resonance Characterization of the Complex of Photosystem I with its Natural Electron Acceptor Ferredoxin. *Biophysical Journal, 83*, 1760–1773.

Fromme, P., & Witt, H. T. (1998). Improved isolation and crystallization of photosystem I for structural analysis. *Biochimica et Biophysica Acta-Bioenergetics, 1365*, 175–184.

Garcia-Ruiz, J. M., Gonzalez-Ramirez, L. A., Gavira, J. A., & Otalora, F. (2002). Granada Crystallisation Box: A new device for protein crystallisation by counter-diffusion techniques. *Acta Crystallographica. Section D, Biological Crystallography, 58*, 1638–1642.

Gerdts, C. J., Elliott, M., Lovell, S., Mixon, M. B., Napuli, A. J., Staker, B. L., et al. (2008). The plug-based nanovolume Microcapillary Protein Crystallization System (MPCS). *Acta Crystallographica. Section D, Biological Crystallography, 64*, 1116–1122.

Jekow, P., Fromme, P., Witt, H. T., & Saenger, W. (1995). Photosystem I from Synechococcus elongatus: Preparation and crystallization of monomers with varying subunit compositions. *Biochimica et Biophysica Acta-Bioenergetics, 1229*, 115–120.

Jekow, P., Schubert, W. D., Fromme, P., Kruip, J., Chitnis, P. R., Rogner, M., et al. (1996). *Zeitschrift Fur Naturforschung C-Journal of Biosciences, 51*, 195–199.

Jordan, P., Fromme, P., Witt, H. T., Klukas, O., Saenger, W., & Krauss, N. (2001). Three-dimensional structure of cyanobacterial photosystem I at 2.5 Å resolution. *Nature, 411*, 909–917.

Kamiya, N., & Shen, J. R. (2003). Crystal structure of oxygen-evolving photosystem II from Thermosynechococcus vulcanus at 3.7-Å resolution. *Proceedings of the National Academy of Sciences of the United States of America, 100*, 98–103.

Katona, G., Andreasson, U., Landau, E. M., Andreasson, L. E., & Neutze, R. (2003). Lipidic Cubic Phase Crystal Structure of the Photosynthetic Reaction Centre from Rhodobacter sphaeroides at 2.35 Å Resolution. *Journal of Molecular Biology, 331*, 681–692.

Kern, J., Loll, B., Luneberg, C., DiFiore, D., Biesiadka, J., Irrgang, K. D., et al. (2005). Purification, characterisation and crystallisation of photosystem II from Thermosynecho coccus elongatus cultivated in a new type of photobioreactor. *Biochimica et Biophysica Acta-Bioenergetics, 1706*, 147–157.

Klukas, O., Schubert, W. D., Jordan, P., Krauss, N., Fromme, P., Witt, H. T., et al. (1999a). Localization of Two Phylloquinones, QK and QK', in an Improved Electron Density Map of Photosystem I at 4-Å Resolution. *Journal of Biological Chemistry, 274*, 7361–7367.

Klukas, O., Schubert, W. D., Jordan, P., Krauss, N., Fromme, P., Witt, H. T., et al. (1999b). Photosystem I, an Improved Model of the Stromal Subunits PsaC, PsaD, and PsaE. *Journal of Biological Chemistry, 274*, 7351–7360.

Koepke, J., Hu, X. C., Muenke, C., Schulten, K., & Michel, H. (1996). The crystal structure of the light-harvesting complex II (B800–850) from Rhodospirillum molischianum. *Structure, 4*, 581–597.

Kohlstadt, M., Dorner, K., Labatzke, R., Koc, C., Hielscher, R., Schiltz, E., et al. (2008). Heterologous Production, Isolation, Characterization and Crystallization of a Soluble Fragment of the NADH:Ubiquinone Oxidoreductase (Complex I) from Aquifex aeolicus. *Biochemistry, 47*, 13036–13045.

Krauss, N., Schubert, W. D., Klukas, O., Fromme, P., Witt, H. T., & Saenger, W. (1996). Photosystem I at 4 Å resolution represents the first structural model of a joint photosynthetic reaction centre and core antenna system. *Nature Structural Biology, 3*, 965–973.

Kurisu, G., Zhang, H., Smith, J. L., & Cramer, W. A. (2003). Structure of the Cytochrome b6f Complex of Oxygenic Photosynthesis: Tuning the Cavity. *Science, 302*, 1009–1014.

Kuta Smatanova, I., Gavira, J. A., Rezacova, P., Vacha, F., & Garcia-Ruiz, J. M. (2006). *Photosynthesis Research, 90*, 255–259.

Liu, Z., Yan, H., Wang, K., Kuang, T., Zhang, J., Gui, L., et al. (2004). Crystal structure of spinach major light-harvesting complex at 2.72 Å resolution. *Nature, 428*, 287–292.

Loll, B., Broser, M., Kos, P. B., Kern, J., Biesiadka, J., Vass, I., et al. (2008). Modeling of variant copies of subunit D1 in the structure of photosystem II from Thermosynechococcus elongatus. *Biological Chemistry, 389*, 609–617.

Loll, B., Kern, J., Saenger, W., Zouni, A., & Biesiadka, J. (2005). Towards complete cofactor arrangement in the 3.0 Å resolution structure of photosystem II. *Nature, 438*, 1040–1044.

Luecke, H., Richter, H. T., & Lanyi, J. K. (1998). Proton transfer pathways in bacteriorhodopsin at 2.3 angstrom resolution. *Science, 280*, 1934–1937.

Luecke, H., Schobert, B., Richter, H. T., Cartailler, J. P., & Lanyi, J. K. (1999). Structure of bacteriorhodopsin at 1.55 Å resolution. *Journal of Molecular Biology, 291*, 899–911.

Misquitta, Y., Cherezov, V., Havas, F., Patterson, S., Mohan, J. M., Wells, A. J., et al. (2004). Rational design of lipid for membrane protein crystallization. *Journal of Structural Biology, 148*, 169–175.

Ng, J. D., Gavira, J. A., & Garcia-Ruiz, J. M. (2003). Protein crystallization by capillary counterdiffusion for applied crystallographic structure determination. *Journal of Structural Biology, 142*, 218–231.

Nogi, T., Fathir, I., Kobayashi, M., Nozawa, T., & Miki, K. (2000). Crystal structures of photosynthetic reaction center and high-potential iron-sulfur protein from Thermochromatium tepidum: Thermostability and electron transfer. *Proceedings of the National Academy of Sciences of the United States of America, 97*, 13561–13566.

Nollert, P. (2005). Membrane protein crystallization in amphiphile phases: Practical and theoretical considerations. *Progress in Biophysics and Molecular Biology, 88*, 339–357.

Nollert, P., Navarro, J., & Landau, E. M. (2002). Crystallization of membrane proteins in cubo. *Methods in Enzymology, 343*, 183–199.

Nollert, P., Qiu, H., Caffrey, M., Rosenbusch, J. P., & Landau, E. M. (2001). Molecular mechanism for the crystallization of bacteriorhodopsin in lipidic cubic phases. *FEBS Letters, 504*, 179–186.

Papiz, M. Z., Prince, S. M., Howard, T., Cogdell, R. J., & Isaacs, N. W. (2003). The structure and thermal motion of the B800-850 LH2 complex from Rps.acidophila at 2.0 Å resolution and 100 K: New structural features and functionally relevant motions. *Journal of Molecular Biology, 326*, 1523–1538.

Pascal, A., Gastaldelli, M., Ceoldo, S., Bassi, R., & Robert, B. (2001). Pigment conformation and pigment-protein interactions in the reconstituted Lhcb4 antenna protein. *FEBS Letters, 492*, 54–57.

Pebay-Peyroula, E., Rummel, G., Rosenbusch, J. P., & Landau, E. M. (1997). X-ray structure of bacteriorhodopsin at 2.5 angstroms from microcrystals grown in lipidic cubic phases. *Science, 277*, 1676–1681.

Roszak, A. W., Howard, T. D., Southall, J., Gardiner, A. T., Law, C. J., Isaacs, N. W., et al. (2003). Crystal structure of the RC-LH1 core complex from Rhodopseudomonas palustris. *Science, 302*, 1969–1972.

Schubert, W. D., Klukas, O., Krauss, N., Saenger, W., Fromme, P., & Witt, H. T. (1997). Photosystem I of Synechococcus elongatus at 4 Å resolution: Comprehensive structure analysis. *Journal of Molecular Biology, 272*, 741–769.

Standfuss, J., Terwisscha van Scheltinga, A. C., Lamborghini, M., & Kuhlbrandt, W. (2005). Mechanisms of photoprotection and nonphotochemical quenching in pea light-harvesting complex at 2.5 Å resolution. *EMBO Journal, 24*, 919–928.

Stolyarova, S., Saridakis, E., Chayen, N. E., & Nemirovsky, Y. (2006). A model for enhanced nucleation of protein crystals on a fractal porous substrate. *Biophysical Journal, 91*, 3857–3863.

Stowell, M. H., McPhillips, T. M., Rees, D. C., Soltis, S. M., Abresch, E., & Feher, G. (1997). Light-induced structural changes in photosynthetic reaction center: Implications for mechanism of electron-proton transfer. *Science, 276*, 812–816.

Stroebel, D., Choquet, Y., Popot, J. L., & Picot, D. (2003). An atypical haem in the cytochrome b(6)f complex. *Nature, 426*, 413–418.

Wadsten, P., Wohri, A. B., Snijder, A., Katona, G., Gardiner, A. T., Cogdell, R. J., et al. (2006). Lipidic sponge phase crystallization of membrane proteins. *Journal of Molecular Biology, 364*, 44–53.

Witt, H. T., Krauss, N., Hinrichs, W., Witt, I., Fromme, P., Pritzkow, W., et al. (1992). *Photosynthesis Research, 34*, 86–86.

Yamashita, E., Zhang, H., & Cramer, W. A. (2007). Structure of the cytochrome b6f complex: Quinone analogue inhibitors as ligands of heme cn. *Journal of Molecular Biology, 370*, 39–52.

Yan, J., Kurisu, G., & Cramer, W. A. (2006). Intraprotein transfer of the quinone analogue inhibitor 2,5-dibromo-3-methyl-6-isopropyl-p-benzoquinone in the cytochrome b6f complex. *Proceedings of the National Academy of Sciences of the United States of America, 103*, 69–74.

Yeates, T. O., Komiya, H., Rees, D. C., Allen, J. P., & Feher, G. (1987). Structure of the reaction center from Rhodobacter sphaeroides R-26: Membrane-protein interactions. *Proceedings of the National Academy of Sciences of the United States of America, 84*, 6438–6442.

Zhang, H., Kurisu, G., Smith, J. L., & Cramer, W. A. (2003). A defined protein-detergent-lipid complex for crystallization of integral membrane proteins: The cytochrome b6f complex of oxygenic photosynthesis. *Proceedings of the National Academy of Sciences of the United States of America, 100*, 5160–5163.

Zouni, A., Jordan, R., Schlodder, E., Fromme, P., & Witt, H. T. (2000). First photosystem II crystals capable of water oxidation. *Biochimica et Biophysica Acta-Bioenergetics, 1457*, 103–105.

Zouni, A., Witt, H. T., Kern, J., Fromme, P., Krauss, N., Saenger, W., et al. (2001). Crystal structure of photosystem II from Synechococcus elongatus at 3.8 Å resolution. *Nature, 409*, 739–743.

CHAPTER 10

A Practical Guide to X-Ray Crystallography of β-barrel Membrane Proteins: Expression, Purification, Detergent Selection, and Crystallization

Mikio Tanabe* and Tina M. Iverson*,†

*Department of Pharmacology, Vanderbilt University Medical Center, Nashville, Tennessee 37232-6600

†Department of Biochemistry, Vanderbilt University Medical Center, Nashville, Tennessee 37232-6600

ABBREVIATIONS

OMP, outer membrane protein; HG, hexyl-glucoside; HpG, heptyl-glucoside; OG, octyl-glucoside; DDG, dodecyl-glucoside; DDM, dodecyl-maltoside; DM, decyl-maltoside; Octyl-POE, octyl-polyoxyethylene; C_8E_4, Octyl-tetra-oxyethylene; $C_{10}E_5$, Decyl-penta-oxyethylene; $C_{12}E_9$, Dodecyl-nona-oxyethylene; LDAO, lauryl-dimethyl-aminoxide; DDAO,

1063-5823/09 $35.00
DOI: 10.1016/S1063-5823(09)63010-6

decyl-dimethyl-aminoxide; HDAO, hexyl-dimethyl-aminoxide; OHESO, octyl-hydroxyethyl-sulfoxide; PEG, polyethylene glycol; MPD, 2-methyl-2,4-pentanediol.

I. OVERVIEW

The outer membranes of prokaryotes, cellular organelles, and chloroplasts contain many proteins that form *β*-barrels. Determination of the structural details of the *β*-barrel outer membrane proteins (OMPs) can contribute to a detailed understanding of their mechanism of function. In the case of OMPs from pathogens, structural studies provide an atomic view of emerging targets to be used for vaccine development.

Through a comprehensive survey of current literature, a general procedure specific for the crystallization of *β*-barrel OMPs has been developed. The procedure begins with overproduction of the target protein in *Escherichia coli* which will lead to expression of properly folded protein into the membrane or require refolding of protein from inclusion bodies. It then outlines techniques to help identify the optimal detergent for extraction from the membrane, refolding from inclusion bodies, and crystallization. A comparison of crystallization conditions from the literature allows the proposal of a new sparse-matrix screen specifically developed to improve the probability of growing diffraction-quality crystals of *β*-barrel OMPs.

II. INTRODUCTION

β-Barrel membrane proteins play a key role in bacterial physiology by rendering the outer membrane selectively permeable to many different substances. The majority of *β*-barrel OMPs of known function facilitate the influx of nutrients and the extrusion of waste between the exterior of the cell and the intermembrane space. However, some OMPs act as enzymes, and some have unknown functions. For OMP channels, also known as porins, the number of *β*-strands comprising the *β*-barrel ranges from 8 to 18 and dictates the pore diameter, which serves as a nonspecific molecular weight cut off for the channel. Some porins are nonselective pores, while others specifically bind substrate. Porins that are selective for anions, sugars, and nucleotides have been identified (Nikaido, 2003; Saier, 2000). OMP transporters usually comprise 20–22 strands and therefore have a larger pore diameter. To prevent nonspecific translocation through the large pore, these transporters contain a specialized, solute-selective plug region. Large solute transporters also contain an extended N-terminus that is required for the interaction with anchoring proteins, such as TonB, on the inner membrane. This substrate-dependent interaction enables transport across both the outer and inner membranes (Moeck & Coulton, 1998; Wiener, 2005).

β-Barrel OMPs have been identified as attractive targets for the prevention and treatment of infection by bacterial pathogens (Lee, Choi, & Xu, 2003; Lin, Huang, & Zhang, 2002). OMPs are exposed at the bacterial cell surface, which makes them accessible to antibodies in the mammalian immune system. As a result, several vaccines on the market and in current clinical trials, use OMPs as an antigen (Baumann, Mansouri, & von Specht, 2004; Girard, Preziosi, Aguado, & Kieny, 2006; Thanassi & Schoen, 2000). Recent indirect evidence suggests that some OMPs may be recognized by the mammalian innate immune system (Massari et al., 2002; Ray, Chatterjee, Bhattacharya, & Biswas, 2003). Thus, revealing how β-barrel OMPs interact with the immune system may be key for the improvement of the efficacy of vaccines against gram-negative bacteria.

OMPs also play a significant role in antibiotic resistance by mediating antibiotic extrusion. All bacteria contain a number of tripartite drug efflux systems. Examples of this include the AcrA (periplasmic adaptor protein), AcrB (inner membrane transporter), and TolC (outer membrane protein) efflux systems from *E. coli,* and the MexA (periplasmic adaptor protein), MexB (inner membrane transporter), and OprM (outer membrane proteins) efflux systems from *Pseudomonas aeruginosa* (Piddock, 2006). These systems include an inner membrane protein (AcrB or MexB) that couples the energy stored in a cation gradient for the transport of molecules across the membrane, and an outer membrane channel (TolC or OprM). The interactions between these two components are believed to be mediated by a periplasmic adaptor protein (AcrA or MexA), typically anchored to the inner membrane by either a single transmembrane helix, or an N-terminal lipid modification (Borges-Walmsley et al., 2003; Eswaran, Koronakis, Higgins, Hughes, & Koronakis, 2004). This type of multidrug resistance (MDR) transporter provides a continuous channel from the cytoplasm to the exterior of the cell. Pharmacological block of this efflux system in combination with traditional antibiotic therapy may improve outcomes and combat the rapid emergence of antibiotic resistance (Pages, Masi, & Barbe, 2005).

While the majority of well-characterized β-barrel membrane proteins are found in the outer membrane of bacteria and intracellular organelles, some viral toxins also form β-barrel membrane proteins (Song et al., 1996). Like many bacterial OMPs, these viral β-barrel membrane proteins can facilitate transmembrane solute translocation or exhibit enzymatic activity. The transport activity contributes to the progression of the viral infection by allowing passage of molecules that interfere with host cell functions by causing actin reorganization, neurotransmitter release, lipase activation, and the Ca^{2+} efflux required to induce cell death (Geny & Popoff, 2006; Heuck, Tweten, & Johnson, 2001).

While the determination of high-resolution three-dimensional structures of β-barrel membrane proteins may provide valuable insight into the mechanisms of function, membrane proteins are difficult to characterize structurally. This is evidenced by the mere 180 unique membrane protein structures available in the Protein Data Bank (PDB) that currently contains over 54,000 entries. This disparity is hypothesized to arise from a combination of poor expression levels, poor stability when detergent solubilized, and a decreased number of potential three-dimensional packing arrangements based on the unique surface hydrophobicity profile (Bannwarth & Schulz, 2003; Granseth, Seppala, Rapp, Daley, & Von Heijne, 2007; Loll, 2003; Raman, Cherezov, & Caffrey, 2006).

While β-barrel membrane proteins are believed to be more stable than their α-helical brethren, only ~25% of the currently available membrane protein structures are of β-barrel membrane proteins (as of September 2008, www.rcsb.org/pdb). Given these statistics, it is not surprising that methods for the crystallization of β-barrel OMPs are lagging behind those developed for their α-helical counterparts. Accordingly, this chapter analyzes the 44 publications reporting successful crystal structure determinations of β-barrel OMPs and uses the compiled data to propose a streamlined protocol for crystallization. This analysis resulted in the development of a novel set of 96 chemical conditions tailored for the screening of initial crystallization conditions for β-barrel membrane proteins by the vapor-diffusion method. In this chapter, we are focusing on the general process of structure determination of β-barrel membrane proteins from expression to crystallization in detergent micelles (Fig. 1). Complementary techniques for the crystallization of both α-helical and β-barrel membrane proteins are the use of the cubic lipidic phase or bicelles for crystallization trials. These techniques are discussed in detail in the chapter by Caffrey and Lyons titled "Monoacylglycerols: The Workhorse Lipids for Crystallizing Membrane Proteins in Mesophases" and the chapter by Faham et al. entitled "Practical Aspects of Membrane Protein Crystallization in Bicelles."

III. Expression and Purification

Determination of the X-ray crystal structure of any macromolecule requires a large quantity of pure, stable, conformationally homogeneous sample. For crystallization of membrane proteins, the quantity of protein required is often greater than for their soluble counterparts, but the expression levels tend to be significantly lower. For example, during the

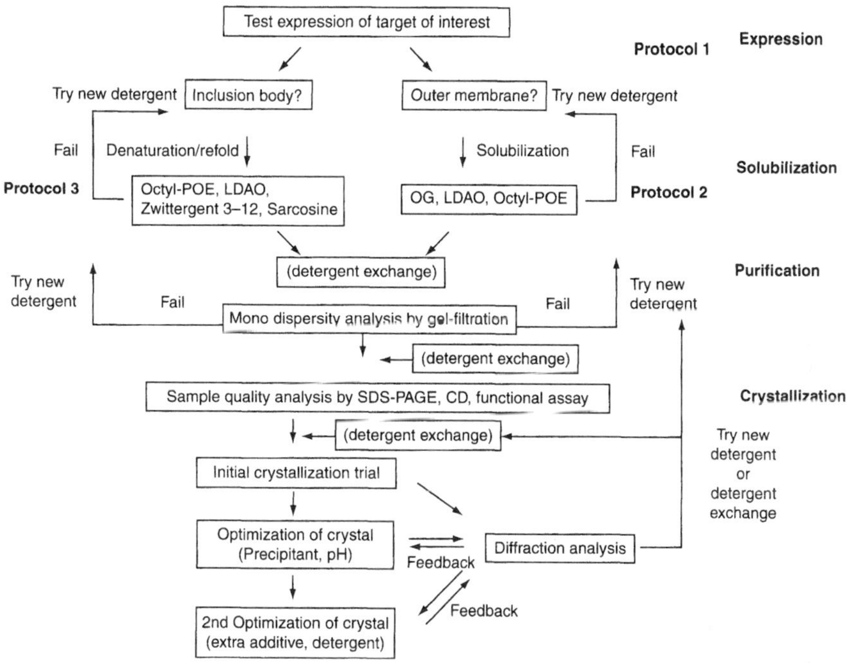

FIGURE 1 A flow chart for expression, purification, and crystallization of β-barrel membrane proteins. This flow chart outlines the general procedure to determine an X-ray crystal of a β-barrel membrane protein. The detailed protocols for expression (Protocol 1), purification from outer membrane (Protocol 2), and refolding from inclusion bodies (Protocol 3) are provided. The most successful detergents for protein solubilization are discussed in Section III.B.

crystallization and phasing of the *Neisseria meningitidis* outer membrane protein PorB, an estimated 300–400 mg of purified protein was required before interpretable experimental electron density maps could be calculated (M. Tanabe and T. M. Iverson, unpublished observation). In this structure determination, nanoliter robotic pipetting technology was employed for all preliminary crystallization trials—in the absence of robotics, an even larger amount of protein might have been required. This example highlights the requirement for careful optimization of protein expression.

A. Plasmid-Driven Expression of β-Barrel Membrane Proteins

Purification of a protein from the native source guarantees protein folding through the organism's native protein-folding machinery; however, few proteins have sufficiently high endogenous expression levels for X-ray crystallographic studies. In addition, a large number of purification steps is required to isolate the protein, and heterogeneous posttranslational modification may hinder crystallization trials. The advantages of heterologous expression include dramatic improvement in protein expression levels, the ability to introduce affinity tags, express protein subdomains, and perform site-directed mutagenesis. These advantages are the reason that the majority of crystal structures are determined from plasmid-expressed proteins. Heterologous overexpression of a membrane protein in *E. coli*, however, frequently does not result in robust expression without the benefit of careful optimization of the expression conditions. Some membrane proteins are only expressed at low levels because of cellular toxicity, an inability to fold correctly in the absence of their native protein-folding machinery, signal sequences that are mismatched to their heterologous host, usage of rare codons, or a combination of any of these factors. If heterologous expression results in low expression, identifying the underlying cause of this low expression is key to improving the protein yield.

A common difficulty with the overexpression of β-barrel membrane proteins in *E. coli* is that increased quantities of OMPs in the outer membrane results in cellular toxicity. Altering the induction conditions for toxic proteins can dramatically improve the yield. The main methods to improve the expression of toxic proteins are to silence low-level expression prior to induction and to decrease the induction time. In one example, high-level heterologous expression of sucrose-specific porin ScrY from *Salmonella typhimurium* was lethal in *E. coli*. However, alteration of the expression protocol to decrease the induction to 1.5 h allowed a maximal yield 8 mg/l of cells (Forst et al., 1993).

The commonly used IPTG-inducible T7 promoter system transcribes low levels of gene product even in the absence of inducing agent. When even low levels of protein expression are toxic, the promoter can be switched, or BL21 (*DE3*) pLys cells may be used. These pLys cells have an increased basal expression level of T7 lysozyme, which suppresses the activity of the T7 polymerase prior to induction (Studier, 1991). The pLys strains can be used for any protein that is toxic to the cells. There are additionally several strains of *E. coli* lacking endogenous OMPs (including *ompA*, *ompC*, and *lamB*) developed specifically for membrane protein overexpression (Derouiche et al., 1996). Presumably porin-limited strains reduce the toxic effects of the overexpression of β-barrel OMPs by reducing the number of other pore-forming β-barrel proteins in the outer membrane. A porin-deletion strain improved the overexpression of the ferrichrome-iron siderophone transporter FhuA from *E. coli* (Locher & Rosenbusch, 1997), the sucrose-specific porin ScrY from *S. typhimurium* (Forst et al., 1993), and the porin PorB from *Neisseria meningitides* (M. Tanabe and T. M. Iverson, unpublished observation). Each of these resulted in successful structure determinations (Forst, Welte, Wacker, & Diederichs, 1998; Locher et al., 1998).

In addition to cellular toxicity, many proteins express at low levels because of unusual codon usage in the gene. Several strains of BL21 (*DE3*) cells include tRNAs encoding for these rare codons (Stratagene). While these strains were initially developed for the expression of mammalian proteins, many bacteria also have altered codon usage, which may cause expression difficulties in *E. coli* (Makrides, 1996; Saier, 1995). In one example, a battery of *Mycobacterium tuberculosis* membrane proteins was subjected to expression trials and successful overexpression required BL21-CodonPlus (*DE3*)-RP strains (Korepanova et al., 2005). Similarly, *Bordetella pertussis* autotransporters, Prn and BrkA were also overexpressed using BL21-CodonPlus (*DE3*)-RIL strain (Dautin, Barnard, Anderson, & Bernstein, 2007).

Misfolding of β-barrel membrane proteins into inclusion bodies is common during overexpresion in *E. coli*. In order to be properly folded, a β-barrel membrane protein must be translocated from the cytoplasm, where it is expressed, to the bacterial outer membrane, where it is to function. In *E. coli*, the SecYEG secretion system is responsible for transport across the inner membrane. Substrates for SecYEG translocation are marked by an N-terminal signal sequence (Driessen & Nouwen, 2008). Following translocation across the inner membrane, the signal sequence is cleaved and the secreted polypeptide chain is folded into a β-barrel and inserted into the outer membrane with assistance from the periplasmic chaperone proteins SurA, Skp, and Omp85 (Bos & Tommassen, 2004; Schleiff & Soll, 2005; Voulhoux & Tommassen, 2004).

It is often speculated that mismatch in signal sequence between *E. coli* and a plasmid-expressed OMP results in the inability of the SecYEG system to recognize an overexpressed β-barrel membrane protein. If initial translocation of the protein across the inner membrane fails, intracellular inclusion bodies form. Alteration of the N-terminal sequence to an ideal *E. coli* signal sequence can improve the translocation of plasmid-expressed OMPs—including those from *E. coli* that have low endogenous expression levels (Ahn, Hwang, Lee, Choi, & Kim, 2007; Ghrayeb et al., 1984). In one example, the *E. coli* autotransporter EspP containing an additional *E. coli* OmpA signal sequence at the N-terminus targeted the location of the protein to the outer membrane (Barnard, Dautin, Lukacik, Bernstein, & Buchanan, 2007). A similar technique was used to target the expression of the filamentous hemagglutinin transporter, FhaC from *B. pertussis*, into outer membrane via the Sec system (Clantin et al., 2007). While the OmpA signal sequence is the most commonly used, and is even included in several commercially available expression vectors such as pASK (IBA) and FLAG (Sigma), signal sequences from other organisms may improve the targeting of β-barrel OMPs to the outer membrane. For example, during the expression of the basic amino acid uptake channel in *P. aeruginosa*, OprD, the *Pseudomonas* signal sequence was removed from the N-terminus of the OprD gene and was replaced with the *Serratia marcescens* ShlB signal sequence. Full-length ShlB is expressed at high levels within the *E. coli* and this technique resulted in successful expression of OprD into the membrane (Biswas, Mohammad, Patel, Movileanu, & van den Berg, 2007).

Formation of inclusion bodies can be influenced by factors other than the signal sequence. During the optimization of expression of the outer membrane iron-uptake transporter PiuA from *Yersinia pestis*, both temperature and the location of the histidine affinity tag influenced the propensity for inclusion body formation. When the protein expression was induced at 37 °C, the protein expressed entirely into inclusion bodies. However, upon reducing temperature to 30 °C, PiuA began to show expression into the membrane (Tanabe et al., 2007, and unpublished observation).

While the majority of β-barrel OMP structures are determined from proteins expressed into the membrane, the use of OMPs refolded from inclusion bodies for structural studies is common: 20% of the reported crystal structures of β-barrel membrane proteins used refolded protein from inclusion bodies. By comparison, 25% of the structures were determined using protein purified from the native organism and 55% used plasmid-driven overexpression that successfully targeted the protein to the outer membrane (Fig. 2). Although the methods discussed above are designed to decrease protein expression into inclusion bodies, the ease of refolding of many β-barrel proteins—and the high quantity of protein that can be expressed into inclusion bodies—can actually make this a method of expression beneficial for structural studies. In one example, when nonspecific porin OmpG from *E. coli* was overexpressed and properly inserted into the outer membrane,

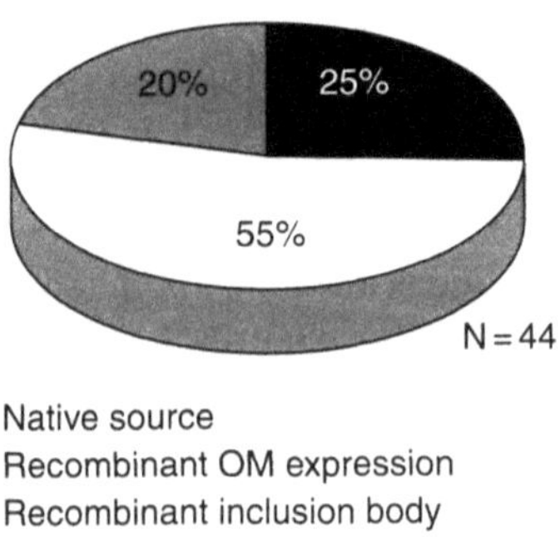

FIGURE 2 Summary of successful expression techniques used for β-barrel outer membrane proteins. Of the 44 published OMP crystal structures, protein was obtained from native source through classical biochemical methods in 11 of the 44 determined structures (25%; *black*), produced by recombinant OM expression in 24 of the 44 determined structures (55%; *white*), and refolded from inclusion bodies in 9 cases (20%; *gray*).

the yield was approximately 0.4 mg protein per liter of bacterial culture (Subbarao & van den Berg, 2006). In comparison, expression of OmpG into inclusion bodies resulted in a yield of 20-30 mg protein per liter of bacterial culture (Yildiz, Vinothkumar, Goswami, & Kuhlbrandt, 2006), a more than 50-fold improvement. A second advantage of expression into inclusion bodies is that purification can be streamlined. Because overexpressed β-barrel OMPs usually comprise more than 95% of the inclusion body, affinity purification is generally unnecessary.

Refolded OMPs from both prokaryotes and eukaryotes are structurally similar to OMPs expressed into the membrane. This is best evidenced by comparison of the crystal structures of the same OMP determined from protein solubilized from the membrane or refolded from inclusion bodies. OmpG form serves as a good example of this. Both OmpG solubilized from the membrane (Subbarao & van den Berg, 2006) and refolded from inclusion bodies (Yildiz et al., 2006) adopt a similar overall fold with a RMSD between Cα atoms of 0.4 Å. Most refolded OMPs are assessed for the formation of the classical β-form architecture using circular dichroism (CD) spectroscopy prior to crystallization.

A general scheme and protocol for initial approaching and troubleshooting the overproduction of β-barrel membrane proteins in *E. coli* is presented in Protocol 1.

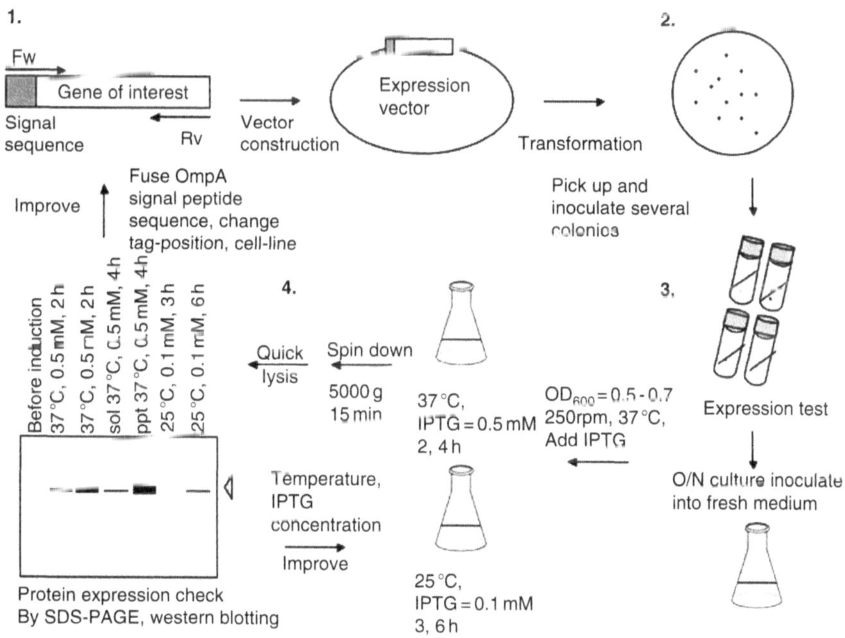

(*continued*)

PROTOCOL 1 Small-scale overexpression trials of outer membrane proteins.

1. *Cloning*. The gene of interest including the predicted N-terminal signal sequence is amplified with PCR and cloned into an expression vector, such as pET (Novagen) or pBAD (Invitrogen). The expression vector is transformed into competent expression cells. The most commonly used strains are any of a series of BL21(DE3). In this protocol, the pET expression system will be used for simplicity. Plasmid uptake is verified by plating onto LB agar containing a selective antibiotic.
2. *Expression trials*. A single colony is used to inoculate 5 ml of LB medium that is incubated overnight at 37 °C with shaking at 250 rpm. Two 250 ml flasks containing 100 ml of freshly prepared LB medium are each inoculated with 0.5 –1 ml of the overnight culture. The cultures are incubated at 37 °C with shaking (230–250 rpm). Once the cells reach log phase, OD_{600} = 0.5–0.7, the expression of OMPs is induced. For the pET expression system, this is accomplished by adding 0.5 mM to one flask, leaving it at 37 °C, and 0.1 mM IPTG to the second flask and moving it to 25 °C. Subsequent (or parallel!) expression trials can use other induction temperatures including 30, 18, and 12 °C. Postinduction time points are collected. At 37 °C, take time points at 2 and 4 h, at 25 °C, 3 and 6 h. If expression trials are performed at other temperatures, the time points are adjusted accordingly so that cooler temperatures are allowed to induce longer. At 30 °C, time points are taken at 2.5 and 5 h, at 18 °C, time points are taken at 12 and 18 h, and at 12 °C, time points are taken at 24 and 36 h. For each time point, cells are harvested by centrifugation 4000 × *g*, 20 min at 4 °C and the medium removed. At this point, the pellet may be stored at –20 °C for later analysis.
3. *Analysis of protein trafficking*. The cell pellet from step 2 is resuspended in lysis buffer (10 mM Tris, 5 mM EDTA (pH 8.0), 1 mg/ml DNaseI, and 0.25 mM PMSF). Quick cell lysis is performed using by sonication for 3–5 min on ice. Cell debris is removed by low-speed centrifugation (20,000 × *g*, 20 min at 4 °C). The supernatant (containing the cytoplasmic fraction and the membranes) and pellet (containing unbroken cells and inclusion bodies) are separated and each is suspended in SDS-PAGE sample buffer (typically final concentration 25 mM Tris-HCl (pH 6.8), 50 mM DTT, 1% SDS, 0.05%, bromophenol blue, and 5% glycerol), and then separated by SDS-PAGE. The estimated expression level of outer membrane integrated or inclusion body formed protein is determined by western blotting for the affinity tag or with a protein specific antibody.

Tips: If proteins have been expressed into inclusion bodies and this is not desirable, the N-terminal signal sequence may be altered (see main text for suggestions). The protein containing the new signal sequence can be subjected to a new round of expression trials (return to step 1). In addition, alteration of expression temperature or reducing the IPTG concentration to as low as 0.05 mM may improve the expression into the membrane.

B. Extraction of OMPs from the Outer Membrane

If a β-barrel OMP is expressed into the outer membrane, the next step is extraction from the membrane and purification in detergent micelles. For any membrane protein, the purification scheme may include separation of the membranes from the cytosol using high-speed centrifugation. For β-barrel proteins, additional separation of the outer membrane from the inner membrane fraction is postulated to improve the purity of the final sample, especially when the protein has not been overexpressed using a plasmid. However, this is not an absolute requirement for structure determination, and the majority of protocols do not include this step.

Whether or not the membranes have been isolated, extraction from the membrane requires the use of a detergent at relatively high concentrations. There are over a thousand detergents commercially available in the U.S., but only a small subset have been recognized as useful for maintaining the fold and biological activity of membrane proteins. Commercially available detergents can be subdivided into four main classes based on the nature of the hydrophilic head group (1) ionic, (2) nonionic, (3) zwitterionic, and (4) bile acid salts (Prive, 2007; Seddon, Curnow, & Booth, 2004). These detergent classes can be further subdivided into different groupings based upon the chemical structure of their head and tail groups. A review of the literature indicates that a subset of commercially available detergents have had a higher success rate for the crystallization of β-barrel membrane proteins (Table I). This analysis reveals that bile acid salts have never been successfully used for structure determination of a β-barrel membrane protein (Table II) and will not be discussed in this context.

1. Ionic Detergents

Ionic detergents can be either cationic or anionic and contain a net charge on the molecule. Sodium dodecyl sulfate (SDS) is an example of an anionic detergent. Like SDS, many ionic detergents are harsh, which may account for the fact that ionic detergents have had relatively little success in the X-ray crystallography of β-barrel membrane proteins. Ionic detergents have been used during extraction in only four examples (9%) of all β-barrel membrane protein structures determined. These ionic detergents have always been exchanged prior to crystallizations and have never been used as the primary detergent during crystallization.

2. NonIonic Detergents

Nonionic detergents are perhaps the most frequently used for both membrane extraction and crystallization of membrane proteins. Nonionic detergents often contain a sugar head group, such as the maltosides and glucosides (e.g., dodecyl maltoside (DDM), octyl glucoside (OG)), but also include the polyoxyethylenes (given as C_mE_n where m is carbon atom number and n is oxyethelene number).

3. Zwitterionic Detergents

Zwitterionic detergents may contain charged species as a part of their chemical structure, but they are electrically neutral. Zwitterionic detergents include amine oxides (e.g., lauryl-dimethyl-aminoxide; LDAO, decyl-dimethyl-

TABLE I

Detergents Successfully Used for β-Barrel Membrane Protein Structure Determination and Their Classification

Class	Ionic detergent		Nonionic detergent				Zwitterionic detergent		
Sub-class	Sarcosine	Sulfuric acid	Alkyl maltoside	Alkyl glucoside	Poly oxyethylene	Glucamide	Amine oxide	Zwittergent	Sulfoxide
	Lauroyl-sarcosine	Sodium-dodecyl-sulfate (SDS)	Dodecyl-maltoside (DDM)	Dodecyl-glucoside (DDG)	Octyl-POE	Decanoyl-hydro-xyrthylglucamide (HEGA-10)	Lauryl-dimethyl-aminoxide (LDAO)	Dodecyl-dimethyl-ammonio-propanesulfonate (Zwittergent 3-12)	Octyl-hydroxyethyl-sulfoxide (OHESO)
			Decyl-maltoside (DM)	Octyl-glucoside (OG)	Octyl-tetra-oxyethylenoxide (C_8E_4)		Decyl-dimethyl-aminoxide (DDAO)	Tetradecyl-dimethyl-ammonio-propanesulfonate (Zwittergent 3-14)	
			Cyclohexyl-methyl-maltoside (Cymal-1)	Heptyl-glucoside (HpG)	Decyl-penta-oxyethylenoxide ($C_{10}E_5$)		Hexyl-dimethy-aminoxide (HDAO)		
			Cyclohexyl-propyl-maltoside (Cymal-3)	Hexyl-glucoside (HG)	Dodecyl-nona-oxyethlenoxide ($C_{12}E_9$)				
					Tetramethylbutyl-phenyl-polyeth-ylene glycol (Triton X-100)				

Detergents are classified here in accordance with the system described in this chapter. The abbreviated detergent name utilized in this chapter is shown in parentheses. Elugent is not on this list, as it is not clear which nonionic detergent is in the actual solution.

Crystallization Conditions of All Determined Unique β-Barrel Membrane Protein

Protein	Reference	PDB ID	Resolution	Expression	Solubilization	Crystallization	Buffer and pH	Precipitants	Salts	Additives
Porin (*R. capsulatus*)	Weiss and Schulz (1992)	2POR	1.8	Native source	2% SDS	0.6% C_8H_4	20 mM Tris (7.2)	23–30% PEG 600	300 mM LiCl	3 mM NaN_3, 10 mM EDTA
OmpF	Cowan et al. (1992)	2OMF	2.4	Vector-OM	3% Octyl-POE	0.6% OHESO, 0.1% Octyl-POE	50 mM Tris (9.8)	8.5–10.5% PEG 2000, 700 mM $MgCl_2$		
PhoE	Cowan et al. (1992)	1PHO	3	Vector-OM	3% Octyl-POE	0.5% OG, 0.1% C_8E_4	25 mM Tris (8.0)	7.5–9.5% PEG 2000	250 mM $MgCl_2$	
Porin (*R. blastica*)	Kreusch, Neubuser, Schiltz, Weckesser, and Schulz (1994)	1PRN	1.96	Native source	2% LDAO	0.6% C_8H_4	20 mM Tris (6.8)	30–38% PEG 600	300-400 mM LiCl	
LamB (*E. coli*)	Schirmer, Keller, Wang, and Rosenbusch (1995)	1MAL	3.1	Vector-OM	3% Octyl-POE	0.1% $C_{12}E_9$, 0.4% DM	20 mM HEPES (7.0)	15–18% PEG 2000	100 mM $MgCl_2$	3 mM NaN_3
LamB (*S. typhimurium*)	Meyer et al. (1997)	2MPR	2.4	Native source	2% LDAO	0.3% C_8E_4, 0.8% HDAO		28–32% PEG 1500		1 mM $MgCl_2$, 1 mM $CaCl_2$, 0.02% NaN_3
Porin (*P. denitrificans*)	Hirsch et al. (1997)	N.D.	3.1	Native source	2% LDAO	1% OG	20 mM Tris (7.5)	28–32% PEG 600, 10% PEG 200	200 mM KCl	10 mM $CaCl_2$, 1 mM NaN_3

(*continued*)

TABLE II (*continued*)

Protein	Reference	PDB ID	Resolution	Expression	Solubilization	Crystallization	Buffer and pH	Precipitants	Salts	Additives
ScrY	Forst, Welte, Wacker, and Diederichs (1998)	1A0T	2.4	Vector-OM	0.6% LDAO	1.2% OG, 1% HG, 1% HDAO	20 mM Tris (7.7)	12–15% PEG 2000	100 mM LiCl	20 mM $MgSO_4$
OmpA	Pautsch and Schulz (1998)	1BXW	2.5	Inclusion body	5% Octyl-POE	0.6% C_8E_4	25 mM Pottasium-phosphate (5.1)	12% PEG 8000, 10% MPD		
FhuA	Locher et al. (1998)	1BY3	2.74	Vector-OM	2% Octyl-POE, 0.5% DG, 1% OG	0.5% OHESO	150 mM Sodium-phosphate (6.2)	33% PEG 2000	450 mM NaCl	
FhuA	Ferguson, Hofmann, Coulton, Diederichs, and Welte (1998)	2FCP	2.5	Vector-OM	1% LDAO	0.8% DDAO	100 mM Cacodylate (6.4)	11% PEG 2000, 3% PEG 200		20% glycerol, 1% *cis*-inositol
OmpK36	Dutzler et al. (1999)	1OSM	3.2	Native source	2% SDS	0.6% OHESO, 0.1% Octyl-POE	50 mM Tris (9.8)	15% PEG 2000	500 mM $MgCl_2$	
OmpLA	Snijder et al. (1999)	1QD5	2.17	Native source	0.65% Triton X-100	1.5% OG	100 mM Bis-Tris (6.0)	25–29% MPD		1 mM $CaCl_2$, 1 mM NaN_3
OmpX	Vogt and Schulz (1999)	1QJ8	1.9	Inclusion body	5% Octyl-POE	0.6% C_8E_4	100 mM Na-acetate (4.6)	30% 2-propanol	200 mM $CaCl_2$	20% glycerol

FepA	Buchanan et al. (1999)	1FEP	2.4	Vector-OM	2% Triton X-100	0.055% LDAO	100 mM Tricine (8.0)	28–32% PEG 1000	350 mM NaCl	10% glycine, 1 mM NaN_3, 1.75% heptanetriol
Omp32	Zeth et al. (2000)	1E54	2.1	Native source	10% Octyl-POE	2% OG	100 mM HEPES (7.5)	1.3–1.4 M Li_2So_4		
TolC	Koronakis et al. (2000)	1EK9	2.1	Vector-OM	5% Triton X-100	0.6% DDG, HG, HpG, OG	20 mM Tris (7.4)	10% PEG 400, 12.5% PEG 2000	400 mM NaCl	10 mM NaCl, 20 mM $MgCl_2$, 1.5% heptanetriol
OmpT	Vandeputte-Rutten et al. (2001)	1I78	2.6	Inclusion body	31.25 mM Zwittergent 3-12	1% OG	100 mM Na-citrate (5.5)	28% MPD	500 mM NaCl	
OpcA	Prince et al. (2002)	1K24	2.03	Inclusion body	5% LDAO	1% $C_{10}E_5$, 0.25% HpG	50 mM Tris (7.5)	20% PEG 4000	150 mM Zn $(CH_3COO)_2$, 50 mM $ZnCl_2$	
FecA	Ferguson et al. (2002)	1KMO	2	Vector-OM	2% Triton X-100	0.055% LDAO	100 mM Tricine (8.0)	28–32% PEG1000	350 mM NaCl	10% glycerol, 1m M NaN_3, 1.75% heptanetriol
BtuB	Chimento et al. (2003)	1NQE	2	Vector-OM	150 mM C_8E_4	20 mM C_8H_4	50 mM Cacodylate (6.6)	4–7% PEG 3350	200–400 mM $Mg(CH_3COO)_2$	
NspA	Vandeputte-Rutten Bos, Tommassen, and Gros (2003)	1P4T	2.55	Inclusion body	1% Zwittergent 3-12	0.06% $C_{10}E_5$	100 mM ADA (6.6)	12% PEG 3000	100 mM Li_2SO_4	2% 2-propanol

(*continued*)

TABLE II *(continued)*

Protein	Reference	PDB ID	Resolution	Expression	Solubilization	Crystallization	Buffer and pH	Precipitants	Salts	Additives
MspA	Faller, Niederweis, and Schulz (2004)	1UUN	2.5	Vector-OM	0.6% C_8E_4	0.2% C_8E_4, 35mM HEGA-10	60 mM Na-citrate (5.6)	12% PEG 4000	100 mM Li_2SO_4	
OprM	Akama et al. (2004)	1WP1	2.56	Native source	2.5% OG	30 mM Cymal-3, 0.1% Octyl-POE	50 mM Tris, 100mM imidazole (8.0)	22–28% MPD	450–600 mM Na-acetate	
NalP	Oomen et al. (2004)	1UYN	2.6	Inclusion body	0.5% Zwittergent 3-12	0.06% $C_{10}E_5$, 0.5% HpG	100 mM Na-citrate (4.0)	9% PEG 1000	200 mM Li_2SO_4	
PagP	Ahn et al. (2004)	1THQ	1.9	Inclusion body	0.5% LDAO	0.05% LDAO	100 mM Na-citrate (5.6)	30% MPD	200 mM NH_4 (CH_3COO)	
Tsx	Ye and van den Berg (2004)	1TLY	3.01	Vector-OM	1% LDAO, 1% OG	0.45% C_8E_4	50 mM Na-acetate (4.3)	27-32% PEG 550		
FadL	van den Berg et al. (2004)	1T16	2.6	Vector-OM	1% LDAO, 1% OG	0.45% C_8E_4	50 mM Cacodylate (5.3)	27–32% PEG 4000		2–5 mM $CuSO_4$
VceA	Federici et al. (2005)	1YC9	1.8	Vector-OM	2% DDM	0.1% DDM, 82 mM OG	100 mM HEPES (7.0)	45% MPD	200 mM NaCl	
FptA	Cobessi, Celia, and Pattus (2005a)	1XKW	2	Native source	1% Octyl-POE	0.05% LDAO	100 mM Na-acetate (4.6)	4.5–9% PEG 8000 or PEG 10000	150 mM $(NH_4)_2SO_4$	25% ethylene glycol
FpvA	Cobessi et al. (2005b)	1XKH	3.6	Native source	1% Zwittergent 3-14	0.75% C_8E_5	100 mM Na-citrate (5.6)	13–16% PEG 4000		20–25% ethylene glycol

OmpC	Basle, Rummel, Storici, Rosenbusch, and Schirmer (2006)	2J1N	2	Vector-OM	3% Octyl-POE	1.8% OHESO, 0.3% Octyl-POE	25 mM Tris (8.0)	28% PEG 2000	500 mM NaCl	
OmpG	Subbarao and van den Berg (2006)	2F1C	2.3	Vector-OM	1% LDAO	0.4% C_8E_4	50 mM Cacodylate (5.5)	28–32% PEG 400, 1.0–1.5 M Na(COOH)		
OmpW	Hong et al. (2006)	2F1V	2.7	Vector-OM	1% LDAO	0.45% C_8E_4	50 mM Na-citrate (5.6)	27–33% MPD	200 mM $NH_4(CH_3COO)$	
Hia	Meng, Surana, St Geme, and Waksman (2006)	2GR8	2	Vector-OM	5% Elugent	0.6% C_8E_4	100 mM CHES (10.0)	12% PEG 4000	200 mM $(NH_4)_2SO_4$	
OprP	Moraes et al. (2007)	2O4V	1.94	Native source	3% C_8E_4	0.6% C_8E_4	100 mM HEPES (6.5)	5% MPD, 7.5–10% PEG 8000	50 mM $KHPO_4$	
OprD	Biswas et al. (2007)	2ODJ	2.9	Vector-OM	1% LDAO	0.4% C_8E_4	100 mM Na-citrate (4.0)	30% PEG 400	100 mM Li_2SO_4, 100 mM NaCl	
Cir collinin Ia	Buchanan et al. (2007)	2HDI	2.5	Vector-OM	5% Elugent	0.05% LDAO, 0.45% C_8E_4	100 mM MES (6.2)	22% PEG 2000	200 mM NaCl	10% glycerol
EspP	Barnard et al. (2007)	2QOM	2.66	Vector-OM	5% Elugent	0.82% OG, 0.34 M Cymal-1	100 mM Cacodylate (6.4)	33% PEG 1000	200 mM NaCl	5% glycerol

(continued)

TABLE II *(continued)*

Protein	Reference	PDB ID	Resolution	Expression	Solubilization	Crystallization	Buffer and pH	Precipitants	Salts	Additives
FhaC	Clantin et al. (2007)	2QDZ	3.15	Vector-OM	1.5% OG	1% OG	400 mM imidazole (6.5)	32% PEG 1000		
OpdK	Biswas, Mohammad, Movileanu, and van den Berg (2008)	2QTK	2.8	Vector-OM	1% LDAO	0.5% C_8E_4	100 mM Na-acetate (6.0)	30% PEG 4000	200 mM $NH_4(CH_3COO)$	
TodX	Hearn, Patel, and van den Berg (2008)	3BS0	2.6	Inclusion body	1% Lauroyl-sarcosine	0.45% C_8E_4	100 mM Na-acetate (4.6)	25% PEG 4000	200 mM $(NH_4)_2SO_4$	
TbuX	Hearn et al. (2008)	3BRY	3.2	Inclusion body	1% Lauroyl-sarcosine	0.45% C_8E_4	50 mM Na-acetate (4.5)	20% PEG 4000	100 mM $MgSO_4$	
PapC	Remaut et al. (2008)	2VQI	3.2	Vector-OM	1% DDM	0.8% C_8E_4, 2 mM LDAO	50 mM Na-citrate (5.6)	5–20% MPD, 6–12% PEG 4000	100 mM $(NH_4)_2SO_4$	

This analysis was based on the information reported for the structure determinations of the 44 available β-barrel membrane protein structures. If a protein has been crystallized with different crystal forms in the same journal issue, the conditions from the best-diffracted crystals were used. Two independent laboratories crystallized FhuA and determined structures by using different strategies at the same time, therefore both conditions were included in the analysis.

aminoxide; DDAO), sulfoxides (e.g., octyl-hydroxyethyl-sulfoxide, OHESO), and propane sulfonates (e.g., zwittergent 3–12). LDAO in particular has been used frequently during the crystallization of both α-helical and β-barrel membrane proteins (Newstead et al., 2008).

It has been proposed that the hydrophilic head group of a detergent largely influences the interaction between the detergent and the protein. While the length of the hydrophobic chain affects the critical micelle concentration (CMC) and aggregation number. Combined together, these two properties dictate the micelle size. Detergents with highly polar or charged hydrophilic ends (i.e., SDS), or shorter hydrophobic chains (i.e., C_6-hexyl glucoside (HG)) and a small micelle size have an increased propensity to denature proteins. In contrast, detergents with only slightly polar head groups and long hydrophobic tails (i.e., C_{12}-DDM) form large micelles that cover the majority of the protein surface (Prive, 2007). While this has a higher likelihood of retaining protein stability and activity, it may also mask the small, soluble regions of protein important for mediation of crystal contacts in the lattice. Intermediate detergents with a weakly polar head group and a hydrophobic tail between 8 and 12 in length, such as OG or LDAO, may help to expose the polar surfaces of the protein and allow an increased surface area for crystal packing while retaining the native fold of the protein.

OMPs may be extracted from the membrane and purified in one detergent, but exchanged into a second detergent or mixture for crystallization. Most β-barrel membrane proteins whose structures have been determined to date have been extracted and purified using nonionic or zwitterionic detergents, with only four exceptions where ionic detergents were used for extraction (Fig. 3A). The most successful detergents in terms of the extraction step are the zwitterionic aminine oxide LDAO (used in 29% of reported structures) and nonionic polyoxyethylene group (39% of reported structures); these are therefore excellent detergents to start with for extraction of novel OMPs. The relatively low cost of these detergents is also advantageous. The nonionic alkylmaltosides (4%) and alkylglucosides (14%) are somewhat less frequently used for the extraction step; however this may reflect the consideration of cost. In addition, successful extraction detergents tend to group into two hydrophobic tail lengths, C_8 (41% of reported structures) and C_{12} (47%); detergents with these two chain lengths were used for extraction just over 88% of all of the successfully determined β-barrel protein structures (Fig. 3B). In some cases more than one detergent was used for solubilization, which brings our total percentage of successfully used detergents for solublization to over 100%. An example of this is the solubilization of the OMP FadL, which was performed by using mixture of 1% of LDAO and OG (van den Berg, Black, Clemons, & Rapoport, 2004).

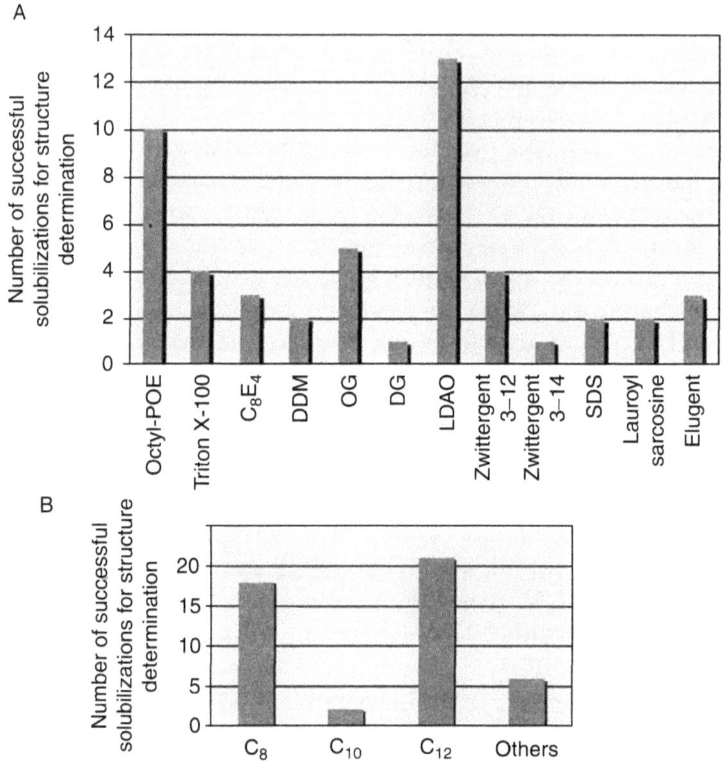

FIGURE 3 Detergents successfully used for solubilization of β-barrel membrane proteins for structure determination. (A) The number citations using a particular detergent for solubilization of a unique β-barrel protein is shown graphically. Using Octyl-POE and LDAO for extraction was the most successful. In some cases more than one detergent was used for solubilization, but they were counted as individual instances here. (B) The length of the hydrophobic tails of detergents successfully used for solublization was analyzed. Detergents with a C_{12} or C_8 tail were the most usefully. Elugent (Calbiochem) has been removed from this analysis since it is a detergent mixture and the company does not report the chain lengths.

Data analysis indicates that 2–3% Octyl-POE (nonionic), 1–2% OG (nonionic), or 1–2% LDAO (zwitterionic) are the most frequently used detergents for β-barrel protein extraction. During extraction, isolated membranes are incubated with buffer solution containing detergent, and incubated for 30 min to 2 h. This step is followed by removal of insoluble material by high-speed centrifugation. The extracted protein is then purified according to a standard protocol with detergents included in all buffers at concentrations 2–10 times the CMC. Detergent exchange prior to crystallization frequently occurs on a chromatography column during purification, usually during the final size exclusion chromatography step, which allows the monodispersity of the sample in the new detergent to be monitored. A protocol detailing extraction, including detergent choice, is provided in Protocol 2.

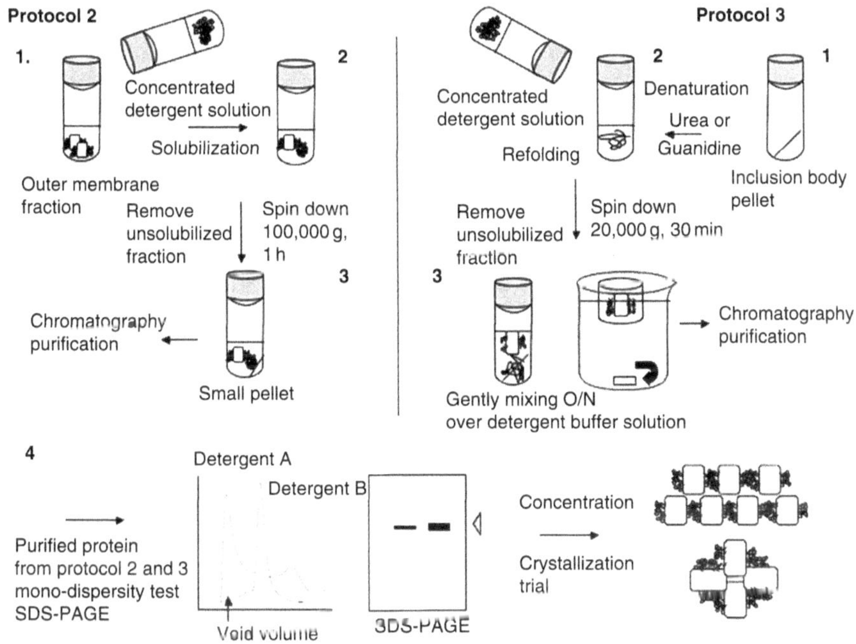

PROTOCOL 2 Solubilization of OMPs from membranes.

1. *Membrane resuspension.* The cell pellet of 200 ml of cells is collected by low speed centrifugation and resuspended in lysis buffer (10 mM Tris, 5 mM EDTA, 1 mg/mL DNaseI, and 0.25 mM PMSF). The lysate is cleared of cellular debris by low-speed centrifugation (20,000 × g 20 min, 4 °C) and membrane fraction is separated from the cytosolic fraction using high-speed centrifugation (100,000 × g, 1.5 h at 4 °C). The supernatent is discarded and the purified membranes are resuspended in 20 ml of buffer (20 mM Tris, 100 mM NaCl, pH 8.0).
2. *Extraction.* The test detergent for extraction is added to the solubilized membranes (e.g., 1–2% LDAO, 2–5% Octyl-POE, 1–2% OG). The detergent-membrane mixture solution is then incubated for 30 min-2 h at 4 °C with gentle agitation.
3. *Clarification.* Insoluble material and nonsolubilized membranes are removed from the detergent extracted membranes using high-speed centrifugation (100,000 × g, 30 min). The detergent solubilized fraction (supernatant) is filtered through a 0.2 μm filter and is subjected to the first purification step to remove the major contaminants. For plasmid-expressed OMPs, this is usually an affinity column. The affinity chromatography is performed according to the standard manufacturers' protocol with 2–10× the CMC of the detergent supplemented into all buffers.

(*continued*)

(*continued*)

4. *Assessment of monodispersity.* The detergent extracted protein is assessed for monodispersity using an analytical gel filtration column, such as a Superdex 200–10/300, equilibrated in a buffer containing 2–10× the CMC of the extraction detergent. A monodisperse sample contains a single peak with a molecular weight similar to that anticipated (see Protocol summary figure, step 4). If the protein is not monodisperse, the extraction detergent is incompatible with the correct fold of the OMP and a new extraction detergent must be identified (go to step 2). If the protein is monodisperse in the extraction detergent, the gel filtration column can be reequilibrated in a battery of other detergents and the protein injected serially to assess for monodispersity. Prior to crystallization trials, the protein is concentrated to 5–50 mg/ml using a molecular cut-off filtration centrifuge tube (such as Amicon (Millipore), vivaspin (GE healthcare)). The purified concentrated protein (>10 mg/ml) is ready to start initial crystallization trials.

Tips: Try several different detergents until single detergent or combination of detergents is identified that yields mono-disperse protein. There are many detergents that support the native fold of the protein, but cannot extract the OMP from the membrane. As a result, it may be necessary to extract the protein from the membrane using one detergent and replace this detergent during chromatography. The exact protein concentration necessary for crystallization trials is protein dependent. The Precrystallizaiton test kit (PCT, Hampton) may gauge a starting value. In each step, the solubilization efficiency and protein purity can be assessed by SDS-PAGE analysis.

C. Refolding of OMPs from Inclusion Bodies

Refolding from inclusion bodies requires four steps (Protocol 3): (1) isolation of the inclusion body, (2) denaturation, (3) refolding, and (4) separation of folded and aggregated proteins. Cells are first lysed, and the inclusion bodies and cellular debris are separated from soluble material by centrifugation. The resultant pellet is washed extensively using a buffer that includes detergent to obtain the pure inclusion bodies. These inclusion bodies are solubilized using a denaturing solution that usually consists of either 8 M urea or 6 M guanidine-HCl. Refolding can be performed by quick dilution, where the clarified inclusion body is diluted 5-100 times into a buffer solution supplemented with detergent, or by slow buffer exchange, where the solution containing the denaturant is gradually replaced by overnight dialysis. During the refolding process, the addition of a high concentration (~0.5 M) of L-arginine is often used as an additive to suppress protein aggregation, and may improve the efficiency of the refolding process. Finally, the refolded OMP is separated from unfolded aggregated sample by size exclusion chromatography which additionally removes any remaining denaturant. A further modification of the quick dilution method was employed during the recent structure determination of the *N. meningitidis* PorB (M. Tanabe and T. M. Iverson, in preparation). After clarification of the urea-solubilized inclusion bodies by centrifugation, urea-solubilized PorB was directly injected onto a size exclusion column. In this example, the remaining urea was exchanged on the gel filtration column and the modified procedure required less than 1 day. The decreased amount of time required to purify proteins refolded from inclusion bodies using this method makes it an attractive alternative that may be used more frequently as a standard protocol in the future.

PROTOCOL 3 The denatunation and refolding of OMPs from inclusion bodies.

1. *Inclusion body isolation.* Cells are lysed, and the inclusion bodies and cellular debris are separated from soluble material by centrifugation at 20,000 × *g* for 20 min at 4 °C. The resultant pellet is washed extensively using a buffer that includes detergent (e.g., 20 mM Tris, 200 mM NaCl, supplement with 1% Triton X-100, or 0.5% Lauroyl-sarcosine) to remove contaminant proteins from unbroken cells and obtain the pure inclusion bodies.
2. *Clarification of the inclusion bodies.* The isolated, pure inclusion bodies are solubilized using a denaturing solution that consists of either 8 M urea or 6 M guanidine-HCl. The suspension is then incubated at 4 °C or room temperature for 10–30 min. Clarification may be enhanced using mild sonication in a bath sonicator.
3. *Dilution and refolding.* To refold by quick dilution, the clarified inclusion body fraction is diluted immediately in a buffer solution supplemented with concentrated detergent (5–100 times higher concentrated than CMC). This can be up to 5% Octyl-POE, 10% LDAO, 2% Zwitergent 3–12, and 1% Lauroyl-sarcosine. To refold slow buffer exchange, the solution containing the denaturant is gradually diluted by overnight dialysis.
4. *Purification.* The solution containing refolded OMP is separated from unfolded material by centrifugation at 20,000 × *g* for 20 min, at 4 °C. The dispersed, refolded OMPs are isolated using gel filtration chromatography and are assessed for monodispersity. This final step additionally removes any remaining denaturant solution. A successful refolding protocol usually results in 20–30% of the inclusion body being refolded correctly. Inclusion bodies that do not contain a peak at the correct molecular weight can be refolded in a different detergent (go to step 3). The protein purity is analyzed by SDS-PAGE. β-Barrel secondary structure is estimated by circular dichroism (CD) spectroscopy to verify proper folding. Detergent exchange and crystallization proceeds similarly to Protocol 2, step 4.

Tips: Some membrane proteins will be unstable when in detergent at 4 °C. If aggregation is a chronic problem, attempt performing the entire purification at room temperature. During the refolding process, the addition of a high concentration (~0.5 M) of L-arginine is often used as an additive to suppress protein aggregation and to improve the efficiency of the refolding process.

As with OMPs expressed into the membrane, the selection of detergent for refolding OMPs from inclusion bodies is empirical. The majority of β-barrel proteins purified from inclusion bodies for structural studies used relatively small alkyl chain and zwitterionic detergents, including 5% Octyl-POE, 0.5–5% LDAO, 0.5–1% Zwittergent 3–12, and 1% Lauroyl-sarcosine. To assess if the detergent selected for refolding supports the native architecture of the OMP, the monodispersity and β-barrel fold can be assessed using multiple complementary methods. The most common is size exclusion chromatography, since this is usually a part of the purification protocol. This can be complemented by CD spectroscopy, dynamic light scattering, analytical ultracentrifugation, and, most importantly, functional analysis.

IV. PRELIMINARY CRYSTALLIZATION AND DIFFRACTION-BASED OPTIMIZATION

A. Detergent Selection

A survey of the methods reported for β-barrel OMP crystal structures shows that the same detergent was used during solubilization and crystallization in only 9% (four examples) of structurally characterized β-barrel proteins (Ahn et al., 2004; Chimento, Kadner, & Wiener, 2003; Clantin et al., 2007; Moraes, Bains, Hancock, & Strynadka, 2007). In the remaining 91%, the detergent used during solubilization differed from the detergent used for crystallization, and a detergent exchange step was included in the protocol. This survey shows that surprisingly few detergents have been successful for growth of diffraction quality crystals, which provides a focused set of starting detergents for future studies. Only nonionic and zwitterionic detergents have been used as the crystallization detergent in β-barrel membrane protein structure determinations. Nonionic detergents currently have the highest success rate for the crystallization of β-barrel membrane proteins, representing 86% of the available crystal structures. Zwitterionic detergents have been used in the 29% of structure determinations (Table II).

The nonionic polyoxyethylenes have been used for more than half of β-barrel crystallizations (in 26 of 44 examples, or 59% of reported structures; Fig. 4). Of the polyoxyethylenes, C_8E_4 has produced diffraction quality crystals the most frequently, having been used to crystallize 41% of structurally characterized β-barrel proteins. Alkyl-glucoside (34%) and amine-oxides (20%) also have good records. The use of alkyl-maltoside (9%) and sulfoxide (9%) and glucamide (2%) has been minor. This analysis suggests that if no prior knowledge of optimum purification techniques is available for a target β-barrel protein, 0.4–0.6% C_8E_4, 1% OG, and 0.05–0.1% LDAO should be used as a first-choice detergents since these three detergents combined account for ~80% of all successful projects.

Reports of crystallizations of β-barrel OMPs using parallel screening in a variety of detergents are consistent with these three detergents being the most likely to form diffraction quality crystals. In one example, the nonspecific porin OmpG was solubilized with four different detergents (OG, DDM, LDAO, and C_8E_4), and subjected to crystallization trials. Diffraction quality crystals were only grown from protein solubilized in either OG or LDAO

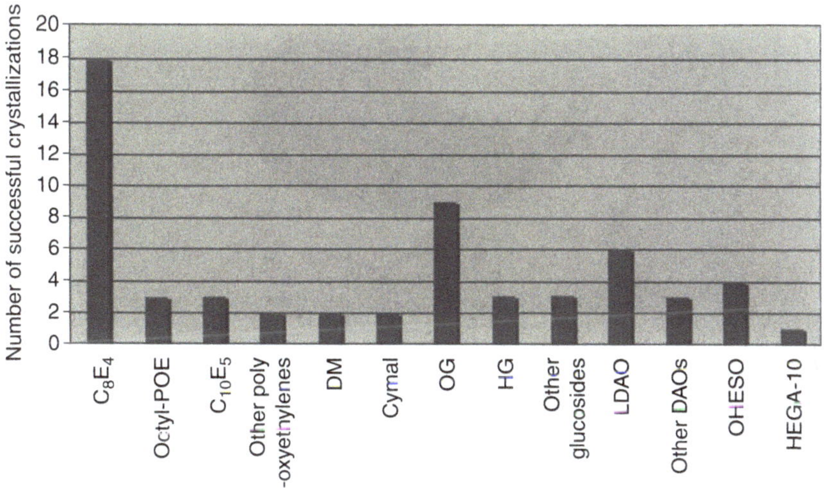

FIGURE 4 Summary of detergents used in crystallization conditions for β-barrel membrane proteins. The various detergents and their classification were tallied for use during crystallization of β-barrel OMP's. Some detergents were used in combination with another in a single publication but were counted as individual instances here.

(Yildiz et al., 2006). Similarly, the small hydrophobic transport channel OmpW was simultaneously purified in LDAO, OG, and C_8E_4; diffraction quality crystals were only obtained from protein purified using C_8E_4 (Hong, Patel, Tamm, & van den Berg, 2006). An expanded battery of detergents was used during crystallization of the human mitochondrial voltage dependent anion channel VDAC-1. This structure determination tested 16 similar subclasses of detergent for crystallization. The best VDAC-1 crystal grew from Cymal-5, an exception to what has been observed for most β-barrel proteins (Meins, Vonrhein, & Zeth, 2008).

Once an initial crystallization lead is identified, switching from a single detergent to a detergent mixture can improve the diffraction quality. Thirty-six percentage of β-barrel protein structures used more than one detergent in the crystallization solution (Fig. 5). This is usually achieved by doping the crystallization reaction with a small amount of the secondary detergent; however, the protein purification protocol can also be altered such that a detergent mixture is used during the last step. Detergents with tail lengths equal to or longer than eight (i.e., OG) are most commonly the "main" detergent, and these are doped with detergents that have a smaller hydrophobic tail, such as Cymal, HG, and hexyl-dimethyl-aminoxide (HDAO). These small detergents are predicted to decrease the micelle size, which may be more conducive to crystal formation. In one example, crystals of the adhesin protein OpcA were dramatically improved by addition of heptyl

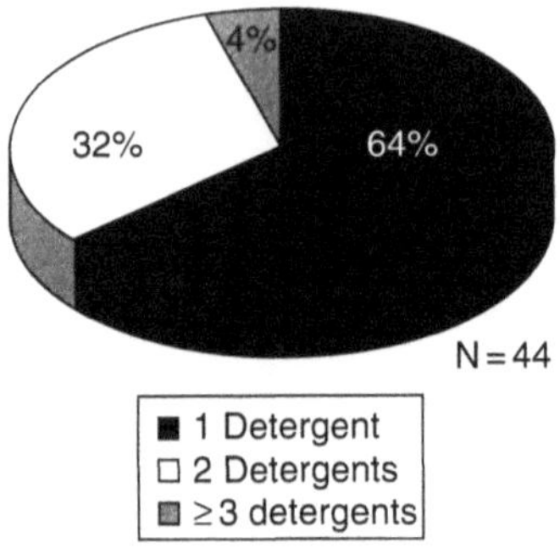

FIGURE 5 Analysis of the number of detergents used in the crystallization of β-barrel membrane proteins. Although most use only one detergent in the crystallization mother liquor, more than 1/3 of the known structures were determined from crystals grown in a solution containing multiple detergents.

glucoside (HpG) to the protein solubilized in $C_{10}E_5$ as a main detergent (Prince, Achtman, & Derrick, 2002). In addition to the effect on micelle size, HDAO was also crucial for suppression of phase separation induced by the precipitant in the crystallization reactions of *S. typhimurium* maltoporin LamB (Meyer, Hofnung, & Schulz, 1997). In an extreme example, TolC used a mixture of four detergents [dodecyl glucoside (DDG), HpG, HG, and OG] in the crystallization reactions (Koronakis, Sharff, Koronakis, Luisi, & Hughes, 2000).

B. Chemical Conditions for β-Barrel Membrane Protein Crystallization

1. Precipitants for Crystallization

Using a literature survey, different precipitants, buffers, salts, and additives were analyzed for their success rate in β-barrel membrane protein crystallization. Important differences between optimal crystallization conditions for β-barrel membrane proteins and soluble and α-helical membrane proteins are observed in this comparison (Kimber et al., 2003; Newstead et al., 2008; Page & Stevens, 2004). Polyethylene glycol (PEG) polymers were used as the precipitating agent in more than 80% of β-barrel OMP crystallizations while methyl pentanediol (MPD) was used in 20% of all β-barrel OMP crystallization (Fig. 6). Only one protein, the *E. coli* virulence-associated porin OmpX, did not use PEGs or MPD, and was instead crystallized using isopropanol as the precipitant (Vogt & Schulz, 1999). In a departure from what is observed for soluble proteins, salts are rarely successful as precipitants of β-barrel OMPs unless they are used in combination with other polymers. Only a single example, the anion-selective channel Omp32 from *Delftia*

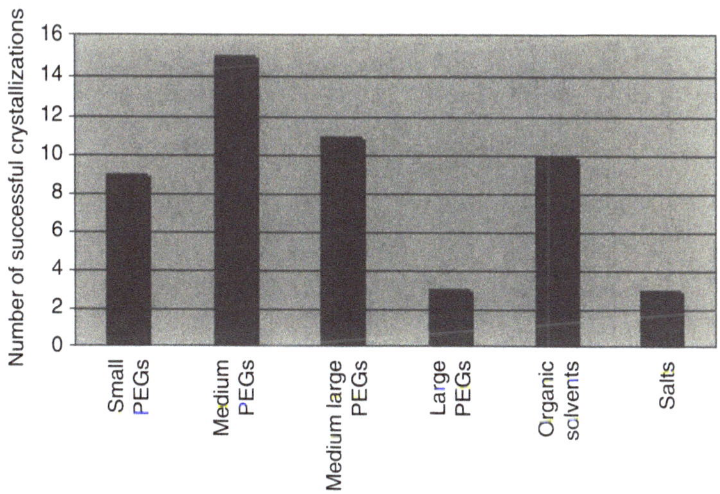

FIGURE 6 Summary of precipitants used for crystallization of β-barrel membrane proteins. The number of different precipitant types used for crystallizations are listed. All instances of precipitant use were counted even if an individual mother liquor contained more than one. Nine cases of the 44 determined structures used small PEGs (MW 200–600), 15 cases—medium PEGs (MW 1000–2500), 11 cases—medium large PEGs (MW 3000-5000), 3 cases-large PEGs (MW > 5000), 10 cases—organic solvents (2-propanol, MPD), and 3 cases used high concentrations (>500 mM) of salts ($MgCl_2$, Li_2SO_4).

acidovorans, has been crystallized using salt (1.3–1.4 M Li_2SO_4) as the lone precipitant (Zeth, Diederichs, Welte, & Engelhardt, 2000). PEGs of low molecular weight (MW 200–600) and medium molecular weight (MW 1000–2500) are more successfully used with β-barrel proteins, whereas high molecular weight PEGs have had a better success rate with soluble proteins (Kimber et al., 2003; Page & Stevens, 2004). The low molecular weight PEGs (MW 200–600) and MPD are both commonly used at a concentration of about 25%. For intermediate molecular weight PEGs (MW 1000–2500), a bimodal distribution of optimal concentrations of the precipitant is observed, with increased probabilities of crystal formation at both 30–35% and 15–20%. For high molecular weight PEG (MW > 4000) the optimal concentration appears to be around 10%. A high concentration (>0.5 M) of $MgCl_2$ and sodium formate is often paired with other precipitants such as PEG 400 (Cowan et al., 1992) and PEG 2000 (Subbarao & van den Berg, 2006).

2. Buffers and Salt Concentrations for Crystallization Conditions

Buffers and pH can greatly influence the crystallization of soluble proteins. Figure 7 shows the pH ranges that have been used for successful crystallization of β-barrel OMPs to date. Most of β-barrel proteins (64% of reported

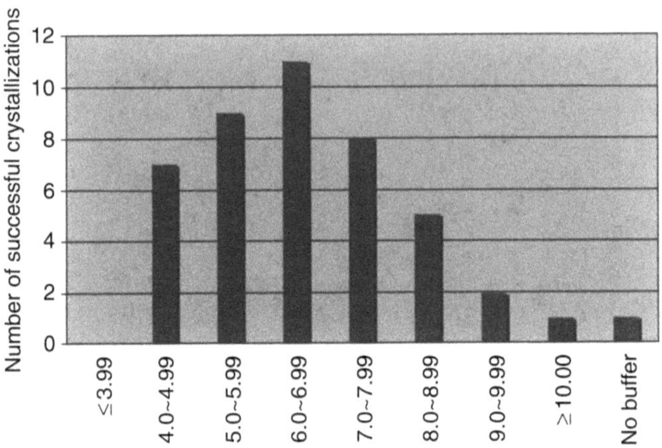

FIGURE 7 Distribution of crystallization buffer pH for β-barrel membrane protein structures. Successful crystallization conditions used pH values that ranged from 4.0 to 10.0 with one exception where no buffer was present in the crystallization mother liquor (Meyer et al., 1997). The distribution centered at pH 6–6.99.

structures) have been crystallized between pH 5 and 8. Eight percent of proteins have been crystallized at extreme pH levels (below 4.5, above 9.5). Most of the commercially available screening solutions cover the pH range from 4.6 to 9.5. The data presented here suggests that a slightly wider pH range between pH 4.0 and 10.5 should cover the broadest crystallization space. This analysis also shows no tight correlation between different buffer types and successful crystallization. As a note, there is no significant difference in this range between membrane and soluble proteins.

Of primary salts, the most commonly used is NaCl, which accounts for 9 of 44 cases (20%) of all β-barrel protein crystallizations. Successful uses of both monovalent (45%) and multivalent cationic ions (18%) are catalogued in the database (Fig. 8). Ammonium sulfate or acetate have also been useful (9% and 7%, respectively).

Small molecules, additional detergents, multivalent cations, salts, and amphiphiles can sometimes contribute to the crystallization process as they can stabilize proteins within the crystal. In this analysis, we have found that 30% of all β-barrel crystallizations use additives. The most common salts used as additives are NaN_3 (16%), $CaCl_2$ (4%), and $MgCl_2$ (4%) (Table II). Small amphiphiles (e.g., heptanetriol, dioxane) and small peptides, which are often critical to for α-helical protein crystallization, can affect detergent micelle size (Timmins, Hauk, Wacker, & Welte, 1991) and are also capable (7%) of improving crystal quality for β-barrel proteins (Buchanan et al.,

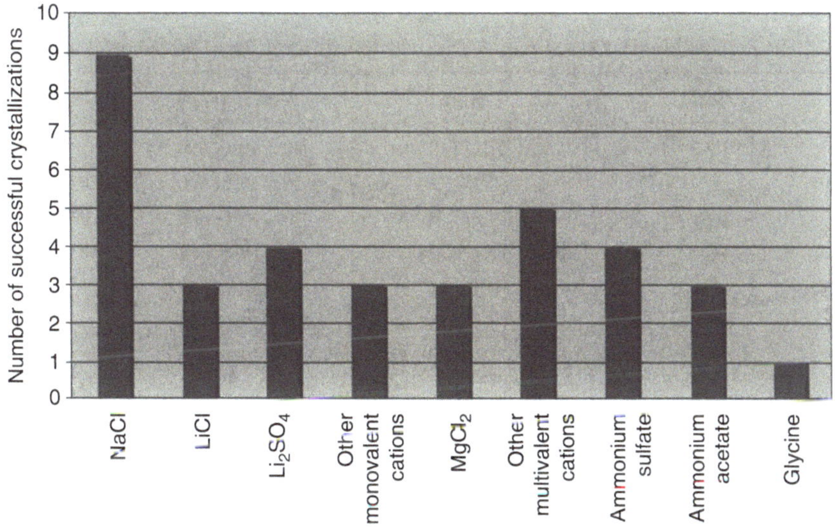

FIGURE 8 Summary of primary salts used during crystallization of β-barrel membrane proteins. The various salts used in the crystallization mother liquor are tallied. If multiple salts were used, each was counted individually. Salts were considered as precipitants if their concentration was large than 500 mM, and considered as additives if their concentration was lower than 20 mM.

1999, Ferguson et al., 2002; Koronakis et al., 2000). Interestingly, 11% of β-barrel membrane protein crystals were grown without the addition of any salt to the crystallization reaction.

The differences in optimal precipitant type and concentration suggest that current commercially available sparse-matrix screening kits may not be the most appropriate starting points in the search for crystallization conditions for β-barrel membrane proteins. The analysis presented here suggests that initial screening should focus on low- to intermediate molecular weight PEGs or MPD at select concentrations. These should be supplemented with a few of the most common salts, such as NaCl, LiCl, Li_2SO_4, $MgCl_2$, ammonium acetate, and ammonium sulfate, at concentrations between 50 and 200 mM. A small number of trials that have ammonium or lithium sulfate or acetate as the main precipitant supplemented with a low concentration of PEGs should be adequate for searching all of crystallization space. With this in mind, we have developed a sparse matrix screen specifically for β-barrel membrane proteins (Table III). Like any sparse-matrix screen, this is designed to provide initial crystallization conditions that need to be optimized through grid screening and the use of additives and detergents.

TABLE III

The Proposed Initial Crystallization Screening Matrix for β-Barrel Membrane Protein

	Salt	Buffer	Precipitant 1	Precipitant 2
1.	0.2 M LiCl	50 mM Tris (7.0)	30% PEG 600	
2.			20% PEG 600	10% PEG 8000
3.	0.2 M KCl	50 mM Tris (7.5)	30% PEG 600	10% PEG 200
4.	0.2 M $MgCl_2$	50 mM Tris (9.8)	15% PEG2000	
5.		50 mM HEPES (7.5)	1.5 M Li_2SO_4	
6.		50 mM Tris (8.0)	15% PEG 2000	
7.		50 mM Cacodylate (6.0)	30% PEG 400	1.5 M Na (COOH)
8.		50 mM Tris (9.8)	10% PEG 200	1.0 M $MgCl_2$
9.	0.2 M LiCl	50 mM HEPES (7.0)	10% PEG 2000	
10.	0.2 M LiCl	50 mM Tris (7.5)	15% PEG 2000	10% MPD
11.		50 mM Na-citrate (6.0)	15% PEG 4000	
12.	0.2 M Li_2SO_4	50 mM HEPES (7.0)	10% PEG 8000	10% MPD
13.	0.2 M Li_2SO_4	50 mM Na-citrate (4.0)	30% PEG 400	
14.	0.2 M $NH_4(CH_3COO)$	50 mM Na-acetate (6.0)	30% PEG 4000	
15.	0.2 M $MgCl_2$	50 mM Tris (7.5)	10% PEG 1500	10% PEG 400
16.	0.2 M NaCl	50 mM HEPES (7.5)	30% MPD	
17.	0.2 M $NH_4(CH_3COO)$	50 mM Tris (8.0)	30% MPD	
18.		50 mM Tris (7.5)	35% PEG 2000	
19.		50 mM potassium phosphate (5.0)	15% PEG 8000	10% MPD
20.	0.2 M $NH_4(CH_3COO)$	50 mM Na-acetate (5.6)	30% MPD	
21.	0.2 M $CaCl_2$	50 mM Glycine (9.0)	30% PEG 400	
22.	0.2 M $CaCl_2$	50 mM Na-acetate (5.6)	30% 2-propanol	10% Glycerol
23.		50 mM MES (6.0)	30% MPD	
24.	0.2 M Zn $(CH_3COO)_2$	50 mM Tris (7.5)	20% PEG 4000	
25.	0.2 M Li_2SO_4	50 mM ADA (6.5)	15% PEG 3000	
26.	0.2 M Li_2SO_4	50 mM Na-acetate (5.6)	10% PEG 1000	
27.		50 mM Tris (7.0)	10% PEG 1000	20% MPD
28.	0.2 M $(NH_4)_2SO_4$	50 mM CHES (10.0)	30% MPD	15% PEG 4000
29.		50 mM Cacodylate (6.5)	30% PEG 1000	

(continued)

TABLE III (*continued*)

	Salt	Buffer	Precipitant 1	Precipitant 2
30.	0.2 M $(NH_4)_2SO_4$	50 mM Na-citrate (5.6)	20% MPD	
31.		50 mM Cacodylate (5.0)	30% PEG 4000	
32.	0.2 M $(NH_4)_2SO_4$	50 mM Na-acetate (4.5)	30% PEG 4000	
33.	0.2 M $MgSO_4$	50 mM Na-acetate (4.6)	20% PEG 4000	
34.		50mM imidazole (6.5)	30% PEG 1000	
35.		50 mM Cacodylate (6.5)	15% PEG 6000	10% PEG 2000
36.		50 mM sodium phosphate (6.0)	30% PEG 2000	
37.		50 mM MES (6.0)	15% PEG 4000	10% Dioxane
38.	0.2 M NaCl	50 mM Tricine (8.0)	30% PEG 1000	
39.	0.2 M NaCl	50 mM Na-citrate (5.6)	15% PEG 1000	
40.		50 mM Na-citrate (6.5)	15% PEG 4000	15% MPD
41.	0.2 M $MgSO_4$	50 mM Na-citrate (5.6)	15% PEG 4000	
42.	0.2 M $(NH_4)_2SO_4$	50 mM Na-citrate (4.6)	10% PEG 10000	
43.		50 mM Cacodylate (6.5)	10% PEG 3350	
44.	0.2 M $(NH_4)_2SO_4$	50 mM Cacodylate (6.0)	30% PEG 5000	
45.			20% PEG 1500	
46.	0.2 M NaCl	50 mM Na-citrate (5.6)	15% PEG 3350	
47		50 mM Na-acetate (5.0)	30% PEG 400	
48.		50 mM ADA (6.0)	1.5 M $(NH_4)_2SO_4$	1 M Li_2SO_4
49.		50 mM Na-acetate (4.0)	30% MPD	
50.		50 mM Na-acetate (4.0)	20% PEG 3000	
51.	0.2 M NaCl	50 mM Na-acetate (4.0)	30% PEG 400	
52.	0.2 M NaCl	50 mM Na-acetate (4.0)	15% PEG 2000	
53.	0.2 M NaCl	50 mM Na-acetate (4.6)	30% PEG 1500	
54.		50 mM Na-acetate (4.6)	15% PEG 4000	
55.	0.2 M $NH_4(CH_3COO)$	50 mM Na-acetate (4.6)	30% PEG 3000	
56.	0.2 M NaCl	50 mM Na-acetate (4.6)	10% PEG 8000	1 M $(NH_4)_2SO_4$
57.	0.2 M NaCl	50 mM Na-citrate (5.6)	30% MPD	
58.		50 mM Na-citrate (5.6)	30% Jeffamine M-600	
59.	0.2M NaCl	50 mM Na-citrate (5.6)	30% PEG 600	
60.	0.2 M Li_2SO_4	50 mM Na-citrate (5.6)	30% PEG 2000	

(*continued*)

TABLE III (*continued*)

	Salt	Buffer	Precipitant 1	Precipitant 2
61.		50 mM Na-citrate (5.6)	30% PEG 4000	
62.	0.2 M $NH_4(CH_3COO)$	50 mM Na-citrate (5.6)	5% PEG 10000	10% PEG 600
63.	0.2 M $(NH_4)_2SO_4$	50 mM Na-citrate (5.6)	30% PEG 3000	
64.		50 mM Cacodylate (6.5)	30% PEG 400	
65.		50 mM Cacodylate (6.5)	15% PEG 2000	
66.			30% PEG 4000	
67.	0.2 M NaCl	50 mM Cacodylate (6.5)	15% PEG 4000	
68.	0.2 M LiCl	50 mM Cacodylate (6.5)	30% PEG 4000	
69.		50 mM Cacodylate (6.5)	10% PEG 8000	1 M $(NH_4)_2SO_4$
70.		50 mM Cacodylate (6.5)	30% MPD	
71.		50 mM Cacodylate (6.5)	30% Ethanol	
72.	0.2 M $MgCl_2$	50 mM HEPES (7.5)	30% PEG 400	
73.	0.2 M Li_2SO_4	50 mM HEPES (7.5)	30% PEG 1000	
74.		50 mM HEPES (7.5)	25% PEG 4000	
75.	0.2 M $(NH_4)_2SO_4$	50 mM HEPES (7.5)	30% MPD	
76.	0.2 M NaCl	50 mM HEPES (7.5)	20% PEG 3000	
77.		50 mM HEPES (7.5)	30% PEG 400	
78.	0.2 M NaCl	50 mM HEPES (7.5)	30% Jeffamine M-600	
79.		50 mM Tris (8.0)	30% PEG 1500	
80.	0.2 M LiCl	50 mM Tris (8.0)	15% PEG 4000	
81.	0.2 M $MgSO_4$	50 mM Tris (8.0)	30% PEG 4000	
82.	0.2 M $NH_4(CH_3COO)$	50 mM Tris (8.0)	10% PEG 8000	10% MPD
83.	0.2 M NaCl		30% MPD	
84.	0.2 M NaCl	50 mM Tris (8.5)	30% PEG 400	
85.	0.2 M Li_2SO_4	50 mM Tris (8.5)	30% PEG 1500	
86.	0.2 M $(NH_4)_2SO_4$	50 mM Tris (8.5)	30% MPD	
87.		50 mM Tris (9.5)	30% PEG 400	
88.		50 mM Tris (9.5)	30% PEG 1000	
89.	0.2 M $MgSO_4$	50 mM Bicine (9.5)	30% PEG 4000	
90.		50 mM Bicine (9.5)	10% PEG 8000	10% MPD
91.	0.2 M NaCl	50 mM Bicine (9.5)	30% MPD	
92.		50 mM Bicine (9.5)	2.0 M $(NH_4)_2SO_4$	10% MPD

(*continued*)

TABLE III (*continued*)

	Salt	Buffer	Precipitant 1	Precipitant 2
93.	0.2 M LiCl	50 mM CHES (10.5)	30% PEG 400	
94.	0.2 M Li_2SO_4	50 mM CHES (10.5)	30% PEG 1000	
95	0.2 M NaCl	50 mM CHES (10.5)	25% PEG 4000	
96		50 mM CHES (10.5)	30% MPD	

All precipitant solution concentrations are shown by percentage weight per volume (%, w/v).

C. Diffraction and Optimization of Crystals

Initial crystals grown from a sparse-matrix screen often suffer from numerous pathologies. Even after optimization of the crystallization conditions to improve the visual quality of the crystals, membrane proteins may have intrinsic disorder, which is believed to result from large micelles interfering with tight, regular crystal packing. As a result, visual quality is less likely to correlated with diffraction quality than in soluble proteins. While membrane protein crystals usually benefit from diffraction-based feedback combined with iterative additive and detergent screening, some initial crystallization leads may result in beautiful crystals that never break the 5 Å barrier. In a recent example from our laboratory, the OMP PorB from *N. meningitidis* was grown in three distinct crystal forms (Fig. 9). The first crystal form grew in space group C222 and these crystals were large (0.2 mm × 0.2 mm × 0.1 mm) and visually stunning (Fig. 9A). However, the best diffraction observed from these crystals was between 6 and 7 Å resolution. By comparison, rhombohedral crystals had poorer optical quality and were significantly smaller,

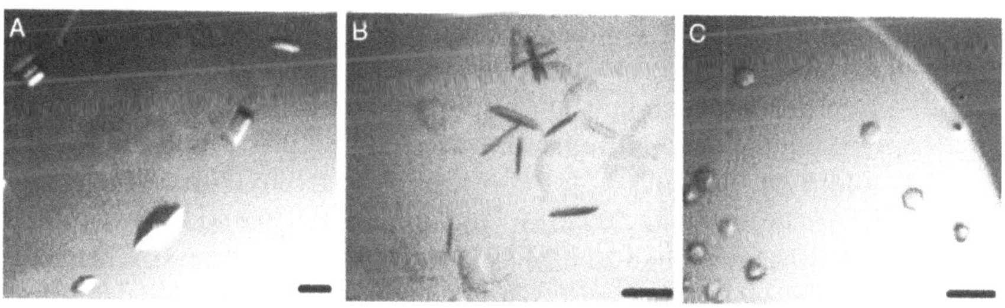

FIGURE 9 Three crystal forms of *N. meningitidis* PorB: (A) crystal form 1 (space group C222) of PorB was grown using PEG 1500 as the precipitant; (B) crystal form 2 (space group R32) of PorB was grown using Jeffamine M-600 as the precipitant; (C) crystal form 3 (space group $P6_3$) of SeMet PorB was grown also using Jeffamine M-600 as the precipitant. The scale bars shown represent 100 μm.

measuring 0.1 mm × 0.1 mm × 0.02 mm (Fig. 9B) but diffracted to 2.9 Å resolution. Se-Met incorporation resulted in another crystal form, $P6_3$ (Fig. 9C). These crystals were smaller than 0.05 mm × 0.05 mm × 0.1 mm, but the diffraction extended to better than 2.0 Å resolution, and a complete data set merging to 2.3 Å resolution was collected. This example is one of many, and suggests that even when initial crystals have been grown, continued broad screening may provide a faster avenue to successful structure determination than optimization of a crystal that diffracts poorly.

1. Conclusion

This chapter represents an in depth analysis of successful examples of β-barrel protein expression, purification, and crystallization conditions. While there is no specific protocol that can be applied to all OMP structural studies, the analysis of successful examples may considerably reduce the time and effort required for future structural studies. This analysis provides practical information that can streamline the process of obtaining high quality X-ray diffraction data of this important class of proteins.

V. MATERIALS AND METHODS

This analysis was based on the crystallization information from the 44 available β-barrel membrane protein structures in the PDB (2008; www.rcsb.org/pdb). To avoid biasing this analysis, conditions from the same β-barrel proteins (including different reaction stages or substrate present or absent forms) and mutant structures were excluded. In these cases, the crystallization conditions from the initial published structure were used. If a protein has been crystallized with different crystal forms in the same journal issue, the conditions from the best-diffracted crystals were cited. Two independent laboratories crystallized FhuA and its structure was solved by using different strategies at the same time (Ferguson, Hofmann, Coulton, Diederichs, & Welte, 1998; Locher et al., 1998), therefore both conditions were included in the analysis.

The concentration of crystallization solution was defined as reservoir well solution. Crystallization conditions were divided into precipitant, buffer pH, salt, and additives. Salts were not considered as precipitants if their concentration was lower than 500 mM, and considered as additives if their concentration was lower than 20 mM (−10% for organic solvents). Chemicals that were only present in the crystal drops not reservoir were also defined as additives. Proteinase inhibitors were not counted as additives. If several detergents or additives are in the crystallization solution, these were counted individually. The percentages of detergents and salts in the crystallization condition were calculated using the number of successful solution in the crystallization condition as a denominator ($n = 44$).

Acknowledgments

This work was supported by National Institutes of Health Grant GM079419 (T.M.I.). M.T. was supported by Uehara memorial foundation fellowship. We are grateful to J. Vey and T. Panosian for critical reading of the manuscript.

References

Akama, H., Kanemaki, M., Yoshimura, M., Tsukihara, T., Kashiwagi, T., Yoneyama, H., et al. (2004). Crystal structure of the drug discharge outer membrane protein, OprM, of *Pseudomonas aeruginosa*: Dual modes of membrane anchoring and occluded cavity end. *Journal of Biological Chemistry, 279*, 52816–52819.

Ahn, J. H., Hwang, M. Y., Lee, K. H., Choi, C. Y., & Kim, D. M. (2007). Use of signal sequences as an *in situ* removable sequence element to stimulate protein synthesis in cell-free extracts. *Nucleic Acids Research, 35*, e21.

Ahn, V. E., Lo, E. I., Engel, C. K., Chen, L., Hwang, P. M., Kay, L. E., et al. (2004). A hydrocarbon ruler measures palmitate in the enzymatic acylation of endotoxin. *EMBO Journal, 23*, 2931–2941.

Bannwarth, M., & Schulz, G. E. (2003). The expression of outer membrane proteins for crystallization. *Biochimica et Biophysica Acta, 1610*, 37–45.

Barnard, T. J., Dautin, N., Lukacik, P., Bernstein, H. D., & Buchanan, S. K. (2007). Autotransporter structure reveals intra-barrel cleavage followed by conformational changes. *Nature Structural and Molecular Biology, 14*, 1214–1220.

Basle, A., Rummel, G., Storici, P., Rosenbusch, J. P., & Schirmer, T. (2006). Crystal structure of osmoporin OmpC from *E. coli* at 2.0 Å. *Journal of Molecular Biology, 362*, 933–942.

Baumann, U., Mansouri, E., & von Specht, B. U. (2004). Recombinant OprF-OprI as a vaccine against *Pseudomonas aeruginosa* infections. *Vaccine, 22*, 840–847.

Biswas, S., Mohammad, M. M., Movileanu, L., & van den Berg, B. (2008). Crystal structure of the outer membrane protein OpdK from *Pseudomonas aeruginosa. Structure (London, England: 1993), 16*, 1027–1035.

Biswas, S., Mohammad, M. M., Patel, D. R., Movileanu, L., & van den Berg, B. (2007). Structural insight into OprD substrate specificity. *Nature Structural and Molecular Biology, 14*, 1108–1109.

Borges-Walmsley, M. I., Beauchamp, J., Kelly, S. M., Jumel, K., Candlish, D., Harding, S. E., et al. (2003). Identification of oligomerization and drug-binding domains of the membrane fusion protein EmrA. *Journal of Biological Chemistry, 278*, 12903–12912.

Bos, M. P., & Tommassen, J. (2004). Biogenesis of the gram-negative bacterial outer membrane. *Current Opinion in Microbiology, 7*, 610–616.

Buchanan, S. K., Lukacik, P., Grizot, S., Ghirlando, R., Ali, M. M., Barnard, T. J., et al. (2007). Structure of colicin I receptor bound to the R-domain of colicin Ia: Implications for protein import. *EMBO Journal, 26*, 2594–2604.

Buchanan, S. K., Smith, B. S., Venkatramani, L., Xia, D., Esser, L., Palnitkar, M., et al. (1999). Crystal structure of the outer membrane active transporter FepA from *Escherichia coli. Nature Structural Biology, 6*, 56–63.

Chimento, D. P., Kadner, R. J., & Wiener, M. C. (2003). The *Escherichia coli* outer membrane cobalamin transporter BtuB: Structural analysis of calcium and substrate binding, and identification of orthologous transporters by sequence/structure conservation. *Journal of Molecular Biology, 332*, 999–1014.

Clantin, B., Delattre, A. S., Rucktooa, P., Saint, N., Meli, A. C., Locht, C., et al. (2007). Structure of the membrane protein FhaC: A member of the Omp85-TpsB transporter superfamily. *Science, 317*, 957–961.

Cobessi, D., Celia, H., Folschweiller, N., Schalk, I. J., Abdallah, M. A., & Pattus, F. (2005a). The crystal structure of the pyoverdine outer membrane receptor FpvA from *Pseudomonas aeruginosa* at 3.6 angstroms resolution. *Journal of Molecular Biology, 347*, 121–134.

Cobessi, D., Celia, H., & Pattus, F. (2005b). Crystal structure at high resolution of ferric-pyochelin and its membrane receptor FptA from *Pseudomonas aeruginosa. Journal of Molecular Biology, 352*, 893–904.

Cowan, S. W., Schirmer, T., Rummel, G., Steiert, M., Ghosh, R., Pauptit, R. A., et al. (1992). Crystal structures explain functional properties of two *E. coli* porins. *Nature, 358*, 727–733.

Dautin, N., Barnard, T. J., Anderson, D. E., & Bernstein, H. D. (2007). Cleavage of a bacterial autotransporter by an evolutionarily convergent autocatalytic mechanism. *EMBO Journal, 26*, 1942–1952.

Derouiche, R., Gavioli, M., Benedetti, H., Prilipov, A., Lazdunski, C., & Lloubes, R. (1996). TolA central domain interacts with *Escherichia coli* porins. *EMBO Journal, 15*, 6408–6415.

Driessen, A. J., & Nouwen, N. (2008). Protein translocation across the bacterial cytoplasmic membrane. *Annual Review of Biochemistry, 77*, 643–667.

Dutzler, R., Rummel, G., Alberti, S., Hernandez-Alles, S., Phale, P., Rosenbusch, J., et al. (1999). Crystal structure and functional characterization of OmpK36, the osmoporin of *Klebsiella pneumoniae. Structure (London, England: 1993), 7*, 425–434.

Eswaran, J., Koronakis, E., Higgins, M. K., Hughes, C., & Koronakis, V. (2004). Three's company: Component structures bring a closer view of tripartite drug efflux pumps. *Current Opinion in Structural Biology, 14*, 741–747.

Faller, M., Niederweis, M., & Schulz, G. E. (2004). The structure of a mycobacterial outer-membrane channel. *Science, 303*, 1189–1192.

Federici, L., Du, D., Walas, F., Matsumura, H., Fernandez-Recio, J., McKeegan, K. S., et al. (2005). The crystal structure of the outer membrane protein VceC from the bacterial pathogen *Vibrio cholerae* at 1.8 Å resolution. *Journal of Biological Chemistry, 280*, 15307–15314.

Ferguson, A. D., Chakraborty, R., Smith, B. S., Esser, L., van der Helm, D., & Deisenhofer, J. (2002). Structural basis of gating by the outer membrane transporter FecA. *Science, 295*, 1715–1719.

Ferguson, A. D., Hofmann, E., Coulton, J. W., Diederichs, K., & Welte, W. (1998). Siderophore-mediated iron transport: Crystal structure of FhuA with bound lipopolysaccharide. *Science, 282*, 2215–2220.

Forst, D., Schulein, K., Wacker, T., Diederichs, K., Kreutz, W., Benz, R., et al. (1993). Crystallization and preliminary X-ray diffraction analysis of ScrY, a specific bacterial outer membrane porin. *Journal of Molecular Biology, 229*, 258–262.

Forst, D., Welte, W., Wacker, T., & Diederichs, K. (1998). Structure of the sucrose-specific porin ScrY from *Salmonella typhimurium* and its complex with sucrose. *Nature Structural Biology, 5*, 37–46.

Geny, B., & Popoff, M. R. (2006). Bacterial protein toxins and lipids: Pore formation or toxin entry into cells. *Biology of the Cell, 98*, 667–678.

Ghrayeb, J., Kimura, H., Takahara, M., Hsiung, H., Masui, Y., & Inouye, M. (1984). Secretion cloning vectors in *Escherichia coli. EMBO Journal, 3*, 2437–2442.

Girard, M. P., Preziosi, M. P., Aguado, M. T., & Kieny, M. P. (2006). A review of vaccine research and development: Meningococcal disease. *Vaccine, 24*, 4692–4700.

Granseth, E., Seppala, S., Rapp, M., Daley, D. O., & Von Heijne, G. (2007). Membrane protein structural biology—how far can the bugs take us? *Molecular Membrane Biology, 24*, 329–332.

Hearn, E. M., Patel, D. R., & van den Berg, B. (2008). Outer-membrane transport of aromatic hydrocarbons as a first step in biodegradation. *Proceedings of the National Academy of Sciences of the United States of America, 105*, 8601–8606.

Heuck, A. P., Tweten, R. K., & Johnson, A. E. (2001). Beta-barrel pore-forming toxins: Intriguing dimorphic proteins. *Biochemistry, 40*, 9065–9073.

Hirsch, A., Breed, J., Saxena, K., Richter, O. M., Ludwig, B., Diederichs, K., et al. (1997). The structure of porin from *Paracoccus denitrificans* at 3.1 Å resolution. *FEBS Letters, 404*, 208–210.

Hong, H., Patel, D. R., Tamm, L. K., & van den Berg, B. (2006). The outer membrane protein OmpW forms an eight-stranded beta-barrel with a hydrophobic channel. *Journal of Biological Chemistry, 281*, 7568–7577.

Kimber, M. S., Vallee, F., Houston, S., Necakov, A., Skarina, T., Evdokimova, E., et al. (2003). Data mining crystallization databases: Knowledge-based approaches to optimize protein crystal screens. *Proteins, 51*, 562–568.

Korepanova, A., Gao, F. P., Hua, Y., Qin, H., Nakamoto, R. K., & Cross, T. A. (2005). Cloning and expression of multiple integral membrane proteins from *Mycobacterium tuberculosis* in *Escherichia coli. Protein Science, 14*, 148–158.

Koronakis, V., Sharff, A., Koronakis, E., Luisi, B., & Hughes, C. (2000). Crystal structure of the bacterial membrane protein TolC central to multidrug efflux and protein export. *Nature, 405*, 914–919.

Kreusch, A., Neubuser, A., Schiltz, E., Weckesser, J., & Schulz, G. E. (1994). Structure of the membrane channel porin from *Rhodopseudomonas blastica* at 2.0 Å resolution. *Protein Science, 3*, 58–63.

Lee, S. Y., Choi, J. H., & Xu, Z. (2003). Microbial cell-surface display. *Trends in Biotechnology, 21*, 45–52.

Lin, J., Huang, S., & Zhang, Q. (2002). Outer membrane proteins: Key players for bacterial adaptation in host niches. *Microbes and Infection, 4*, 325–331.

Locher, K. P., Rees, B., Koebnik, R., Mitschler, A., Moulinier, L., Rosenbusch, J. P., et al. (1998). Transmembrane signaling across the ligand-gated FhuA receptor: Crystal structures of free and ferrichrome-bound states reveal allosteric changes. *Cell, 95*, 771–778.

Locher, K. P., & Rosenbusch, J. P. (1997). Oligomeric states and siderophore binding of the ligand-gated FhuA protein that forms channels across *Escherichia coli* outer membranes. *European Journal of Biochemistry, 247*, 770–775.

Loll, P. J. (2003). Membrane protein structural biology: The high throughput challenge. *Journal of Structural Biology, 142*, 144–153.

Makrides, S. C. (1996). Strategies for achieving high-level expression of genes in *Escherichia coli. Microbiological Reviews, 60*, 512–538.

Massari, P., Henneke, P., Ho, Y., Latz, E., Golenbock, D. T., & Wetzler, L. M. (2002). Cutting edge: Immune stimulation by neisserial porins is toll-like receptor 2 and MyD88 dependent. *Journal of Immunology, 168*, 1533–1537.

Meins, T., Vonrhein, C., & Zeth, K. (2008). Crystallization and preliminary X-ray crystallographic studies of human voltage-dependent anion channel isoform I (HVDAC1). *Acta Crystallographica. Section F, Structural Biology and Crystallization Communications, 64*, 651–655.

Meng, G., Surana, N. K., St Geme, J. W., III, & Waksman, G. (2006). Structure of the outer membrane translocator domain of the *Haemophilus influenzae* Hia trimeric autotransporter. *EMBO Journal, 25*, 2297–2304.

Meyer, J. E., Hofnung, M., & Schulz, G. E. (1997). Structure of maltoporin from *Salmonella typhimurium* ligated with a nitrophenyl-maltotrioside. *Journal of Molecular Biology, 266*, 761–775.

Moeck, G. S., & Coulton, J. W. (1998). TonB-dependent iron acquisition: Mechanisms of siderophore-mediated active transport. *Molecular Microbiology, 28*, 675–681.

Moraes, T. F., Bains, M., Hancock, R. E., & Strynadka, N. C. (2007). An arginine ladder in OprP mediates phosphate-specific transfer across the outer membrane. *Nature Structural and Molecular Biology, 14*, 85–87.

Newstead, S., Ferrandon, S., & Iwata, S. (2008). Rationalizing alpha-helical membrane protein crystallization. *Protein Science, 17*, 466–472.

Nikaido, H. (2003). Molecular basis of bacterial outer membrane permeability revisited. *Microbiology and Molecular Biology Reviews, 67*, 593–656.

Oomen, C. J., van Ulsen, P., van Gelder, P., Feijen, M., Tommassen, J., & Gros, P. (2004). Structure of the translocator domain of a bacterial autotransporter. *EMBO Journal, 23*, 1257–1266.

Page, R., & Stevens, R. C. (2004). Crystallization data mining in structural genomics: Using positive and negative results to optimize protein crystallization screens. *Methods, 34*, 373–389.

Pages, J. M., Masi, M., & Barbe, J. (2005). Inhibitors of efflux pumps in gram-negative bacteria. *Trends in Molecular Medicine, 11*, 382–389.

Pautsch, A., & Schulz, G. E. (1998). Structure of the outer membrane protein A transmembrane domain. *Nature Structural Biology, 5*, 1013–1017.

Piddock, L. J. (2006). Multidrug-resistance efflux pumps—not just for resistance. *Nature Reviews Microbiology, 4*, 629–636.

Prince, S. M., Achtman, M., & Derrick, J. P. (2002). Crystal structure of the OpcA integral membrane adhesin from *Neisseria meningitidis*. *Proceedings of the National Academy of Sciences of the United States of America, 99*, 3417–3421.

Prive, G. G. (2007). Detergents for the stabilization and crystallization of membrane proteins. *Methods, 41*, 388–397.

Raman, P., Cherezov, V., & Caffrey, M. (2006). The membrane protein data bank. *Cellular and Molecular Life Sciences, 63*, 36–51.

Ray, A., Chatterjee, N. S., Bhattacharya, S. K., & Biswas, T. (2003). Porin of *Shigella dysenteriae* enhances mRNA levels for Toll-like receptor 2 and MyD88, up-regulates CD80 of murine macrophage, and induces the release of interleukin-12. *FEMS Immunology and Medical Microbiology, 39*, 213–219.

Remaut, H., Tang, C., Henderson, N. S., Pinkner, J. S., Wang, T., Hultgren, S. J., et al. (2008). Fiber formation across the bacterial outer membrane by the chaperone/usher pathway. *Cell, 133*, 640–652.

Saier, M. H., Jr. (1995). Differential codon usage: A safeguard against inappropriate expression of specialized genes? *FEBS Letters, 362*, 1–4.

Saier, M. H., Jr. (2000). Families of proteins forming transmembrane channels. *Journal of Membrane Biology, 175*, 165–180.

Schirmer, T., Keller, T. A., Wang, Y. F., & Rosenbusch, J. P. (1995). Structural basis for sugar translocation through maltoporin channels at 3.1 Å resolution. *Science, 267*, 512–514.

Schleiff, E., & Soll, J. (2005). Membrane protein insertion: Mixing eukaryotic and prokaryotic concepts. *EMBO Reports, 6*, 1023–1027.

Seddon, A. M., Curnow, P., & Booth, P. J. (2004). Membrane proteins, lipids and detergents: Not just a soap opera. *Biochimica et Biophysica Acta, 1666*, 105–117.

Snijder, H. J., Ubarretxena-Belandia, I., Blaauw, M., Kalk, K. H., Verheij, H. M., Egmond, M. R., et al. (1999). Structural evidence for dimerization-regulated activation of an integral membrane phospholipase. *Nature, 401*, 717–721.

Song, L., Hobaugh, M. R., Shustak, C., Cheley, S., Bayley, H., & Gouaux, J. E. (1996). Structure of *staphylococcal* alpha-hemolysin, a heptameric transmembrane pore. *Science, 274*, 1859–1866.

Studier, F. W. (1991). Use of bacteriophage T7 lysozyme to improve an inducible T7 expression system. *Journal of Molecular Biology, 219*, 37–44.

Subbarao, G. V., & van den Berg, B. (2006). Crystal structure of the monomeric porin OmpG. *Journal of Molecular Biology, 360*, 750–759.

Tanabe, M., Mirza, O., Bertrand, T., Atkins, H. S., Titball, R. W., Iwata, S., et al. (2007). Structures of OppA and PstS from *Yersinia pestis* indicate variability of interactions with transmembrane domains. *Acta Crystallographica. Section D, Biological Crystallography, 63*, 1185–1193.

Thanassi, W. T., & Schoen, R. T. (2000). The Lyme disease vaccine: Conception, development, and implementation. *Annals of Internal Medicine, 132*, 661–668.

Timmins, P. A., Hauk, J., Wacker, T., & Welte, W. (1991). The influence of heptane-1,2,3-triol on the size and shape of LDAO micelles. Implications for the crystallisation of membrane proteins. *FEBS Letters, 280*, 115–120.

van den Berg, B., Black, P. N., Clemons, W. M., Jr., & Rapoport, T. A. (2004). Crystal structure of the long-chain fatty acid transporter FadL. *Science, 304*, 1506–1509.

Vandeputte-Rutten, L., Bos, M. P., Tommassen, J., & Gros, P. (2003). Crystal structure of *Neisserial* surface protein A (NspA), a conserved outer membrane protein with vaccine potential. *Journal of Biological Chemistry, 278*, 24825–24830.

Vandeputte-Rutten, L., Kramer, R. A., Kroon, J., Dekker, N., Egmond, M. R., & Gros, P. (2001). Crystal structure of the outer membrane protease OmpT from *Escherichia coli* suggests a novel catalytic site. *EMBO Journal, 20*, 5033–5039.

Vogt, J., & Schulz, G. E. (1999). The structure of the outer membrane protein OmpX from *Escherichia coli* reveals possible mechanisms of virulence. *Structure, 7*, 1301–1309.

Voulhoux, R., & Tommassen, J. (2004). Omp85, an evolutionarily conserved bacterial protein involved in outer-membrane-protein assembly. *Research in Microbiology, 155*, 129–135.

Weiss, M. S., & Schulz, G. E. (1992). Structure of porin refined at 1.8 Å resolution. *Journal of Molecular Biology, 227*, 493–509.

Wiener, M. C. (2005). TonB dependent outer membrane transport: Going for Baroque? *Current Opinion in Structural Biology, 15*, 394–400.

Ye, J., & van den Berg, B. (2004). Crystal structure of the bacterial nucleoside transporter Tsx. *EMBO Journal, 23*, 3187–3195.

Yildiz, O., Vinothkumar, K. R., Goswami, P., & Kuhlbrandt, W. (2006). Structure of the monomeric outer-membrane porin OmpG in the open and closed conformation. *EMBO Journal, 25*, 3702–3713.

Zeth, K., Diederichs, K., Welte, W., & Engelhardt, H. (2000). Crystal structure of Omp32, the anion-selective porin from *Commamonas acidovorans*, in complex with a periplasmic peptide at 2.1 Å resolution. *Structure (London, England: 1993), 8*, 981–992.

CHAPTER 11

Bacterial Membrane Proteins: The New Soluble Proteins?

Hubing Lou,* Konstantinos Beis,[†] and James H. Naismith*

*Centre for Biomolecular Sciences, The University of St Andrews, Fife KY16 9ST, United Kingdom

[†]Imperial College London, South Kensington Campus, London SW7 2AZ, United Kingdom

I. OVERVIEW

In recent years, the successful structure determination of membrane proteins has become accelerated. Advances in protein production and crystallization have underpinned the structure determination of novel membrane proteins. The outer and inner membrane proteins of bacteria have been the most intensively studied. In this review, we will discuss the structure of outer membrane proteins (Omps), and highlight the recent advances in protein production. Omps have a variety of different functions but at their core they allow compounds in and out the cell. Omps are relatively easy to produce and crystallize and this has resulted in the publication of many studies. The defining characteristic of these proteins is the β-barrel transmembrane region. It has assumed that all Omps are β-barrels. The publication of the first α-helical

Current Topics in Membranes, Volume 63

1063-5823/09 $35.00
DOI: 10.1016/S1063-5823(09)63011-8

Omps, Wza shows this is not the case. The structural determination of eukaryotic integral and bacterial inner membrane proteins has been more challenging. In most cases production of stable protein has been the principal difficulty. Several research groups have developed new technologies to overcome these challenges. This review will highlight some of the most recent ones. The advances in protein production and crystallization technology also resulted in the long sought structure of a human G-protein coupled receptors (GPCR).

II. STRUCTURES OF OMPS

Gram-negative bacteria have two specialized membranes. The OM is a complex mixture of protein, lipid, and carbohydrate. It provides the bacterial cell with a robust physical barrier and corresponding resilience to the environmental conditions or bactericidal toxins or host immune system during infection. The composition of this lipid bilayer is asymmetrical in contrast to the symmetrical nature of eukaryotic membranes and the inner membrane of bacteria. Omps were considered to uniformly be β-barrel proteins; this belies the diversity in their structure and function and in any event is no longer.

Porins were the first Omps to be studied. The first X-ray structure of a porin from *Rhodobacter capsulatus* was determined in 1990 (Weiss, Wacker, Weckesser, Welte, & Schulz, 1990) quickly followed by OmpF and PhoE from *Escherichia coli* (Cowan et al., 1992), a porin from *Rhodopseudomonas blastica* (Kreusch, Neubuser, Schiltz, Weckesser, & Schulz, 1994), OmpK36 from *Klebsiella pnueumoniae* (Dutzler et al., 1999), Omp32 from *Comamonas acidovorans* (Zeth, Diederichs, Welte, & Engelhardt, 2000), and *Delftia acidovorans* (Zachariae, Kluhspies, De, Engelhardt, & Zeth, 2006), and more recently OmpC from *E. coli* (Basle, Rummel, Storici, Rosenbusch, & Schirmer, 2006). These porins have the same striking appearance, 16-stranded hollow β-barrels. These porins have a characteristic trimeric arrangement with a threefold axis normal the membrane plane. A representative structure of OmpC is shown in Fig. 1. The hydrophobic interactions between monomeric barrel surfaces stabilize the trimer. The barrels all show tight turns on the periplasmic side and large, irregular loops on the extracellular side. The extracellular loops have been numbered and are referred to as Loop L1, L2, etc. L2 interacts with the neighboring monomer to stabilize the trimer (Phale et al., 1998). L3 folds into the barrel forming a "constriction zone" at about midway into the barrel. As a result when viewed via a space-filling model, the central pore has an hourglass shape. The constriction site has a transverse electrostatic field caused by acidic

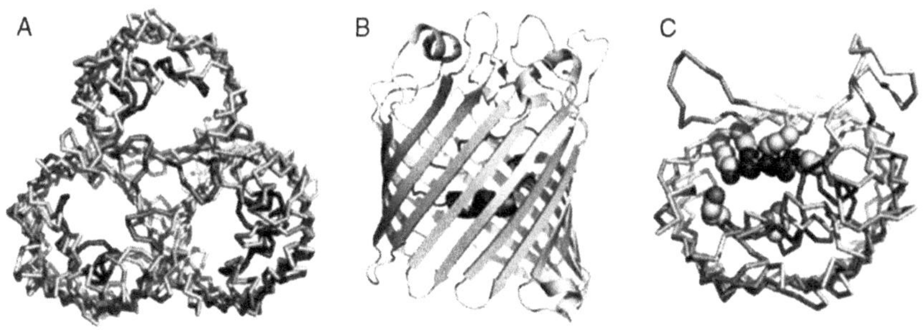

FIGURE 1 Structure of OmpC porin from *E. coli* (PDB code: 2J1N): (A) view of the trimer from the outside of membrane, loop L2 is colored in orange, L3 is in purple; (B) side view of the monomer; and (C) view of the constriction zone of the channel, featuring the opposite-charged residues that form the electrostatic field. (See Color Insert.)

residues in L3 facing a cluster of basic residues at the opposite barrel wall. This charge constellation is thought to make an important contribution to the porin properties (Nikaido, 2003).

These porins are known as "general porins" and are water-filled channels with no particular substrate specificity. Their biological role is to control the passage of hydrophilic solutes based on the substrate molecular size (Nikaido, 2003). In addition to general porins, bacteria have substrate-specific porins which are also trimeric β-barrel proteins. The substrate-specific porins include the maltose-specific channel LamB from *E. coli* (Schirmer, Keller, Wang, & Rosenbusch, 1995) and its homolog from *Salmonella typhimurium* (Meyer, Hofnung, & Schulz, 1997), the sucrose-specific channel ScrY from *S. typhimurium* (Forst, Welte, Wacker, & Diederichs, 1998), from *Pseudomonas aeruginosa* the phosphate-specific transporter OprP (Moraes, Bains, Hancock, & Strynadka, 2007), OprD (Biswas, Mohammad, Patel, Movileanu, & van den Berg, 2007), and OpdK (Biswas, Mohammad, Movileanu, & van den Berg, 2008).

Monomers of LamB and ScrY have 18 β-stranded structures instead of 16 strands as seen in general porins and also form a homotrimer, with the threefold parallel to the membrane normal. Once again, L3 folds back inside the β-barrel (Fig. 2). X-ray analysis of sugar-soaked LamB crystal identified a substrate translocation pathway (Schirmer et al., 1995) consisting of a row of aromatic amino acids, the "greasy slide" (Fig. 2B), lined by polar residues, the "ionic track." The hydrophobic face of sugar makes van der Waals' contacts with the greasy slide while the sugar hydroxyl groups make hydrogen bonds with the ionic track. It has been proposed that movement of the substrate through the channel proceeds by a sequence of hydrogen bond making and breaking. Most of the channel-lining residues are conserved between LamB and ScrY, but the differences confer the specificity.

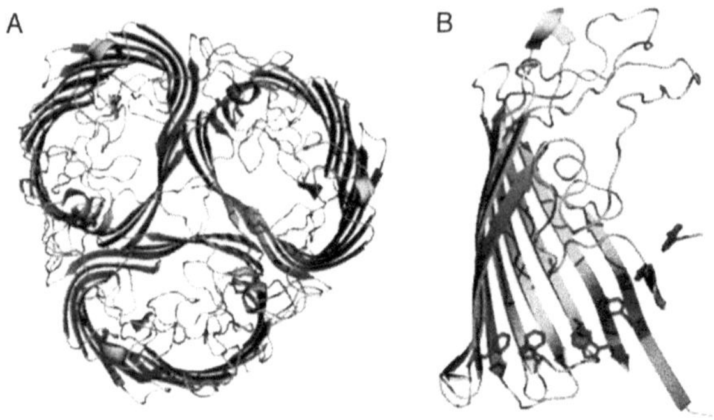

FIGURE 2 Structure of LamB (PDB code: 1MAL): (A) view of the trimer from extracellular side; and (B) aromatic residues that contribute to the greasy slide are shown in purple. (See Color Insert.)

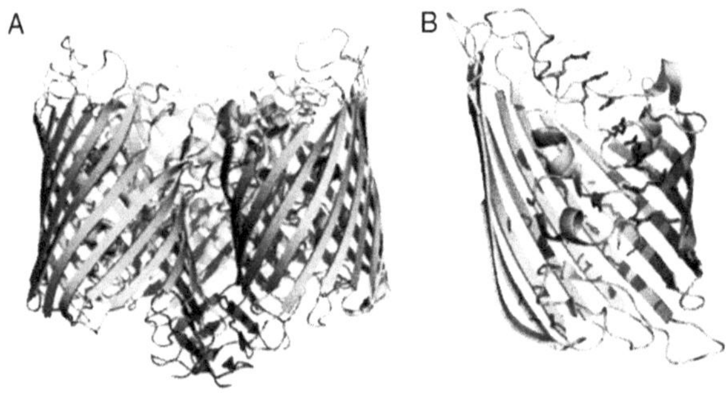

FIGURE 3 Structure of OprP (PDB code: 2O4V): (A) side view of OprP trimer, the tricorn N-terminal strands are colored red; and (B) the arginine ladder located on the upper portion of each monomer of OprP (green stick). (See Color Insert.)

The *P. aeruginosa* OprP is a 16-strand antiparallel β-barrel but unusually, OprP has an extended periplasmic N-terminus involved in stabilizing the trimer though a "tricorn"-like strand exchange (Fig. 3A). Inside the channel, there is a nine-residue arginine "ladder" that spans from the extracellular surface down through the constriction zone (Fig. 3B). This ladder is proposed to control the transit of the phosphate anion (Moraes et al., 2007). Lysine residues coat the inner periplasmic surface, creating an "electropositive sink" that pulls the phosphates through the eyelet and into the cell.

OprD and OpdK are 18 β-strand substrate-specific porins from *P. aeruginosa*; OprD transports basic amino acids such as lysine and arginine (Trias & Nikaido, 1990) and OpdK is responsible for transporting vanillate and related small aromatic acids (Biswas et al., 2008). They both crystallize as monomers as shown by analysis of their crystal packing. However, biochemical and biophysical studies showed that they form trimers in the bacterial OM. Similar to OprP, OprD contains a basic ladder of arginine and lysine residues that may provide a path for substrate diffusion through the membrane. In OpdK, only a portion of the ladder residues are present for substrate transport. The charge distribution of residues lining the pore is very similar between these two, which does not provide an explanation for the different substrate specificities. However, in OprD, the surface of the periplasmic funnel is negatively charged whilst in OpdK it is positively charged. This difference seems to underlie the differing substrate specificities.

Besides trimeric 16–18 β-strand porins, bacteria also have dimeric β-barrel proteins. The first dimeric structure was the outer membrane phospholipase A (OMPLA). OMPLA is an integral membrane enzyme which catalyzes the hydrolysis of acyl ester bonds in phospholipids in a Ca^{2+}-dependent manner (Ubarretxena-Belandia et al., 1999). The monomer of OMPLA is a 12-stranded antiparallel β-barrel but like the porins has long extracellular loops and short periplasmic turns. The interior OMPLA is polar and contains an intricate hydrogen-bonding network; however, the central pore of OMPLA does not apparently function as a channel (Snijder et al., 1999). The dimer interface is formed by the one side of the barrel (Fig. 4) and two Ca^{2+} are adjacent to the interface. The active sites lie at the outer edge of the barrel at the interface between the two monomers with OMPLA dimerization, essential for biological function (Snijder et al., 1999).

PapC is the translocation pore responsible for assembly of adhesive pili on the surface of gram-negative pathogenic bacteria, acting as P pilus "ushers" (Remaut et al., 2008). P pili are complex extended fibers produced by pyelonephritic strains of *E. coli*. PapC, considered to be the prototype for bacterial usher proteins, is found as a dimer arranged similarly to that seen for OMPLA.

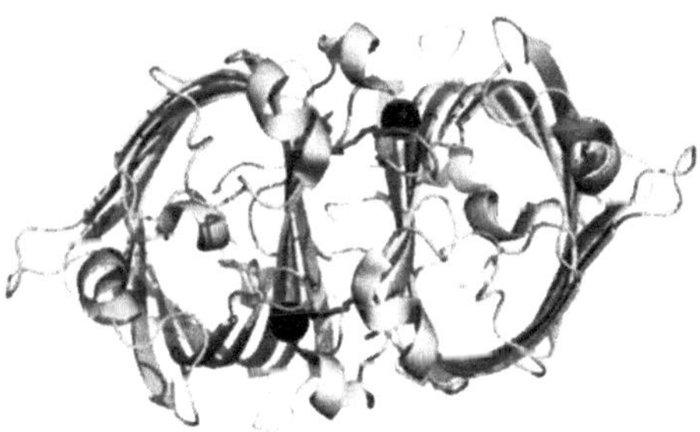

FIGURE 4 Structure of dimeric OMPLA (PDB code: 1QD6) viewed from the extracellular side. The two black spheres represent two calcium ions, the substrate analog, hexadecanesulphonylfluoride, at the active sites are colored purple. (See Color Insert.)

Each PapC monomer is a 24 β-stranded monomer (the largest barrel observed to date). A long sequence connects the two β-strands $\beta6$ and $\beta7$ and is located in the centre of the pore forming a "plug" domain. The plug domain is a six-stranded β-sandwich, the plane of the strands are parallel to the assumed membrane plane. The plug completely occludes the translocation pore and is held in place by a β-hairpin (connecting strands $\beta5$ and $\beta6$) that folds into the channel lumen, the only helix in the structure sits above (on the extracellular side) the β-hairpin (Fig. 5). The inward curvature of the $\beta5$–$\beta6$ hairpin creates a gap in the side of the β-barrel that is believed to extend into the OM bilayer (Fig. 5), this partly "missing stave" is thus far a unique structural feature.

Monomeric β-barrel protein structures also exist and they include structural proteins, enzymes, and transporters. One example, the TonB-dependent active transporters consist of a 22 β-stranded barrel that uses the proton-motive force across the cytoplasmic membrane through the TonB-ExbB-ExbD energy-transducing complex to transport specific substrates across the bacterial OM (Moeck & Coulton, 1998). The iron-siderophore transporters FhuA, FepA, FecA from *E. coli*, FptA and FpvA from *P. aeruginosa*, the cobalamins (e.g., cyanocobalamin, vitamin B_{12}) transporter BtuB from *E. coli*, the colicin I receptor Cir from *E. coli* all belong to this class, despite a low-sequence similarity (Cobessi et al., 2005). Strands of the β-barrel are connected by long loops on the extracellular side and short turns in the periplasmic side as already seen in porins. The N-terminal domain, often referred to as a "plug" or "cork" domain, contains mixed four stranded

FIGURE 5 Structure of PapC translocation domain (PDB code: 2VQI). (A) The PapC translocation channel viewed from the extracellular side. The plug domain, $\beta5$–$\beta6$ hairpin, the α-helix are colored purple, orange, and blue, respectively. (B) β-Barrel viewed from side. Structural elements are colored as in (A), the abnormal $\beta5$–$\beta6$ is labeled. (See Color Insert.)

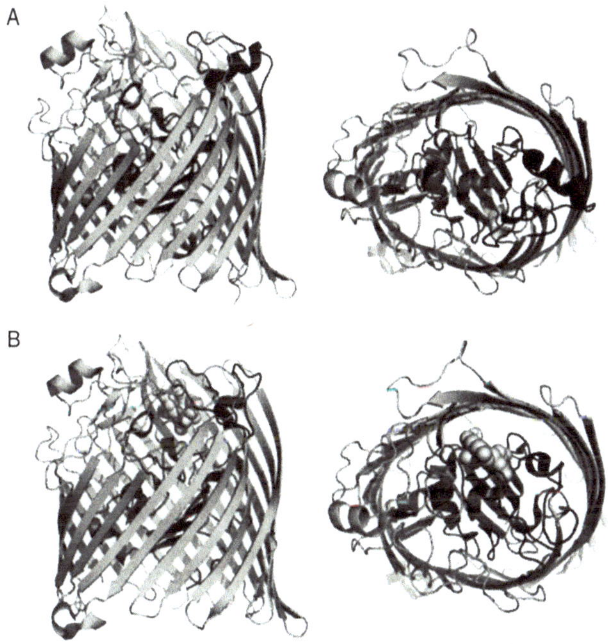

FIGURE 6 Structures of FecA with (A) unliganded (PDB code: 1KMO) and (B) liganded (PDB code: 1KMP). The two extracellular loops L7 and L8, which undergo major conformational changes upon ligand binding, are shown in blue. The plug domain is colored purple. The switch helix, located in the periplasmic pocket of FecA, is colored orange and is only observed in the unliganded conformation. The substrate is shown as yellow spheres. (See Color Insert.)

β-sheets connected by a series of short β-strands, α-helices, and irregular secondary structural elements positioned in the middle of the barrel (Fig. 6A).

Siderophores are secreted by bacteria to acquire iron (Ferguson & Deisenhofer, 2002) with the iron-siderophore complex transported through the OM. All the iron-siderophore transporters characterized to date have essentially identical structures. The barrels are 60–70 Å in height and have an elliptical cross section. The transporters can exist in three states, empty, bound to siderophore, and bound to the iron-siderophore complex. All three binding states have been characterized for FecA (Ferguson et al., 2002; Yue, Grizot, & Buchanan, 2003). Comparison between the unloaded FecA and FecA bound to siderophore (citrate) reveals only minor differences (~1 Å) in the extracellular loops L7 and L8. Major conformational changes occur when FecA binds the ferric citrate complex, including 10 Å movements in the two extracellular loops L7 and L8. Conformational changes also occur in the plug domain and an N-terminal segment located within the periplasmic pocket, termed the switch helix. The helix is thought to unwind and become disordered (Fig. 6B). Structures of FhuA and FhuA bound with iron-ferrichrome have been reported from two independent groups (Ferguson, Hofmann, Coulton, Diederichs, & Welte, 1998; Locher et al., 1998). The barrel domain and the extracellular loops undergo only minor changes and the key difference lies in the plug domain and the "switch helix." Upon ligand binding, the plug domain translates upward toward the ligand and the switch helix completely unwound and bended ~180° in the opposite direction of the former helix axis (Ferguson et al., 1998).

Compared to the structures of iron-siderophore transporters, BtuB and Cir have shorter transmembrane barrels with BtuB around 55 Å and Cir around 40 Å. In general these structures possess shorter extracellular loops, except that in Cir loops L7 and L8 are very long (Buchanan et al., 2007). When BtuB binds the cyanocobalamin, small conformational changes occur in several extracellular loops and once again large changes are observed in the plug domain. The Ton box, a highly conserved stretch of seven amino acid residues near the N-terminal in TonB-dependent transporters (Chimento, Mohanty, Kadner, & Wiener, 2003) also undergoes change. Cir undergoes large and unusual conformational changes upon binding its substrate colicin Ia, the extracellular loops L7 and L8 move as a rigid body to a more open conformation as compared with the same loops in the uncomplexed Cir structure (Fig. 7) (Buchanan et al., 2007).

FIGURE 7 Conformational changes in Cir upon ligand binding. A superposition of uncomplexed Cir (PDB code: 2HDF) and colicin-bound Cir (PDB code: 2HDI) shows the largest change occur in the extracellular loops L7 and L8. The green colored residues represent L7 and L8 from the uncomplexed Cir and the orange colored ones represent that from colicin-bound Cir. (See Color Insert.)

Interestingly, the conformational change in FhuA upon binding TonB is relatively minor (Pawelek et al., 2006) while the conformation of Ton box in BtuB-TonB complex is significant (Shultis, Purdy, Banchs, & Wiener, 2006). In all structures to date, the plug domain obstructs the channel. Therefore, a conformational change of the luminal domain is required to create a substrate path. The mechanistic basis of this change is unknown.

OmpA, OmpX, OmpW, PagP from *E. coli*, and NspA from *Neisseria meningitidis* all belong to the same eight β-stranded barrel family (Fig. 8). The basic architecture, long extracellular loops and short periplasmic turns, is again found; however, there are important differences. For example, PagP

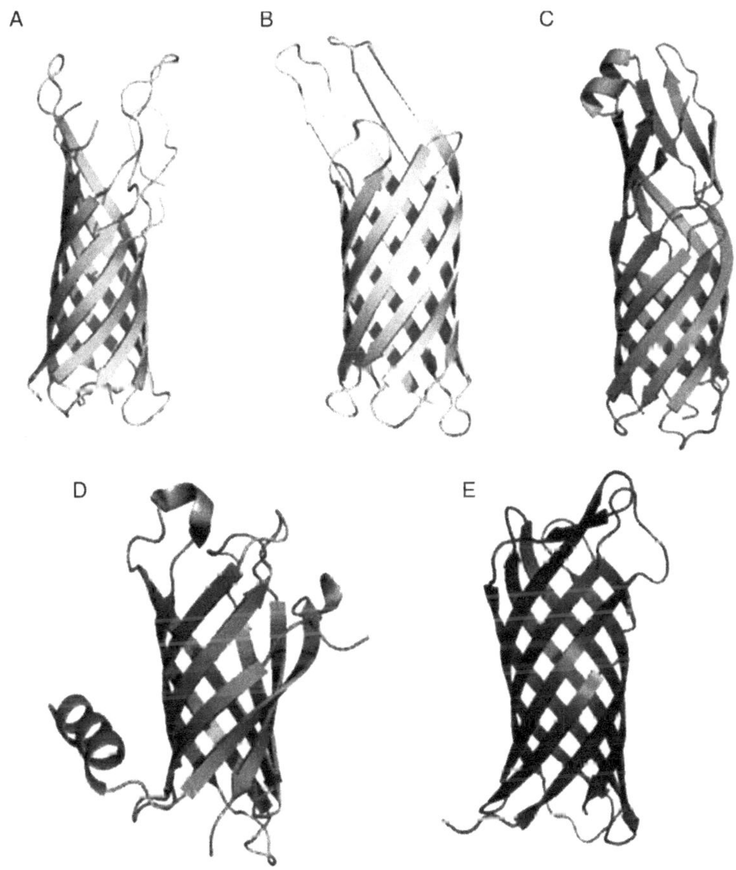

FIGURE 8 A gallery of eight-stranded β-barrel proteins: (A) OmpA (PDB code: 1BXW), (B) OmpX (PDB code: 1QJ8), (C) OmpW (PDB code: 2F1V), (D) PagP (PDB code: 1THQ), and (E) NspA (PDB code: 1P4T).

has a periplasmic N-terminal amphipathic α-helix (Fig. 8D) and in OmpW, the interior of the barrel is a hydrophobic channel possibly involved in the transport of small hydrophobic molecules (Hong, Patel, Tamm, & van den Berg, 2006). The interior of OmpA, OmpX, and NspA contains an extensive hydrogen-bonding network but no channel through which ions or other molecules can be transported (Pautsch & Schulz, 1998; Vandeputte-Rutten, Bos, Tommassen, & Gros, 2003; Vogt & Schulz, 1999). PagP has an unusual interior: the upper half of the barrel core is distinctly hydrophobic and is devoid of interior waters, while the lower half has a typical hydrophilic interior filled with polar side chains (Ahn et al., 2004).

As of the date of writing (September, 2008), there are only two Omps with 10 strands, one is the protease OmpT from *E. coli* (Vandeputte-Rutten et al., 2001) and the other is adhesin OpcA from *N. meningitidis* (Prince, Achtman, & Derrick, 2002). A comparison of these two structures showed that the overall fold was similar: a long transmembrane β-barrel that protrudes far from the lipid bilayer into the extracellular space (Fig. 9).

The overall shape of the OpcA β-barrel was more uniform than OmpT; the top of OmpT barrel is circular whereas in the central part of the molecule the cross-section is elliptical. In contrast, the OpcA barrel has an elliptical cross section along its barrel axis. Both barrels are apparently accessible to water from the periplasmic side, but both are blocked on the extracellular face. A 12-stranded single β-barrel is found in the *E. coli* nucleoside transporter Tsx (Ye & van den Berg, 2004). The *N. meningitidis* autotransporter NalP β-domain (Oomen et al., 2004) is also a 12 stranded β-barrel but has an α-helical peptide (Fig. 10A) in the middle. The autotransporter EspP from *E. coli* has a similar structure as NalP (Barnard, Dautin, Lukacik, Bernstein, & Buchanan, 2007) but unlike NalP, the β-domain of EspP begins with a short α-helix. However, full length EspP is predicted to contain an amphipathic α-helix spanning the length of the barrel pore as seen in NalP (Barnard et al., 2007). The autotransporter Hia from *Haemophilus influenzae* is a 12-stranded β-barrel domain superficially similar to that in NalP (Meng, Surana, St Geme, & Waksman, 2006). Strikingly in Hia the barrel is assembled by three subunits with each contributing four β-strands. Three α-helices, one from each subunit, fill the central pore (Fig. 10B). This is radically different from other bacterial Omps discussed thus far whose barrels are typically formed from a single monomer.

This multisubunit single barrel is also found in TolC (Koronakis, Sharff, Koronakis, Luisi, & Hughes, 2000), VceC (Federici et al., 2005), OprM (Akama et al., 2004), MspA (Faller, Niederweis, & Schulz, 2004), and α-hemolysin (Song et al., 1996). Although very different in sequence, *E. coli* TolC, *Vibrio cholerae* VceC and *P. aeruginosa* OprM share a high degree of structural similarity. They have a 12-stranded β-barrel anchored to the OM

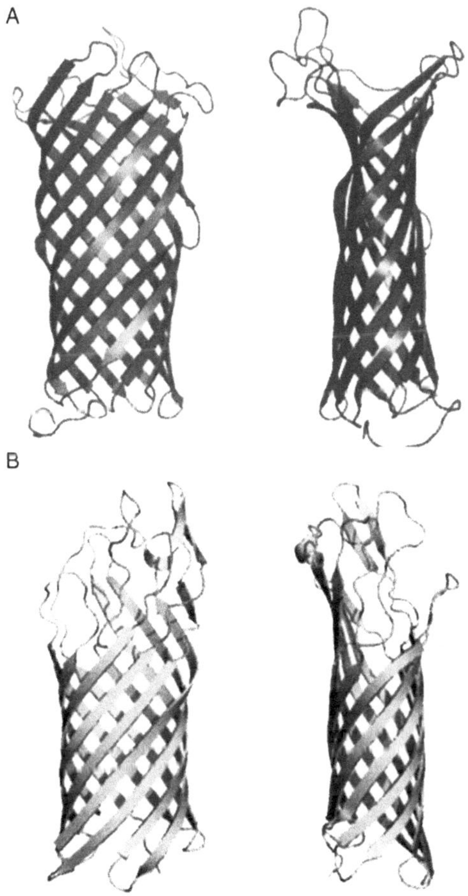

FIGURE 9 Structures of OmpT (PDB code: 1I78) and OpcA (PDB code: 1K24): (A) side view of OmpT with 90° in respect to each other, and (B) side view of OpcA with 90° turn with respect to each other.

and attached to a long α-helical periplasmic barrel; both barrels formed from three protomers (Fig. 11A). The β-barrel is completely open in TolC, while in both OprM and VceC three extracellular loops form a constriction. The α-helical barrel is closed at the periplasmic end in all three proteins and is assumed to represent a "resting state." As these three proteins are responsible for the export of drugs and other toxic compounds from the cytoplasm (Federici et al., 2005) the periplasmic domain must open to support this function.

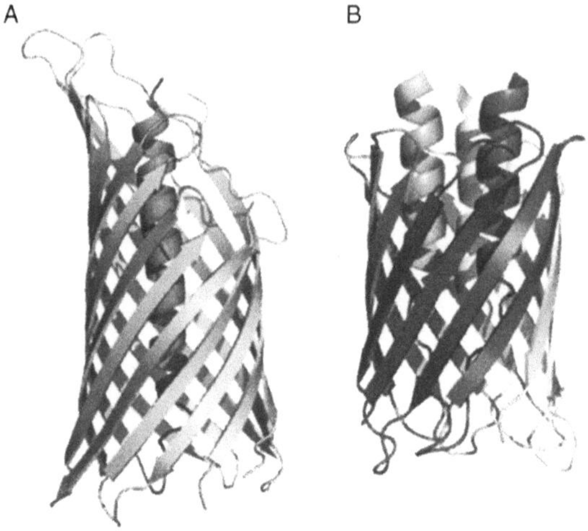

FIGURE 10 Structure of the β-domain from NalP (PDB code: 1UYN) and Hia (PDB code: 2GRB): (A) side view of β-domain of NalP, the N-terminal α-helix is colored red; and (B) side view of Hia β-domain, each subunit is colored in a different color. (See Color Insert.)

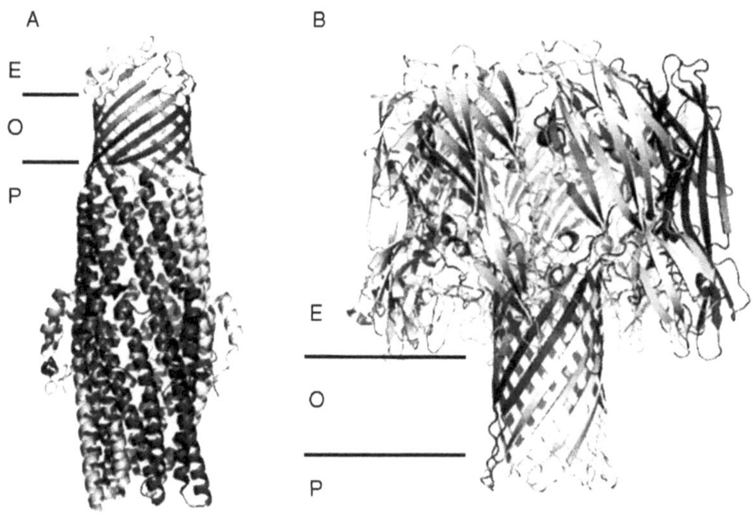

FIGURE 11 Structure of TolC (PDB code: 1EK9) and α-hemolysin (PDB code: 7AHL). (A) Three subunits in structure of TolC are in three different color. (B) In the structure of α-hemolysin, one subunit is colored purple. The approximate position of the outer membrane (O) is indicated by horizontal lines with the extracellular side (E) at the top and the periplasmic side (P) at the bottom. (See Color Insert.)

α-Hemolysin is assembled from seven subunits to form a 14-stranded barrel with a large extracellular domain shaped like a mushroom with a large central hydrophilic channel (Fig. 11B). MspA, a 16-stranded barrel porin from *Mycobacterium smegmatis* has a 16-stranded β-barrel formed from eight subunits (two stands per protomer) (Faller et al., 2004).

OmpG is a single protomer and 14-stranded β-barrel functioning as a porin. It has a large channel and lacks the constriction zone seen in other porins (Subbarao & van den Berg, 2006; Yildiz, Vinothkumar, Goswami, & Kuhlbrandt, 2006). Instead, OmpG has flexible extracellular loops which undergo conformational changes under different pH conditions. At neutral pH the pore is open (Fig. 12A) but at pH 5.6 (or lower) the pore is blocked by loop L6 which folds across and into the channel (Fig. 12B).

The fatty acid transporter FadL from *E. coli* (van den Berg, Black, Clemons, & Rapoport, 2004), the aromatic hydrocarbon transporter TodX from *Pseudomonas putida* and TbuX from *Ralstonia pickettii* (Hearn, Patel, & van den Berg, 2008) each have similar 14-stranded barrels. In each structure the lumen is occluded by an N-terminal "hatch domain" consisting of three short helices. The L3 consists of two antiparallel α-helices which form a hydrophobic cleft thought to bind substrate. In TodX and TbuX, the loop lies flat on the top of the barrel while in FadL it protrudes into the extracellular environment (Fig. 13). Based on structural data, a model has emerged in which substrate is bound by the extracellular loop L3, transits into the central hydrophobic channel. A conformational change of the hatch domain allows substrate to diffuse into the periplasm.

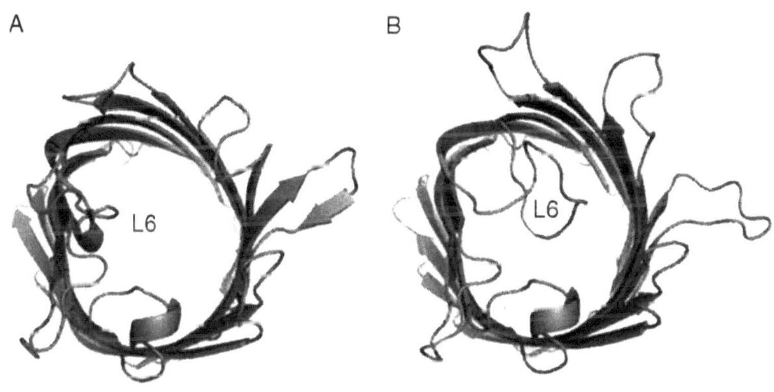

FIGURE 12 Structure of OmpG in two conformation states viewed from the extracellular side of the membrane. Loop L6 which undergoes largest conformational change is labeled: (A) open conformation of OmpG (PDB code: 2IWV); (B) closed conformation of OmpG (PDB code: 2IWW).

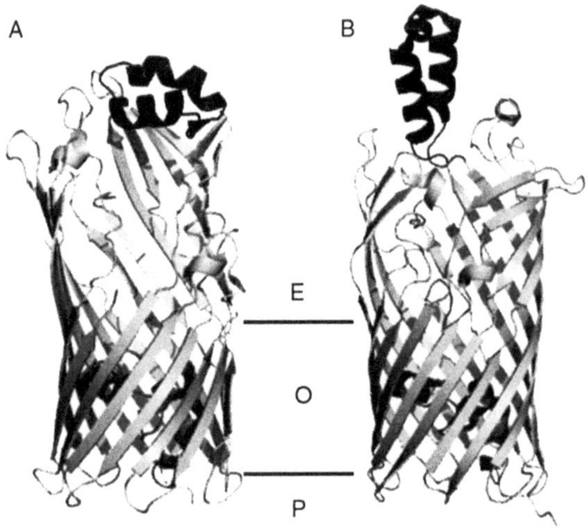

FIGURE 13 Structure features of TodX (A, PDB code: 3BSO) and FadL (B, PDB code: 1T16). The N-terminal hatch domain is colored in purple, the extracellular loop L3 is in blue. The approximate position of the outer membrane (O) is indicated by horizontal lines with the extracellular side (E) at the top and the periplasm (P) at the bottom. (See Color Insert.)

Despite the variability, the dominant feature is the β-barrel architecture, as has been seen for other Omps. β-barrel proteins are transported and assembled by a specialized protein transport machinery involving multiple proteins (Kim et al., 2007). The outer membrane protein 85-two-partner secretion B (Omp85-TpsB) superfamily is at the heart of this system. Omp85-TpsB is thought to contain a conserved C-terminal transmembrane β-barrel and a soluble N-terminal region harboring putative polypeptide-transport-associated (POTRA) domains (Clantin et al., 2007). FhaC, an outer membrane protein from the Omp85-TpsB transporter family, mediates the secretion of *Bordetella pertussis* filamentous hemagglutinin (FHA), an elongated right-handed parallel β-helix (Clantin et al., 2007). The protein is a monomer and comprises a 16-stranded β-barrel with loop L6 forming a hairpin which is inserted into the barrel (Fig. 14). The N-terminus of the protein is located in the extracellular milieu and folds into a long α-helix (H1) that goes through the transmembrane β-barrel. The helix is connected to a periplasmic module consisting of two structurally related POTRA domains that precede the β-barrel. The domains consist of 75 residues that form three-stranded β-sheets and one α-helix. They share the same strand-helix-strand-strand topology and are thought to recognize the N-terminal of FHA providing a template for assembly and export. Translocation is proposed to start with FHA adopting an extended β-hairpin structure during transit which refolds

FIGURE 14 Structure of FhaC (PDB code: 2QDZ). The α-helix H1 is colored red, POTRA 1 yellow, POTRA 2 blue, loop L6 in olive. (See Color Insert.)

at the cell surface. After the C-terminus of FHA has reached the cell surface, the N-terminus of FHA dissociates from POTRA1 and completes translocation (Clantin et al., 2007). This templating function of the POTRA domains is predicted to be the basis by which β-barrels themselves are formed. A stretch of amino acids in the unfolded protein binds to the POTRA domain creating the first β-strand against which the other strands of the protein assemble. This process once started is presumably spontaneous. The β-barrel structure is extremely strong, being held together by extensive main chain hydrogen bonds. In fact many β-barrels can be removed from the membrane by very harsh detergents without unfolding, they are relatively insensitive (at least in the membrane portion) to proteases and can handled quite harshly during purification (e.g., precipitation).

Until recently all Omps were presumed to follow the β-barrel paradigm and the POTRA templating system provided a simple rationale for this. The structure of Wza showed this not to be the case; it has an α-helical barrel that spans the OM (Dong et al., 2006). Wza is assembled from eight protomers with a very large central periplasmic cavity (reminiscent of TolC) (Fig. 15). Although, the α-helical barrel is completely open, the central cavity is closed to the periplasm. Creating an opening into this central cavity would seem essential for function but the trigger for any such an opening is unknown.

The human voltage-dependent anion channel (VDAC) from mitochondria is the first example of a eukaryotic porin structure (Bayrhuber et al., 2008; Hiller et al., 2008). The OM of mitochondria contains three membrane

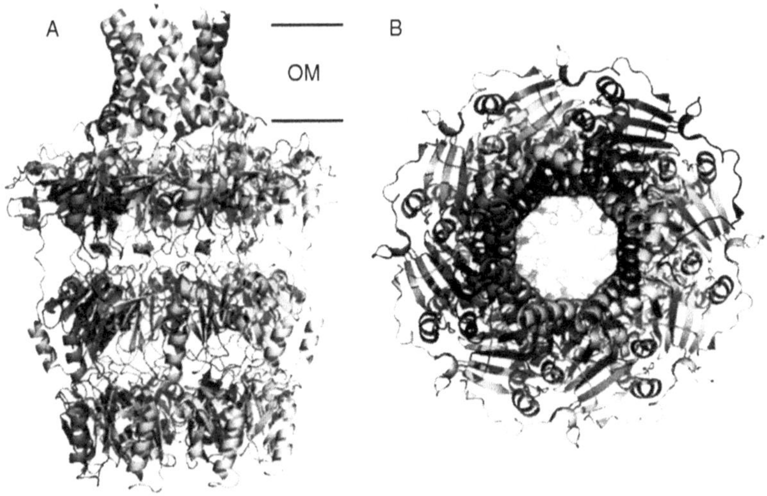

FIGURE 15 Structure of Wza (PDB code: 2J58): (A) side view of Wza. The OM position is marked; and (B) the channel of Wza, viewed from outside of the cell.

protein families; the translocase of the outer mitochondrial membrane (TOM), the sorting and assembly machinery (SAM), and the VDAC (Blachly-Dyson & Forte, 2001; Hill et al., 1998; Wiedemann et al., 2003). Similar to the bacterial porins, the human VDAC adopts a β-barrel architecture composed of 19 β-strands with an α-helix located horizontally, midway within the pore (Bayrhuber et al., 2008; Hiller et al., 2008) (Fig. 16). The VDAC is 30 Å in length and the pore is around 25 Å in diameter. In contrast to the bacterial porins, the N-terminal tail of VDAC is not part of the barrel, but it folds horizontally inside the barrel; this structural feature is similar to the way L3 is folded in bacterial porins. The N-terminus is also involved in voltage gating. The oligomeric state of the VDAC is unclear as it can exist in equilibrium as monomers, dimers, trimers, tetramers, hexamers, and higher oligomers (Malia & Wagner, 2007). Applying symmetry operators, the VDAC forms a dimer (Bayrhuber et al., 2008).

III. RECOMBINANT PRODUCTION OF OMPS

OM can either be produced in a functional state in the OMs or as inclusion bodies. The robust nature of Omps has resulted in both methods being successful for the structural elucidation of porins. All Omps that are transported to the OM contain an N-terminal signal peptide, usually 20 amino

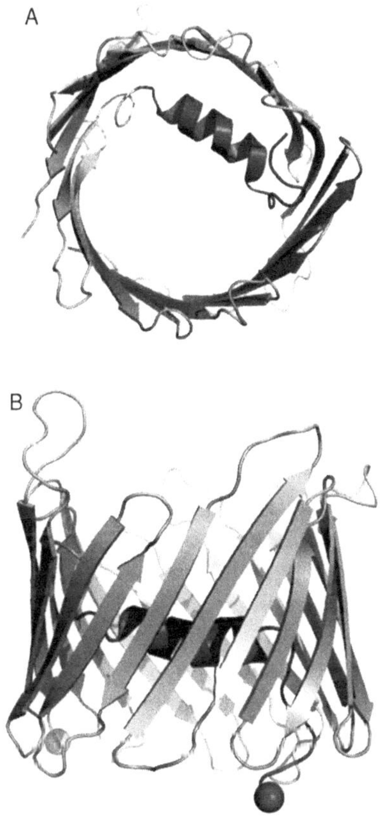

FIGURE 16 Structure of human VDAC: (A) view from outside the mitochondrion; and (B) view along the membrane. The N-terminal tail is colored magenta. The N- and C-terminal residues are shown in ball (PDB code: 2JK4). (See Color Insert.)

acids long, which directs them through the translocon in the IM to the periplasm. The signal peptide is removed during translocation by signal peptidases; it is then folded and inserted in the OM. For the recombinant production of Omps it is important to take into account the signal peptide. *E. coli* is usually used as the host organism for the overexpression of either native or heterologous recombinant Omps; in some cases, the *E. coli* signal peptidase cannot process the signal peptide of the heterologous proteins, resulting in inclusion bodies or no expression. Manipulation of the signal peptide of the target protein can resolve this (Hearn et al., 2008). Absence of the signal peptide from the recombinant protein results in the production of the target protein in the form of inclusion bodies. The most common inducible system is the T7 (Dubendorff & Studier, 1991; Studier, Rosenberg, Dunn, & Dubendorff, 1990); however, the arabinose promoter which allows for very tight regulation of the overexpression conditions (Guzman, Belin, Carson, & Beckwith, 1995), and leaky expression systems are also used. The introduction of affinity tags makes purification straightforward although because of the signal processing these usually are inserted at the C-terminus. Noncleavable affinity tags (or cleaved tags which leave a long tail, common at the C-terminus) can inhibit crystallization, as seen for Wza. Electron microscopy analysis revealed the formation of large aggregates in the presence of a histidine tag. Thus caution is required when tagging the C-terminus. The most common *E. coli* strain is the BL21(DE3) strain; however, this and many other strains have abundant outer membrane porins, OmpF, OmpC, and LamB which can complicate purification. The OmpC deficient *E. coli* porin strain BL21(DE3)Omp8 helps overcome this (Prilipov, Phale, Van Gelder, Rosenbusch, & Koebnik, 1998). The so-called "Walker strains" (C41(DE3) and C43(DE3)) can facilitate the expression of toxic proteins (Miroux & Walker, 1996) and produce high-biomass density.

Omps can be expressed as inclusion bodies in the absence of a signal peptide, extracted and purified in the presence of urea in a relatively straightforward manner. Many can be readily and correctly refolded in the presence of a detergent, although an assay to assess correct functional refolding is desirable. A successful example of this method for protein production is the X-ray structure of OmpG from *E. coli* (Yildiz et al., 2006). To date, there is no successful example of expression of eukaryotic Omps in their functional state via the OM of *E. coli*.

IV. RECOMBINANT BACTERIAL INNER MEMBRANE PROTEINS AND EUKARYOTIC MEMBRANE PROTEINS

The production of inner membrane and eukaryotic proteins is different from their OM counterparts. The proteins have to be produced in a functional form. The host for production of endogenous and heterologous prokaryotic membrane proteins is *E. coli*; this system has been successful for all of the prokaryotic inner membrane structures in the PDB. It is important to know the environment that the eukaryotic protein is targeted to (plasma membrane, vesicles, ER). This makes the production of recombinant eukaryotic proteins particularly challenging (there are no structures of eukaryotic proteins expressed heterologously in *E. coli*). This is probably due to differences in membrane lipid composition for eukaryotic versus *E. coli* cells as well as the absence of posttranslational modifications such as glycosylation in prokaryotic systems. The most commonly used systems for expression of eukaryotic membrane proteins are yeast, *Pichia pastoris* and *Saccharomyces cerevisiae*, followed by insect cells and mammalian cell lines. Each of these techniques has their own advantages and disadvantages (Junge et al., 2008).

One of the bottlenecks towards the successful structure determination of inner membrane proteins is the selection of highly expressed and stable targets after extraction for purification and crystallization trials; this can be an expensive and time consuming procedure. There are some techniques that have been developed in recent years that can facilitate such processes, resulting in increases in the structure determination of inner and eukaryotic membrane proteins. The GFP-method is based on a cleavable GFP protein with an octa-histidine tag fused to the C-terminus of the target protein. This method allows for rapid and cost efficient monitoring of overexpression, solubilization, and purification of IMP. This system has been developed for both prokaryotic (Drew, Lerch, Kunji, Slotboom, & de Gier, 2006; Kawate & Gouaux, 2006), *E. coli*, and eukaryotic (Drew et al., 2008; Newstead, Kim, von Heijne, Iwata, & Drew, 2007), *S. cerevisiae*, expression systems. This also allows in-gel fluorescence and fluorescence size-exclusion chromatography (Kawate & Gouaux, 2006) of GFP fusion proteins without the need for purification in order to determine the best expressed protein construct and the best detergent for extraction and purification that produces monodisperse protein. This method has been used in the structure determination of the Na^+/Cl^--dependent neurotransmitter transporter (Yamashita, Singh, Kawate, Jin, & Gouaux, 2005).

The blot method allows screening of expression and solubilization for multiple constructs and can be performed in a 96-well plate (Eshaghi et al., 2005). Proteins are blotted on filters with antibodies to the purification tag after lysis and solubilization (Cornvik et al., 2005). This method has been used for the structure determination of the CorA (Eshaghi et al., 2006).

V. CRYSTALLIZATION OF MEMBRANE PROTEINS

The most common technique for the crystallization of membrane proteins is vapor diffusion. To obtain high-quality diffracting crystals it is usually important to have monodisperse sample in an active state. Pure sample sometimes does not only mean a single band in a PAGE gel. It is important to also consider other protein modifications other than the obvious ones such as glycosylation or phosphorylation, prior to crystallization. For Omps crystals, the quality depends on removing as much of the protein-bound lipids as possible (Kim, 1998), whereas for α-helical membrane proteins it is necessary to retain more lipids (Guan, Smirnova, Verner, Nagamori, & Kaback, 2006) possibly avoiding size-exclusion chromatography. Removal of lipids from Omps is usually achieved by either binding the protein to an anion-exchange or affinity column, if a tag is present, or alternatively, application to a size-exclusion column followed by washing the protein extensively with buffer-containing detergent. To retain lipids for inner membrane proteins it is necessary to minimize the number of purification steps. In the UK membrane protein laboratory, samples are used that have and have not been through size-exclusion chromatography. Recently a new sparse-matrix screen, MemPlus, has been developed by Molecular Dimensions for the crystallization of Omps; the screen is based around the published crystallization conditions of Omps and constitutes of 48 different conditions. The group of Iwata has developed three screens that are specific for inner membrane protein crystallization, MemStart, MemSys (Iwata, 2003), and MemGold (Newstead, Ferrandon, & Iwata, 2008).

In UK membrane protein laboratory, the initial crystallization screens are conducted in the detergent that the protein has been purified in; typically, this is 1% octyl-glucoside (OG) for Omps and 0.03% dodecyl-β-D-maltopyranoside (DDM) for the others. Crystals that are obtained in DDM usually do not diffract very well since the DDM has a rather large micelle. Therefore, it is essential to screen for additional detergents or even try exchange for a smaller detergent during the purification process. Additives can also help with the optimization of crystal quality. The article by Newstead et al. has a detailed and useful analysis of the most successful additives and detergents that have been used in the past to obtain the crystal structures of membrane proteins (Newstead et al., 2008).

In contrast to Omps that are "easier" to obtain high-quality diffracting crystals for, inner and eukaryotic membrane protein crystals suffer from low-resolution diffraction or no diffraction. Many crystallization methods have been advanced in recent years to overcome these obstacles. On the other hand, lipidic cubic and sponge phases allow the membrane proteins to diffuse in a lipid environment mimicking the native environment more closely (crystals grow free of detergent) (Landau & Rosenbusch, 1996). Even though the lipidic techniques are used more widely for helical membrane proteins, Caffrey and colleagues have also determined the crystal structure of BtuB, an Omps, using the *in meso* method where the protein diffuses in lipid bilayers (Cherezov et al., 2006).

VI. New Strategies for Eukaryotic Proteins from the Study of GPCR

There are over 1000 GPCR in the human genome and they underlie most signaling events. They constitute one of the major protein classes targeted by the pharmaceutical industry. It is difficult to identify a single class of protein molecule that is more important economically and scientifically. Despite the scientific need to obtain structural information from this protein, until recently the community relied solely on the pioneering work for the structure of rhodopsin to create models for these vital proteins. It would be reasonable to conclude that since funding for such an important target was forthcoming over a prolonger period, GPCR's represented a particular challenge to crystallographers. These challenges appear inherent to the molecules themselves, they vary in conformation, possibly having multiple states and they have long flexible regions. Both of these features are known to complicate any attempt to crystallize any protein. A further complication was that of production of a homogeneous batch of active protein.

Five papers in a 12-month period have transformed our understanding of these proteins reporting structure of $\beta 2$ adrenergic receptor (Cherezov et al., 2007; Rasmussen et al., 2007), $\beta 1$ adrenergic receptor (Warne et al., 2008), A2A adenosine receptor (Jaakola et al., 2008), and opsin (Park, Scheerer, Hofmann, Choe, & Ernst, 2008). The $\beta 2$ adrenergic receptor and A2A adenosine receptor are found with ligands whereas the structures for $\beta 1$ adrenergic receptor and opsin were of the apo form. In terms of techniques, the most important and hopeful observation is that the papers report four distinct approaches that led to success in overcoming the particular challenges of GPCRs. Although these approaches seem simple in retrospect, it is important to recognize that they were the culmination of years of careful experimentation and imagination. The utility of these methods suggest we may not need to wait quite so long for future G-protein coupled protein structures.

The use of antibodies, Fab or Fv fragments, to fix structures into more rigid conformers is well known (Hunte, Koepke, Lange, Rossmanith, & Michel, 2000; Zhou, Morais-Cabral, Kaufman, & MacKinnon, 2001), their use with $\beta 2$ adrenergic receptor led to the first structure of a GPCR (Rasmussen et al., 2007). The active protein was expressed using insect cells. The antibody approach used here has an additional element. The antibody was chosen to recognize only the active conformation of the third intracellular loop of the protein and it did not recognize unfolded protein. The antibody did not compromise ligand binding which happens on the other (extracellular) side of the membrane. The antibody served the well-known purpose of locking the protein into a more defined conformation state and by purifying the complex it presumably allowed nonfunctional protein to be removed. For $\beta 2$ adrenergic receptor it also greatly increased the polar surface area available for crystal contacts thus favoring crystal formation. In general, hydrophobic contacts do not favor crystal formation because they lack the strong angular component of hydrogen bonds that gives rise to specific lattice contacts. The antibody comprised around 50% of the scattering.

At 3.4–3.7 Å resolution, the structural detail is limited but none the less important insights into structure were gained (Rasmussen et al., 2007). In particular, the structure revealed the precise helical arrangement and the binding site of carazolol. The following week, the high-resolution structure of the same protein was described in linked papers (Cherezov et al., 2007). Crystals were obtained by a novel protein engineering strategy, which introduced T4 lysozyme into the third cytoplasmic loop as well as removal of the flexible C-terminus (Cherezov et al., 2007; Rosenbaum et al., 2007) (Fig. 17). This chimeric approach was first reported for the K^+ ion channel (Long, Tao, Campbell, & MacKinnon, 2007; Nishida, Cadene, Chait, & MacKinnon, 2007). As with the antibody structure, the protein was expressed in insect cells. The choice of T4 lysozyme was guided by consideration of the likely distance of the residues in the loop which were to be replaced as well as a requirement for a soluble small stable protein. Although the replacement of the third loop disrupts the binding to G-proteins, the engineered protein still binds and responds to ligands. This engineered protein led to a 2.4 Å structure. As would be expected the higher resolution (and order) of the crystals manifests itself in a more complete description of the structure. In describing the

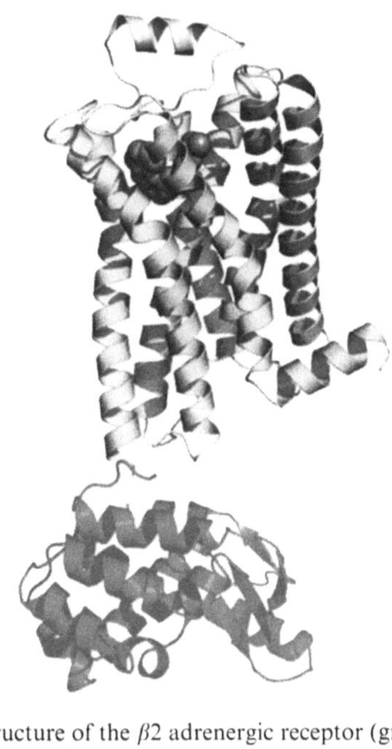

FIGURE 17 Crystal structure of the β2 adrenergic receptor (green) with the L3 replaced by T4 lysozyme (red). Carazolol is shown as orange balls (PDB code: 2RH1). (See Color Insert.)

crystallization of the fusion protein, the authors make the point that the quality of the crystals was dependent on using the lipidic cubic phase crystallization. This method is entirely different from that pursued in traditional crystallization experiments (which are often variants of vapor diffusion). Lipidic cubic phase is based on a formation by lipid, protein, and water a continuous structured array. As the array has no micelle (hence discontinuity) diffusion is thought to be continuous and therefore growth of crystals are more controlled. From a methodological viewpoint it is comforting to see that both strategies which involve quite radical but distinct approaches to the protein gave essentially the similar GPCR structures bound to the identical ligand. Obtaining an apo structure of GPCR was thought to be particularly challenging because many of the proteins aggregate in the absence of the ligand. The structure of opsin (the apo form of rhodopsin) showed that a rapid specially adapted purification protocol allowed production of functional apo receptor from source (bovine eye) (Park et al., 2008). Although clearly successful, the normal concentration of G-proteins in their *in vivo* setting is too small for this likely to be widely applicable but it may be a very powerful approach in some cases.

The fourth approach details a rational site-directed mutagenesis strategy aimed at making stable GPCRs (Serrano-Vega, Magnani, Shibata, & Tate, 2008). Stability is the *sine non qua* of crystal, a protein which aggregates or unfolds is often hard to purify and impossible to crystallize. The protein engineering approach relies on a robust assay for function and an assessment of stability. Most GPCR's can be assayed in some manner, although radioactive binding assays are the least easy to make high throughput. The assessment of stability can be performed by measuring activity versus temperature or more recently using a fluorescent derivative (Alexandrov, Mileni, Chien, Hanson, & Stevens, 2008). This latter approach allows extremely rapid screening for protein stability and requires only that there are cysteines which are exposed upon unfolding. The technique is limited to those proteins containing a buried cysteine; however, one could be engineered into the protein if required.

VII. SUMMARY

Membrane proteins are special from one standpoint alone. In the absence of detergent, they will often aggregate and precipitate. This is because unlike conventional soluble proteins they expose large areas of hydrophobic surface which will bind to other hydrophobic surfaces in a nonspecific manner. The problem of finding a suitable detergent remains trial and error, although new analytical technologies are being developed which help to identify the most likely candidates without the expense and waste of crystallization experiments. The relative uniformity of bacterial Omps suggests the detergent choice is more limited. The expression of membrane proteins is a challenge but this is true for many soluble proteins. It is unclear whether expression of membrane proteins is really a special case or whether problems of being part of a larger protein complex, toxicity, and lack of chaperone system found for soluble proteins are more common for membrane proteins. In one sense membrane proteins are disadvantaged, the volume of membrane in a cell is usually considerably less than soluble compartment, posing a limit on the overexpression. Cell-free technology and other new expression systems may overcome or reduce this limitation. In counterpart, the purification of overexpressed membrane proteins may be more straightforward as the fractionation of the membrane portion of cells reduces contaminants. It is clear that membrane proteins will be the subject of increasing study, given their importance this is to be welcomed and we would conclude with an invitation to other structural biologists "come on in, the water is lovely."

References

Ahn, V. E., Lo, E. I., Engel, C. K., Chen, L., Hwang, P. M., Kay, L. E., et al. (2004). A hydrocarbon ruler measures palmitate in the enzymatic acylation of endotoxin. *EMBO Journal, 23*, 2931–2941.

Akama, H., Kanemaki, M., Yoshimura, M., Tsukihara, T., Kashiwagi, T., Yoneyama, H., et al. (2004). Crystal structure of the drug discharge outer membrane protein, OprM, of *Pseudomonas aeruginosa*: Dual modes of membrane anchoring and occluded cavity end. *The Journal of Biological Chemistry, 279*, 52816–52819.

Alexandrov, A. I., Mileni, M., Chien, E. Y., Hanson, M. A., & Stevens, R. C. (2008). Microscale fluorescent thermal stability assay for membrane proteins. *Structure, 16*, 351–359.

Barnard, T. J., Dautin, N., Lukacik, P., Bernstein, H. D., & Buchanan, S. K. (2007). Autotransporter structure reveals intra-barrel cleavage followed by conformational changes. *Nature Structural and Molecular Biology, 14*, 1214–1220.

Basle, A., Rummel, G., Storici, P., Rosenbusch, J. P., & Schirmer, T. (2006). Crystal structure of osmoporin OmpC from *E. coli* at 2.0 A. *Journal of Molecular Biology, 362*, 933–942.

Bayrhuber, M., Meins, T., Habeck, M., Becker, S., Giller, K., Villinger, S., et al. (2008). Structure of the human voltage-dependent anion channel. *Proceedings of the National Academy of Sciences of the United States of America, 105*, 15370–15375.

Biswas, S., Mohammad, M. M., Movileanu, L., & van den Berg, B. (2008). Crystal structure of the outer membrane protein OpdK from *Pseudomonas aeruginosa. Structure, 16*, 1027–1035.

Biswas, S., Mohammad, M. M., Patel, D. R., Movileanu, L., & van den Berg, B. (2007). Structural insight into OprD substrate specificity. *Nature Structural and Molecular Biology, 14*, 1108–1109.

Blachly-Dyson, E., & Forte, M. (2001). VDAC channels. *IUBMB Life, 52*, 113–118.

Buchanan, S. K., Lukacik, P., Grizot, S., Ghirlando, R., Ali, M. M., Barnard, T. J., et al. (2007). Structure of colicin I receptor bound to the R-domain of colicin Ia: Implications for protein import. *EMBO Journal, 26*, 2594–2604.

Cherezov, V., Rosenbaum, D. M., Hanson, M. A., Rasmussen, S. G., Thian, F. S., Kobilka, T. S., et al. (2007). High-resolution crystal structure of an engineered human beta2-adrenergic G protein-coupled receptor. *Science, 318*, 1258–1265.

Cherezov, V., Yamashita, E., Liu, W., Zhalnina, M., Cramer, W. A., & Caffrey, M. (2006). *In meso* structure of the cobalamin transporter, BtuB, at 1.95 A resolution. *Journal of Molecular Biology, 364*, 716–734.

Chimento, D. P., Mohanty, A. K., Kadner, R. J., & Wiener, M. C. (2003). Substrate-induced transmembrane signaling in the cobalamin transporter BtuB. *Nature Structural Biology, 10*, 394–401.

Clantin, B., Delattre, A. S., Rucktooa, P., Saint, N., Meli, A. C., Locht, C., et al. (2007). Structure of the membrane protein FhaC: A member of the Omp85-TpsB transporter superfamily. *Science, 317*, 957–961.

Cobessi, D., Celia, H., Folschweiller, N., Schalk, I. J., Abdallah, M. A., & Pattus, F. (2005). The crystal structure of the pyoverdine outer membrane receptor FpvA from *Pseudomonas aeruginosa* at 3.6 angstroms resolution. *Journal of Molecular Biology, 347*, 121–134.

Cornvik, T., Dahlroth, S. L., Magnusdottir, A., Herman, M. D., Knaust, R., Ekberg, M., et al. (2005). Colony filtration blot: A new screening method for soluble protein expression in *Escherichia coli*. *Nature Methods, 2*, 507–509.

Cowan, S. W., Schirmer, T., Rummel, G., Steiert, M., Ghosh, R., Pauptit, R. A., et al. (1992). Crystal structures explain functional properties of two *E. coli* porins. *Nature, 358*, 727–733.

Dong, C., Beis, K., Nesper, J., Brunkan-Lamontagne, A. L., Clarke, B. R., Whitfield, C., et al. (2006). Wza the translocon for *E. coli* capsular polysaccharides defines a new class of membrane protein. *Nature, 444*, 226–229.

Drew, D., Lerch, M., Kunji, E., Slotboom, D. J., & de Gier, J. W. (2006). Optimization of membrane protein overexpression and purification using GFP fusions. *Nature Methods, 3*, 303–313.

Drew, D., Newstead, S., Sonoda, Y., Kim, H., von Heijne, G., & Iwata, S. (2008). GFP-based optimization scheme for the overexpression and purification of eukaryotic membrane proteins in *Saccharomyces cerevisiae*. *Nature Protocols, 3*, 784–798.

Dubendorff, J. W., & Studier, F. W. (1991). Controlling basal expression in an inducible T7 expression system by blocking the target T7 promoter with lac repressor. *Journal of Molecular Biology, 219*(1), 45–59.

Dutzler, R., Rummel, G., Alberti, S., Hernandez-Alles, S., Phale, P., Rosenbusch, J., et al. (1999). Crystal structure and functional characterization of OmpK36, the osmoporin of *Klebsiella pneumoniae*. *Structure, 7*, 425–434.

Eshaghi, S., Hedren, M., Nasser, M. I., Hammarberg, T., Thornell, A., & Nordlund, P. (2005). An efficient strategy for high-throughput expression screening of recombinant integral membrane proteins. *Protein Science, 14*, 676–683.

Eshaghi, S., Niegowski, D., Kohl, A., Martinez Molina, D., Lesley, S. A., & Nordlund, P. (2006). Crystal structure of a divalent metal ion transporter CorA at 2.9 angstrom resolution. *Science, 313*, 354–357.

Faller, M., Niederweis, M., & Schulz, G. E. (2004). The structure of a mycobacterial outer-membrane channel. *Science, 303*, 1189–1192.

Federici, L., Du, D., Walas, F., Matsumura, H., Fernandez-Recio, J., McKeegan, K. S., et al. (2005). The crystal structure of the outer membrane protein VceC from the bacterial pathogen *Vibrio cholerae* at 1.8 Å resolution. *The Journal of Biological Chemistry, 280*, 15307–15314.

Ferguson, A. D., Chakraborty, R., Smith, B. S., Esser, L., van der Helm, D., & Deisenhofer, J. (2002). Structural basis of gating by the outer membrane transporter FecA. *Science, 295*, 1715–1719.

Ferguson, A. D., & Deisenhofer, J. (2002). TonB-dependent receptors-structural perspectives. *Biochimica Et Biophysica Acta, 1565*, 318–332.

Ferguson, A. D., Hofmann, E., Coulton, J. W., Diederichs, K., & Welte, W. (1998). Siderophore-mediated iron transport: Crystal structure of FhuA with bound lipopolysaccharide. *Science, 282*, 2215–2220.

Forst, D., Welte, W., Wacker, T., & Diederichs, K. (1998). Structure of the sucrose-specific porin ScrY from *Salmonella typhimurium* and its complex with sucrose. *Nature Structural Biology, 5*, 37–46.

Guan, L., Smirnova, I. N., Verner, G., Nagamori, S., & Kaback, H. R. (2006). Manipulating phospholipids for crystallization of a membrane transport protein. *Proceedings of the National Academy of Sciences of the United States of America, 103*, 1723–1726.

Guzman, L. M., Belin, D., Carson, M. J., & Beckwith, J. (1995). Tight regulation, modulation, and high-level expression by vectors containing the arabinose PBAD promoter. *Journal of Bacteriology, 177*, 4121–4130.

Hearn, E. M., Patel, D. R., & van den Berg, B. (2008). Outer-membrane transport of aromatic hydrocarbons as a first step in biodegradation. *Proceedings of the National Academy of Sciences of the United States of America, 105*, 8601–8606.

Hill, K., Model, K., Ryan, M. T., Dietmeier, K., Martin, F., Wagner, R., et al. (1998). Tom40 forms the hydrophilic channel of the mitochondrial import pore for preproteins [see comment]. *Nature, 395*, 516–521.

Hiller, S., Garces, R. G., Malia, T. J., Orekhov, V. Y., Colombini, M., & Wagner, G. (2008). Solution structure of the integral human membrane protein VDAC-1 in detergent micelles. *Science, 321*, 1206–1210.

Hong, H., Patel, D. R., Tamm, L. K., & van den Berg, B. (2006). The outer membrane protein OmpW forms an eight-stranded beta-barrel with a hydrophobic channel. *The Journal of Biological Chemistry, 281*, 7568–7577.

Hunte, C., Koepke, J., Lange, C., Rossmanith, T., & Michel, H. (2000). Structure at 2.3 A resolution of the cytochrome *bc*(1) complex from the yeast *Saccharomyces cerevisiae* co-crystallized with an antibody Fv fragment. *Structure, 8*, 669–684.

Iwata, S. (2003). Crystallization informatics of membrane proteins. In S. Iwata, (Ed.), *Methods and results in crystallization of membrane proteins*, pp. 281–294. La Jolla, California: International University Line.

Jaakola, V. P., Griffith, M. T., Hanson, M. A., Cherezov, V., Chien, E. Y., Lane, J. R., et al. (2008). The 2.6 Angstrom crystal structure of a human A2A adenosine receptor bound to an antagonist. *Science, 322*(5905), 1211–1217.

Junge, F., Schneider, B., Reckel, S., Schwarz, D., Dotsch, V., & Bernhard, F. (2008). Large-scale production of functional membrane proteins. *Cellular and Molecular Life Sciences, 65*, 1729–1755.

Kawate, T., & Gouaux, E. (2006). Fluorescence-detection size-exclusion chromatography for precrystallization screening of integral membrane proteins. *Structure, 14*, 673–681.

Kim, H. (1998). Crystallization of OmpC osmoporin from *Escherichia coli*. *Acta Crystallographica, 54*, 1399–1400.

Kim, S., Malinverni, J. C., Sliz, P., Silhavy, T. J., Harrison, S. C., & Kahne, D. (2007). Structure and function of an essential component of the outer membrane protein assembly machine. *Science, 317*, 961–964.

Koronakis, V., Sharff, A., Koronakis, E., Luisi, B., & Hughes, C. (2000). Crystal structure of the bacterial membrane protein TolC central to multidrug efflux and protein export. *Nature, 405*, 914–919.

Kreusch, A., Neubuser, A., Schiltz, E., Weckesser, J., & Schulz, G. E. (1994). Structure of the membrane channel porin from *Rhodopseudomonas blastica* at 2.0 A resolution. *Protein Science, 3*, 58–63.

Landau, E. M., & Rosenbusch, J. P. (1996). Lipidic cubic phases: A novel concept for the crystallization of membrane proteins. *Proceedings of the National Academy of Sciences of the United States of America, 93*, 14532–14535.

Locher, K. P., Rees, B., Koebnik, R., Mitschler, A., Moulinier, L., Rosenbusch, J. P., et al. (1998). Transmembrane signaling across the ligand-gated FhuA receptor: Crystal structures of free and ferrichrome-bound states reveal allosteric changes. *Cell, 95*, 771–778.

Long, S. B., Tao, X., Campbell, E. B., & MacKinnon, R. (2007). Atomic structure of a voltage-dependent K+ channel in a lipid membrane-like environment. *Nature, 450*, 376–382.

Malia, T. J., & Wagner, G. (2007). NMR structural investigation of the mitochondrial outer membrane protein VDAC and its interaction with antiapoptotic Bcl-xL. *Biochemistry, 46*, 514–525.

Meng, G., Surana, N. K., St Geme, J. W., III, & Waksman, G. (2006). Structure of the outer membrane translocator domain of the *Haemophilus influenzae* Hia trimeric autotransporter. *EMBO Journal, 25*, 2297–2304.

Meyer, J. E., Hofnung, M., & Schulz, G. E. (1997). Structure of maltoporin from *Salmonella typhimurium* ligated with a nitrophenyl-maltotrioside. *Journal of Molecular Biology, 266*, 761–775.

Miroux, B., & Walker, J. E. (1996). Over-production of proteins in *Escherichia coli*: Mutant hosts that allow synthesis of some membrane proteins and globular proteins at high levels. *Journal of Molecular Biology, 260*, 289–298.

Moeck, G. S., & Coulton, J. W. (1998). TonB-dependent iron acquisition: Mechanisms of siderophore-mediated active transport. *Molecular Microbiology, 28*, 675–681.

Moraes, T. F., Bains, M., Hancock, R. E., & Strynadka, N. C. (2007). An arginine ladder in OprP mediates phosphate-specific transfer across the outer membrane. *Nature Structural and Molecular Biology, 14*, 85–87.

Newstead, S., Ferrandon, S., & Iwata, S. (2008). Rationalizing alpha-helical membrane protein crystallization. *Protein Science, 17*, 466–472.

Newstead, S., Kim, H., von Heijne, G., Iwata, S., & Drew, D. (2007). High-throughput fluorescent-based optimization of eukaryotic membrane protein overexpression and purification in *Saccharomyces cerevisiae. Proceedings of the National Academy of Sciences of the United States of America, 104*, 13936–13941.

Nikaido, H. (2003). Molecular basis of bacterial outer membrane permeability revisited. *Microbiology and Molecular Biology Reviews, 67*, 593–656.

Nishida, M., Cadene, M., Chait, B. T., & MacKinnon, R. (2007). Crystal structure of a Kir3.1-prokaryotic Kir channel chimera. *EMBO Journal, 26*, 4005–4015.

Oomen, C. J., van Ulsen, P., van Gelder, P., Feijen, M., Tommassen, J., & Gros, P. (2004). Structure of the translocator domain of a bacterial autotransporter. *EMBO Journal, 23*, 1257–1266.

Park, J. H., Scheerer, P., Hofmann, K. P., Choe, H. W., & Ernst, O. P. (2008). Crystal structure of the ligand-free G-protein-coupled receptor opsin. *Nature, 454*, 183–187.

Pautsch, A., & Schulz, G. E. (1998). Structure of the outer membrane protein A transmembrane domain. *Nature Structural Biology, 5*, 1013–1017.

Pawelek, P. D., Croteau, N., Ng-Thow-Hing, C., Khursigara, C. M., Moiseeva, N., Allaire, M., et al. (2006). Structure of TonB in complex with FhuA, *E. coli* outer membrane receptor. *Science, 312*, 1399–1402.

Phale, P. S., Philippsen, A., Kiefhaber, T., Koebnik, R., Phale, V. P., Schirmer, T., et al. (1998). Stability of trimeric OmpF porin: The contributions of the latching loop L2. *Biochemistry, 37*, 15663–15670.

Prilipov, A., Phale, P. S., Van Gelder, P., Rosenbusch, J. P., & Koebnik, R. (1998). Coupling site-directed mutagenesis with high-level expression: Large scale production of mutant porins from *E. coli. FEMS Microbiology Letters, 163*, 65–72.

Prince, S. M., Achtman, M., & Derrick, J. P. (2002). Crystal structure of the OpcA integral membrane adhesin from *Neisseria meningitidis*. *Proceedings of the National Academy of Sciences of the United States of America, 99*, 3417–3421.

Rasmussen, S. G., Choi, H. J., Rosenbaum, D. M., Kobilka, T. S., Thian, F. S., Edwards, P. C., et al. (2007). Crystal structure of the human beta2 adrenergic G-protein-coupled receptor. *Nature, 450*, 383–387.

Remaut, H., Tang, C., Henderson, N. S., Pinkner, J. S., Wang, T., Hultgren, S. J., et al. (2008). Fiber formation across the bacterial outer membrane by the chaperone/usher pathway. *Cell, 133*, 640–652.

Rosenbaum, D. M., Cherezov, V., Hanson, M. A., Rasmussen, S. G., Thian, F. S., Kobilka, T. S., et al. (2007). GPCR engineering yields high-resolution structural insights into beta2-adrenergic receptor function. *Science, 318*, 1266–1273.

Schirmer, T., Keller, T. A., Wang, Y. F., & Rosenbusch, J. P. (1995). Structural basis for sugar translocation through maltoporin channels at 3.1 A resolution. *Science, 267*, 512–514.

Serrano-Vega, M. J., Magnani, F., Shibata, Y., & Tate, C. G. (2008). Conformational thermostabilization of the beta1-adrenergic receptor in a detergent-resistant form. *Proceedings of the National Academy of Sciences of the United States of America, 105*, 877–882.

Shultis, D. D., Purdy, M. D., Banchs, C. N., & Wiener, M. C. (2006). Outer membrane active transport: Structure of the BtuB:TonB complex. *Science, 312*, 1396–1399.

Snijder, H. J., Ubarretxena-Belandia, I., Blaauw, M., Kalk, K. H., Verheij, H. M., Egmond, M. R., et al. (1999). Structural evidence for dimerization-regulated activation of an integral membrane phospholipase. *Nature, 401*, 717–721.

Song, L., Hobaugh, M. R., Shustak, C., Cheley, S., Bayley, H., & Gouaux, J. E. (1996). Structure of staphylococcal alpha-hemolysin, a heptameric transmembrane pore. *Science, 274*, 1859–1866.

Studier, F. W., Rosenberg, A. H., Dunn, J. J., & Dubendorff, J. W. (1990). Use of T7 RNA polymerase to direct expression of cloned genes. *Methods in Enzymology, 185*, 60–89.

Subbarao, G. V., & van den Berg, B. (2006). Crystal structure of the monomeric porin OmpG. *Journal of Molecular Biology, 360*, 750–759.

Trias, J., & Nikaido, H. (1990). Protein D2 channel of the *Pseudomonas aeruginosa* outer membrane has a binding site for basic amino acids and peptides. *The Journal of Biological Chemistry, 265*, 15680–15684.

Ubarretxena-Belandia, I., Hozeman, L., van der Brink-van der Laan, E., Pap, E. H., Egmond, H. M., Verheij, H. M., et al. (1999). Outer membrane phospholipase A is dimeric in phospholipid bilayers: A cross-linking and fluorescence resonance energy transfer study. *Biochemistry, 38*, 7398–7405.

van den Berg, B., Black, P. N., Clemons, W. M., Jr., & Rapoport, T. A. (2004). Crystal structure of the long-chain fatty acid transporter FadL. *Science, 304*, 1506–1509.

Vandeputte-Rutten, L., Bos, M. P., Tommassen, J., & Gros, P. (2003). Crystal structure of Neisserial surface protein A (NspA), a conserved outer membrane protein with vaccine potential. *The Journal of Biological Chemistry, 278*, 24825–24830.

Vandeputte-Rutten, L., Kramer, R. A., Kroon, J., Dekker, N., Egmond, M. R., & Gros, P. (2001). Crystal structure of the outer membrane protease OmpT from *Escherichia coli* suggests a novel catalytic site. *EMBO Journal, 20*, 5033–5039.

Vogt, J., & Schulz, G. E. (1999). The structure of the outer membrane protein OmpX from *Escherichia coli* reveals possible mechanisms of virulence. *Structure, 7*, 1301–1309.

Warne, T., Serrano-Vega, M. J., Baker, J. G., Moukhametzianov, R., Edwards, P. C., Henderson, R., et al. (2008). Structure of a beta1-adrenergic G-protein-coupled receptor. *Nature, 454*, 486–491.

Weiss, M. S., Wacker, T., Weckesser, J., Welte, W., & Schulz, G. E. (1990). The three-dimensional structure of porin from *Rhodobacter capsulatus* at 3 A resolution. *FEBS Letters, 267*, 268–272.

Wiedemann, N., Kozjak, V., Chacinska, A., Schonfisch, B., Rospert, S., Ryan, M. T., et al. (2003). Machinery for protein sorting and assembly in the mitochondrial outer membrane. *Nature, 424*, 565–571.

Yamashita, A., Singh, S. K., Kawate, T., Jin, Y., & Gouaux, E. (2005). Crystal structure of a bacterial homologue of Na^{+}/Cl^{-}-dependent neurotransmitter transporters. *Nature, 437*, 215–223.

Ye, J., & van den Berg, B. (2004). Crystal structure of the bacterial nucleoside transporter Tsx. *EMBO Journal, 23*, 3187–3195.

Yildiz, O., Vinothkumar, K. R., Goswami, P., & Kuhlbrandt, W. (2006). Structure of the monomeric outer-membrane porin OmpG in the open and closed conformation. *EMBO Journal, 25*, 3702–3713.

Yue, W. W., Grizot, S., & Buchanan, S. K. (2003). Structural evidence for iron-free citrate and ferric citrate binding to the TonB-dependent outer membrane transporter FecA. *Journal of Molecular Biology, 332*, 353–368.

Zachariae, U., Kluhspies, T., De, S., Engelhardt, H., & Zeth, K. (2006). High resolution crystal structures and molecular dynamics studies reveal substrate binding in the porin Omp32. *The Journal of Biological Chemistry, 281*, 7413–7420.

Zeth, K., Diederichs, K., Welte, W., & Engelhardt, H. (2000). Crystal structure of Omp32, the anion-selective porin from *Comamonas acidovorans*, in complex with a periplasmic peptide at 2.1 A resolution. *Structure, 8*, 981–992.

Zhou, Y., Morais-Cabral, J. H., Kaufman, A., & MacKinnon, R. (2001). Chemistry of ion coordination and hydration revealed by a K+ channel-Fab complex at 2.0 A resolution. *Nature, 414*, 43–48.

Index

P

R

S

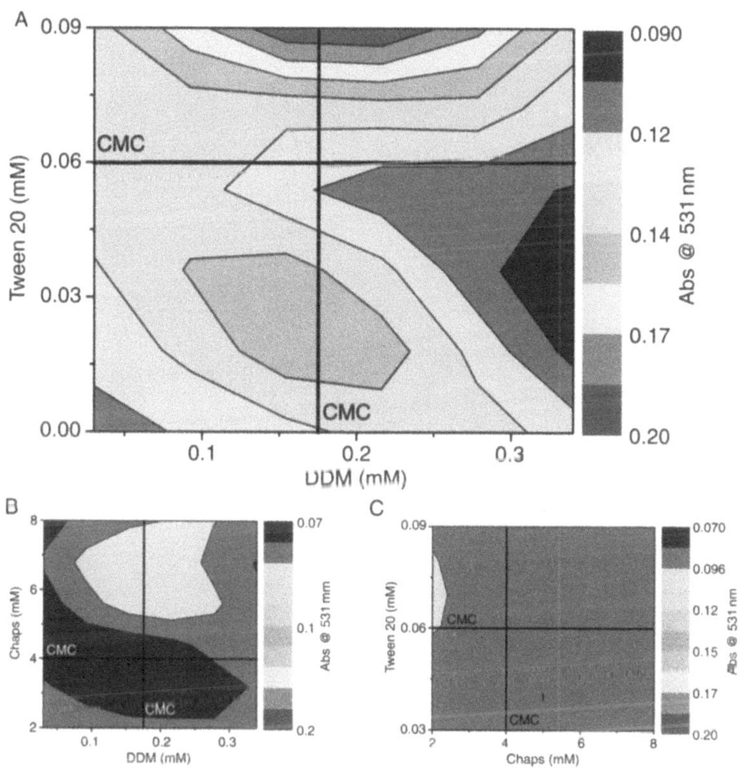

Please refer to Fig. 7.6 in text for figure legend.

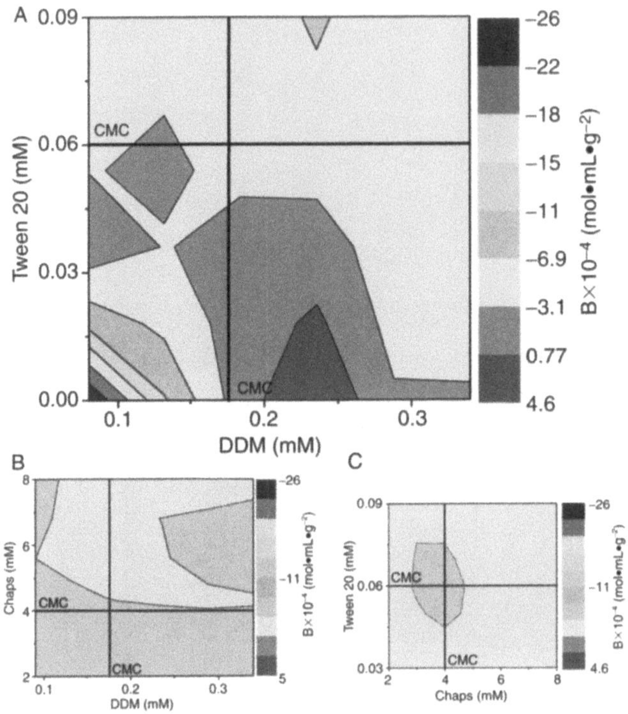

Please refer to Fig. 7.7 in text for figure legend.

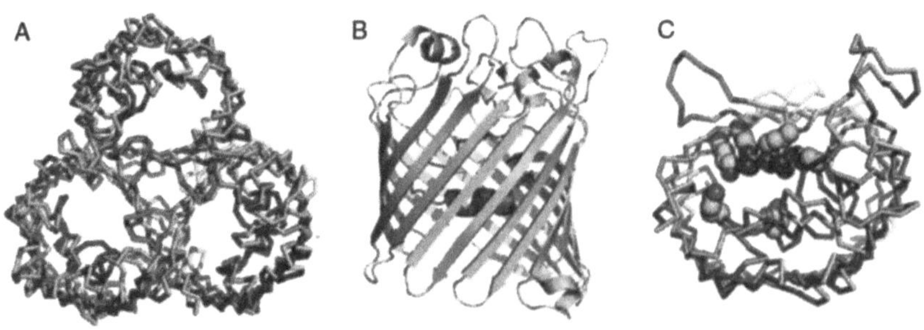

Please refer to Fig. 11.1 in text for figure legend.

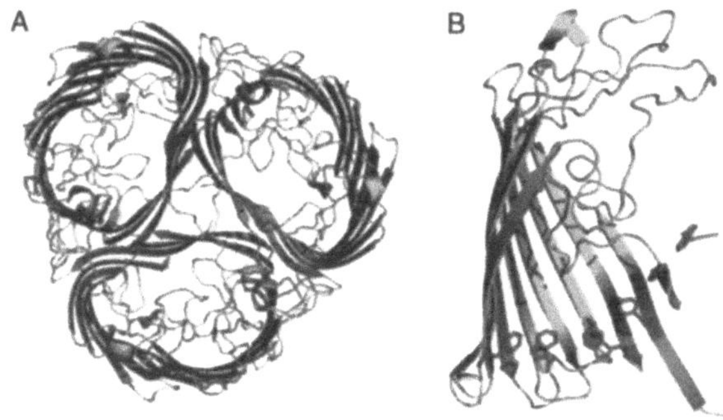

Please refer to Fig. 11.2 in text for figure legend.

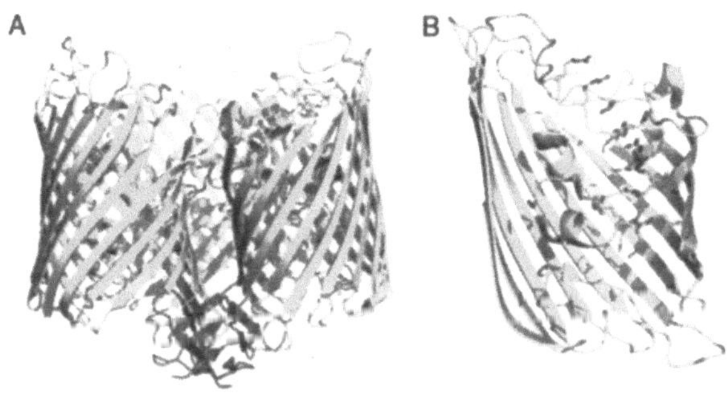

Please refer to Fig. 11.3 in text for figure legend.

Please refer to Fig. 11.4 in text for figure legend.

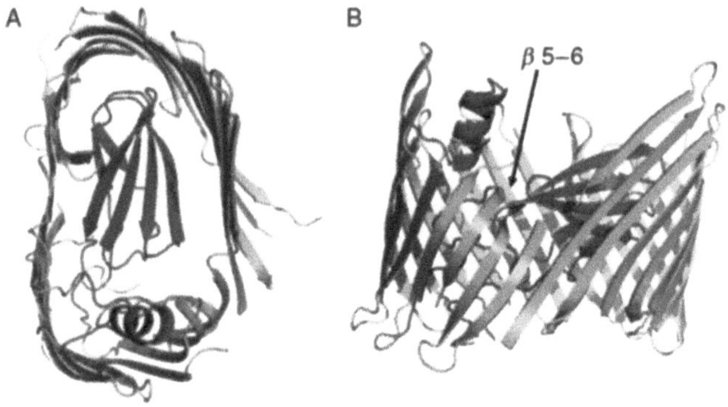

Please refer to Fig. 11.5 in text for figure legend.

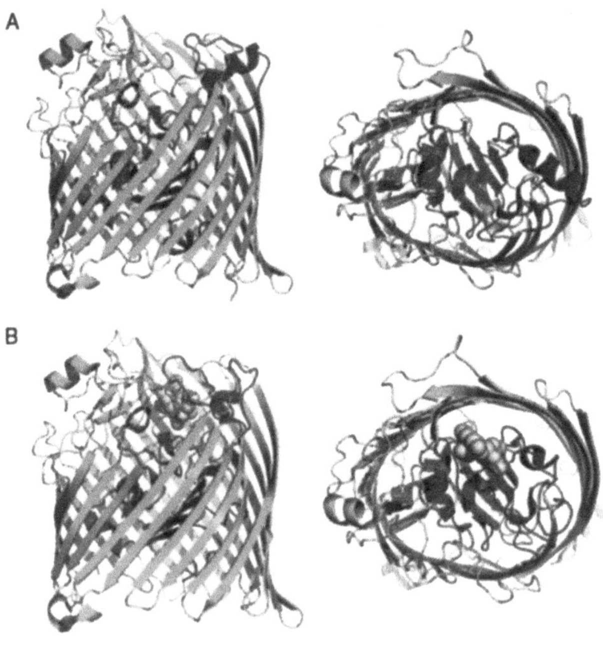

Please refer to Fig. 11.6 in text for figure legend.

Please refer to Fig. 11.7 in text for figure legend.

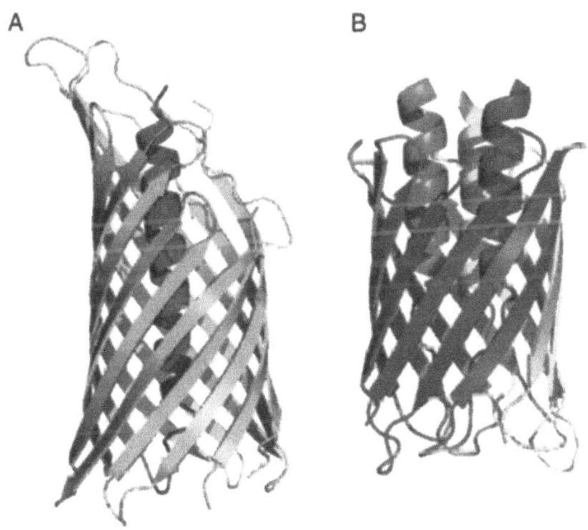

Please refer to Fig. 11.10 in text for figure legend.

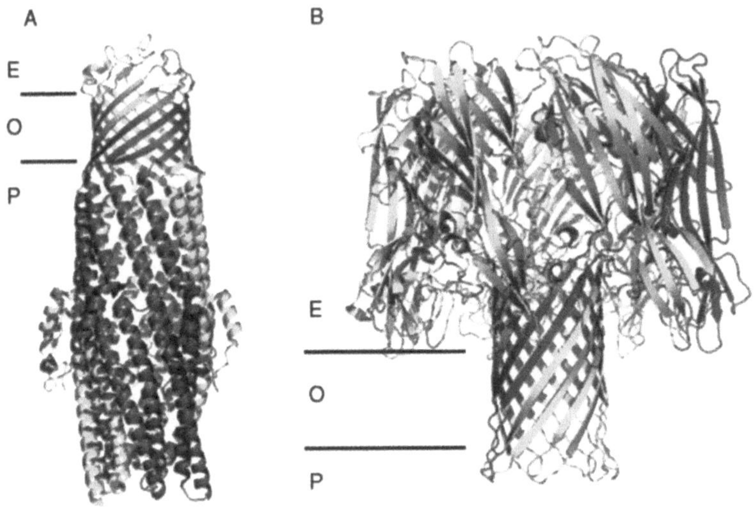

Please refer to Fig. 11.11 in text for figure legend.

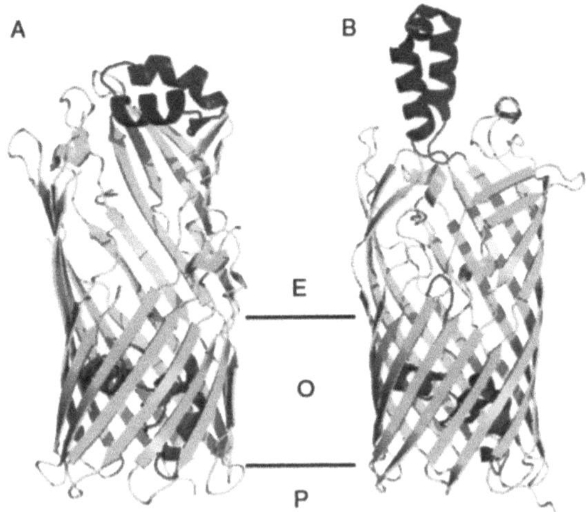

Please refer to Fig. 11.13 in text for figure legend.

Please refer to Fig. 11.14 in text for figure legend.

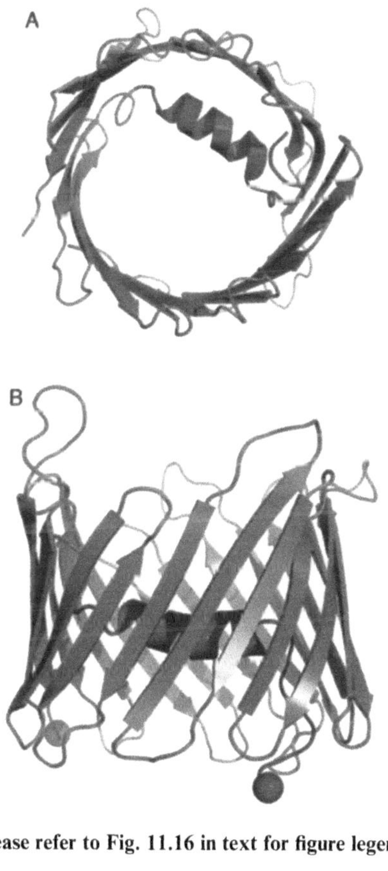

Please refer to Fig. 11.16 in text for figure legend.

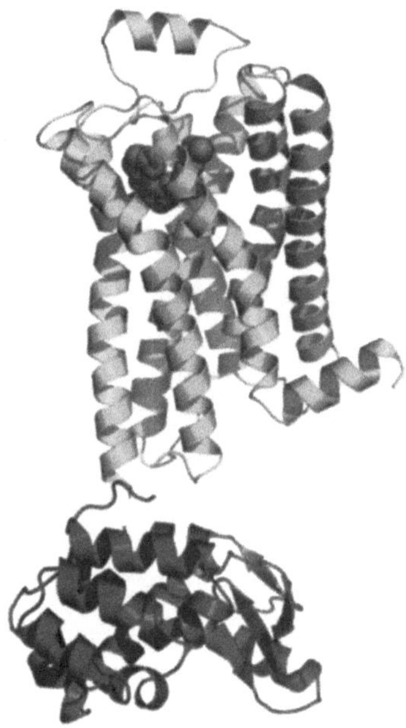

Please refer to Fig. 11.17 in text for figure legend.

FSC
www.fsc.org
MIX
Papier | Fördert
gute Waldnutzung
FSC® C083111

Kolibri 360 GmbH
Ferdinand-Jühlke-Straße 7
99095 Erfurt, Deutschland
produktsicherheit@kolibri360.de